Natural Attenuation of Environmental Contaminants

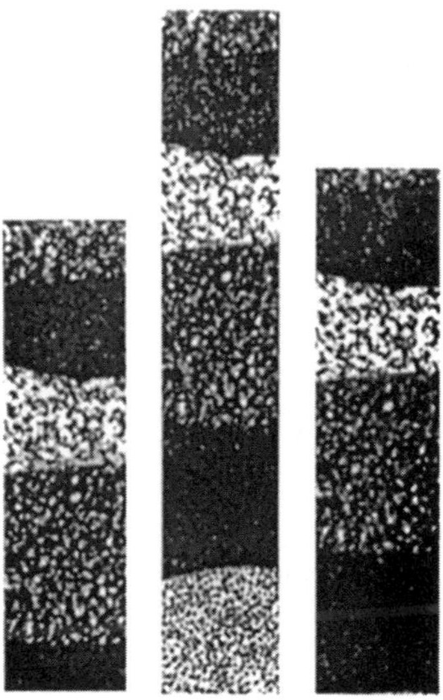

Editors

Andrea Leeson, Mark E. Kelley, Hanadi S. Rifai, and Victor S. Magar

The Sixth International In Situ and On-Site Bioremediation Symposium

San Diego, California, June 4–7, 2001

BATTELLE PRESS
Columbus • Richland

Library of Congress Cataloging-in-Publication Data

International In Situ and On-Site Bioremediation Symposium (6th : 2001 : San Diego, Calif.)
Natural attenuation of environmental contaminants : the Sixth International In Situ and On-Site Bioremediation Symposium : San Diego, California, June 4-7, 2001 / editors, A. Leeson … [et al.].
p. cm.– (The Sixth International In Situ and On-Site Bioremediation Symposium ; 2)
Includes bibliographical references and index.
ISBN 1-57477-112-4 (hc. : alk. paper)
1. Hazardous wastes--Natural attenuation--Congresses. 2. Bioremediation--Congresses. I. Leeson, Andrea, 1962- . II. Title. III. Series: International In Situ and On-Site Bioremediation Symposium (6th : 2001 : San Diego, Calif.). Sixth International In Situ and On-Site Bioremediation Symposium ; 2.
TD192.5.I56 2001 vol. 2
[TD1060]
628.5 s--dc21
[628.5]

2001041196

Printed in the United States of America

Battelle Press
505 King Avenue
Columbus, Ohio 43201, USA
614-424-6393 or 1-800-451-3543
Fax: 1-614-424-3819
Internet: press@battelle.org
Website: www.battelle.org/bookstore

For information on future environmental conferences, write to:
Battelle
Environmental Restoration Department, Room 10-123B
505 King Avenue
Columbus, Ohio 43201-2693
Phone: 614-424-7604
Fax: 614-424-3667
Website: www.battelle.org/conferences

CONTENTS

FOREWORD

The papers in this volume correspond to presentations made at the Sixth International In Situ and On-Site Bioremediation Symposium (San Diego, California, June 4-7 2001). The program included approximately 600 presentations in 50 sessions on a variety of bioremediation and supporting technologies used for a wide range of contaminants.

This volume focuses on *Natural Attenuation of Environmental Contaminants*. Natural attenuation continues to gain increasing attention from site managers and regulators as a viable alternative to more costly engineered site remediation approaches. Natural attenuation was first demonstrated at sites with hydrocarbon contamination, but recent studies at sites contaminated by chlorinated compounds have shown that this approach often has merit for these more challenging contamination problems. Covering natural attenuation in media ranging from deep aquifers to shallow soils, and for contaminants ranging from fuels to solvents to herbicides, this volume offers the reader a comprehensive overview of case studies that represent the current state of the art in natural attenuation approaches to site remediation.

The author of each presentation accepted for the symposium program was invited to prepare an eight-page paper. According to its topic, each paper received was tentatively assigned to one of ten volumes and subsequently was reviewed by the editors of that volume and by the Symposium chairs. We appreciate the significant commitment of time by the volume editors, each of whom reviewed as many as 40 papers. The result of the review was that 352 papers were accepted for publication and assembled into the following ten volumes:

Bioremediation of MTBE, Alcohols, and Ethers — 6(1). Eds: Victor S. Magar, James T. Gibbs, Kirk T. O'Reilly, Michael R. Hyman, and Andrea Leeson.

Natural Attenuation of Environmental Contaminants — 6(2). Eds: Andrea Leeson, Mark E. Kelley, Hanadi S. Rifai, and Victor S. Magar.

Bioremediation of Energetics, Phenolics, and Polycyclic Aromatic Hydrocarbons — 6(3). Eds: Victor S. Magar, Glenn Johnson, Say Kee Ong, and Andrea Leeson.

Innovative Methods in Support of Bioremediation — 6(4). Eds: Victor S. Magar, Timothy M. Vogel, C. Marjorie Aelion, and Andrea Leeson.

Phytoremediation, Wetlands, and Sediments — 6(5). Eds: Andrea Leeson, Eric A. Foote, M. Katherine Banks, and Victor S. Magar.

Ex Situ Biological Treatment Technologies — 6(6). Eds: Victor S. Magar, F. Michael von Fahnestock, and Andrea Leeson.

Anaerobic Degradation of Chlorinated Solvents— 6(7). Eds: Victor S. Magar, Donna E. Fennell, Jeffrey J. Morse, Bruce C. Alleman, and Andrea Leeson.

Bioaugmentation, Biobarriers, and Biogeochemistry — 6(8). Eds: Andrea Leeson, Bruce C. Alleman, Pedro J. Alvarez, and Victor S. Magar.

Bioremediation of Inorganic Compounds — 6(9). Eds: Andrea Leeson, Brent M. Peyton, Jeffrey L. Means, and Victor S. Magar.

In Situ Aeration and Aerobic Remediation — 6(10). Eds: Andrea Leeson, Paul C. Johnson, Robert E. Hinchee, Lewis Semprini, and Victor S. Magar.

In addition to the volume editors, we would like to thank the Battelle staff who assembled the ten volumes and prepared them for printing: Lori Helsel, Carol Young, Loretta Bahn, Regina Lynch, and Gina Melaragno. Joseph Sheldrick, manager of Battelle Press, provided valuable production-planning advice and coordinated with the printer; he and Gar Dingess designed the covers.

The Bioremediation Symposium is sponsored and organized by Battelle Memorial Institute, with the assistance of a number of environmental remediation organizations. In 2001, the following co-sponsors made financial contributions toward the Symposium:

Geomatrix Consultants, Inc.
The IT Group, Inc.
Parsons
Regenesis
U.S. Air Force Center for Environmental Excellence (AFCEE)
U.S. Naval Facilities Engineering Command (NAVFAC)

Additional participating organizations assisted with distribution of information about the Symposium:

Ajou University, College of Engineering
American Petroleum Institute
Asian Institute of Technology
National Center for Integrated Bioremediation Research & Development (University of Michigan)
U.S. Air Force Research Laboratory, Air Expeditionary Forces Technologies Division
U.S. Environmental Protection Agency
Western Region Hazardous Substance Research Center (Stanford University and Oregon State University)

Although the technical review provided guidance to the authors to help clarify their presentations, the materials in these volumes ultimately represent the authors' results and interpretations. The support provided to the Symposium by Battelle, the co-sponsors, and the participating organizations should not be construed as their endorsement of the content of these volumes.

Andrea Leeson & Victor Magar, Battelle
2001 Bioremediation Symposium Co-Chairs

A NOVEL TECHNIQUE TO QUANTIFY TPH REDUCTION DURING NATURAL ATTENUATION STUDIES BY MEASURING NITRATE CONCENTRATIONS

Stewart E. Lehman, Brian E. Dragich, and ***Nirupam Pal***
(California Polytechnic State University, San Luis Obispo, CA)

ABSTRACT: Thousands of petroleum-contaminated sites throughout the continental USA pose a serious threat to humans and the ecosystem. Natural attenuation/intrinsic bioremediation is one of the most preferred remediation technologies. A novel technique has been developed to indirectly measure the biodegradation rate by monitoring the various electron acceptors, namely nitrate, sulfate, ferric ion, and manganese. These ionic species are used in a sequential fashion during anaerobic biodegradation for hydrocarbons under the subsurface. This group developed various stoichiometric balances combining bioenergetics and kinetics. These results can be used to predict the rate of intrinsic bioremediation for petroleum hydrocarbon-contaminated soils and groundwater by accurately measuring the various ionic concentrations upstream and downstream of a contaminated plume.

INTRODUCTION

Contamination of groundwater and soil with fuel hydrocarbons is a widespread environmental problem. Natural attenuation has become a frequently used remediation technique for sites with petroleum-contaminated aquifers. Of all the natural attenuation processes, intrinsic bioremediation is of great importance. Intrinsic bioremediation refers to the component of natural attenuation brought about by biological degradation mechanisms (Wiedemeier et al., 1999).

During intrinsic bioremediation of petroleum-contaminated sites, microorganisms can degrade total petroleum hydrocarbons (TPH) via a number of mechanisms dependent on aquifer conditions. Parameters such as TPH concentration, electron acceptor availability, oxidation-reduction potential (ORP), and pH are critical in determining which biodegradation process will occur. Direct measurement of TPH for evaluating intrinsic bioremediation is difficult due to TPH sorption to the groundwater matrix and the low solubility of TPH in water. Measurement accuracy may vary widely, depending on the products involved and the amount of post-release chemical changes that may have occurred (Calabrese et al., 1993). Specific problems relate to the fact that a significant portion of the more volatile compounds in gasoline and light fuel oil may be lost in the solvent concentration step (Calabrese et al., 1993). Another difficulty in obtaining accurate TPH concentrations is hydrocarbon mixtures used for instrument calibration have a constant composition, whereas the relative composition of petroleum products and their residues are highly variable (Calabrese et al., 1993). Since TPH measurement is often highly variable, the more reliable measurement of electron acceptor concentrations in groundwater could better estimate the extent of TPH biodegradation in the subsurface.

This research investigated a novel approach to monitor TPH biodegradation via analysis of electron acceptor concentrations in groundwater. This paper demonstrates the concept using nitrate as the model electron acceptor. Bioenergetics theory was used to develop a balanced stoichiometric equation with nitrate as the electron acceptor. This equation was used to quantify TPH degradation through nitrate measurements. Bioenergetics is an approach to create chemical redox equations while stoichiometrically balancing chemical elements, electrons, charge, and energy. This approach integrates factors controlling microbial growth and relates it to materials, which cells consume or produce (Rittmann and McCarty, 2001). The notable difference between stoichiometric equations obtained using bioenergetics and traditional redox equations is the inclusion of biomass formation. In traditional redox reactions, all energy extracted from the electron donor is assumed to be converted to energy, or respiration. Microcosm studies were conducted to: 1) support the use of bioenergetically derived equations for a mass balance approach of quantifying TPH reduction by monitoring CO_2 production, 2) determine the relationship between nitrate reduction and TPH reduction using a defined media, and 3) determine the relationship between nitrate reduction and TPH reduction using groundwater samples from a contaminated site.

MATERIALS AND METHODS

Chemicals. All chemicals were obtained from Spectrum Laboratory Products (Gardena, CA). TPH was extracted from UNOCAL's Guadalupe Oil Fields, located in Guadalupe, CA. Free-phase TPH (consisting of heavy California crude oil and a kerosene-like thinning agent) was extracted and rid of water by adding 15 g anhydrous sodium sulfite to 200 mL of TPH.

Microorganisms and Media. Inoculum was prepared by extracting microorganisms from hydrocarbon-affected soil. Indigenous microorganisms were obtained by adding 50 mg hydrocarbon-contaminated soil to 1 L of DI water having a 1% NaCl concentration. The mixture was shaken vigorously and then allowed to settle for a few minutes. The supernatant was decanted and centrifuged at 6,000 rpm for 30 minutes. The supernatant was decanted again and used as inoculum for all experiments. The inoculum tested was negative for the presence of nitrate.

Microcosm media was composed of the following reagents: 2.75 g K_2HPO_4/L; 0.575 g KH_2PO_4/L; 56 mg $MgCl_2$/L; 5 mg $ZnCl_2$/L; 25 mg Na_2MoO_4/L; 14 mg $CaCl_2$/L; and 25 mg $NaNO_3$/L. Reagents were added to 18 MΩDI water, mixed thoroughly, and autoclaved. Fifty milligrams of TPH was added to each bioreactor as the sole carbon source. A solution containing 10,000 mg TPH/L was prepared in a glass vial by adding 35 mL of acetone to 350 mg of TPH. The TPH/acetone solution was used as a stock for adding TPH/acetone mixture aliquots to the bioreactors.

For experiments conducted with field groundwater samples, fresh groundwater was extracted from the field site. The groundwater was unamended for the experimental studies and contained an initial concentration of 23 mg NO_3^-/L and no TPH. Indigenous microorganisms present in the groundwater were used as inoculum for the groundwater experiments.

Biodegradation Studies. Microcosm studies were performed in 250-mL Pyrex screw top flasks (Corning Inc., Corning, NY) attached to a Micro-Oxymax anaerobic respirometer (Columbus Instruments, Columbus, OH). Prior to use, the bioreactors were cleaned, dried, and sterilized for 30 minutes at 121°C in an autoclave. Once cooled, 1 mL of a 10,000 mg TPH/L solution was added to each reactor. The solution was prepared in a glass vial by adding 35 mL of acetone to 350 mg of TPH. The acetone was allowed to completely volatilize, leaving 10 mg TPH in each bioreactor. One hundred ninety-five milliliters of autoclaved media was then added to the bioreactors. Bioreactors were purged with nitrogen to remove any oxygen and measured for dissolved oxygen (DO) using a calibrated DO electrode (Cole – Parmer Instrument Company, Vernon Hills, IL). ORP was then measured in each bioreactor using pH/mV meter (Cole Parmer Instrument Company, Vernon Hills, IL) with a platinum redox electrode (Orion Research, Inc, Beverly, MA), while purged with nitrogen, to ensure proper nitrate utilizing conditions. A poising agent was not required to adjust the ORP in the microcosms because the solution ORP was within the nitrate reduction range. Bioreactors were then immediately connected to the Micro-Oxymax anaerobic respirometer. Once connected to the respirometer and leak tested, 5 mL of inoculum was added to each reactor through the septum. The initial TPH and nitrate concentrations were 50 mg TPH/L and 18 mg NO_3^-/L. The microcosm experiments were run concurrently with a media control, containing 200 mL of autoclaved media, for reference.

Groundwater microcosm experiments were conducted in 500-mL Pyrex media bottles serving as bioreactors. Bioreactors were prepared similarly as above except 2 mL of the 10,000 mg TPH/L solution was added, and 400 mL of groundwater was added to each reactor, for an initial concentration of 50 mg TPH/L.

Analysis. Carbon dioxide production was measured and recorded by the Micro-Oxymax respirometer. Nitrate concentrations were analyzed using a DIONEX DX-120 ion chromatograph (IC) (Dionex Corporation, Sunnyvale, CA). A Dionex ASRS-ULTRA 4mm suppressor, IonPac AS9-HC anion column with an AG9-HC guard column was used with the eluent flow rate set to 1 mL/min, and the detection stabilizer temperature set to 30°C. Additionally, 9 mM Na_2CO_3 eluent was used with a sample injection volume of 25-μL and run time of 28 minutes. Calibration standards were prepared, analyzed by the IC, and their peaks were stored in the PeakNet software program for a calibration curve. Initial and final DO and ORP measurements were made to ensure nitrate-reducing conditions were kept throughout the experiments.

Microbial Stoichiometry. Bioenergetics states a fraction of the electrons transferred (f_e) are used to provide energy while the remaining electrons (f_s) are used for the growth of microbial cells. f_e and f_s are determined using Gibbs free energy values from the electron donor reaction (R_d), electron acceptor reaction (R_a), and the cell synthesis reaction (R_c). The sum of f_e and f_s is 1.0. A detailed derivation and explanation of bioenergetics theory is provided in Rittmann and McCarty (2001). Below is a brief derivation of the stoichiometric equations used for analysis. Hexane was used as an approximate electron donor for all stoichiometry due to TPH both having a complex composition and being an array of compounds. Because of this approximation,

$$-R_d\text{: } 0.0263\ C_6H_{14} + 0.316\ H_2O \rightarrow 0.158\ CO_2\ + H^+\ + e^-$$

$$f_eR_a\text{: } 0.108\ NO_3^- + 0.54\ e^- + 0.648\ H^+ \rightarrow 0.054\ N_2 + 0.324\ H_2O$$

$$f_sR_c\text{: } 0.0821\ CO_2 + 0.0164\ NO_3^- + 0.476\ H^+ + 0.46\ e^- \rightarrow 0.0164\ C_5H_7O_2N\ +\ 0.18\ H_2O$$

$$R\text{: } 0.0263\ C_6H_{14} + 0.1244\ NO_3^- + 0.124\ H^+ \rightarrow 0.0164C_5H_7O_2N + 0.054N_2 + 0.076CO_2 + 0.188H_2O$$

Solving for one mole of Hexane:

$$\boxed{C_6H_{14} + 4.72NO_3^- + 4.71H^+ \rightarrow 0.62C_5H_7O_2N + 2.052N_2 + 2.88CO_2 + 7.14H_2O} \quad (1)$$

RESULTS

Media Controls. A media control experiment was performed to determine the quantity of CO_2 evolved in bioreactors containing only microbial media. The CO_2 evolved from the media control was then subtracted from the total CO_2 produced during biodegradation studies to give the CO_2 produced solely from TPH degradation. During the media control experiment, a minimal amount of CO_2 was observed. The observed CO_2 production was used during analysis.

Verification of TPH/CO_2 Stoichiometry. A relationship between CO_2 production and TPH reduction was previously determined in which nitrate was used as the sole electron acceptor (Kuan and Perkins, 1999). In order to verify the presented stoichiometric relationship between TPH and CO_2, previous experimental CO_2 values and the TPH/CO_2 mass ratio, derived from Equation 1, were used to predict theoretical TPH concentrations. Theoretical TPH concentrations were then compared to experimental TPH concentrations determined by GC analysis from Kuan and Perkins (1999). Results showed an excellent relationship between experimentally obtained and theoretically approximated TPH concentrations (Figure 1) and thus justified the use of a mass balance approach using stoichiometry derived from bioenergetics. The experimental and theoretical TPH/CO_2 ratios obtained were 0.6494 and 0.679 g TPH/g NO_3^- respectively, with an experimental error of 4.6%.

Nitrate Microcosms. Microcosm experiments were conducted using nitrate as the sole electron acceptor. After 16 days, CO_2 increased to approximately 13 mg CO_2/L while nitrate concentration decreased to approximately 10 mg NO_3^-/L, from the initial 20 mg NO_3^-/L, for triplicate runs. Data monitoring ceased after 16 days because low nitrate concentrations caused it to be the limiting reactant in the degradation of TPH. Results from the triplicate runs were all consistent with each other (Figure 2).

Using adjusted carbon dioxide values obtained from the anaerobic respirometer, TPH was then approximated using the stoichiometric relationship between CO_2 and TPH (Equation 1). Figure 3 demonstrates TPH reduction, based on adjusted CO_2 values, for triplicate runs. Approximately 8 mg TPH/L was degraded due to biological reduction. A first-order kinetic model was applied to TPH data and the average TPH first-order rate constant for the nitrate-amended microcosms was 0.011 day^{-1} with a standard deviation of 0.0012 day^{-1} between triplicate runs (Figure 3).

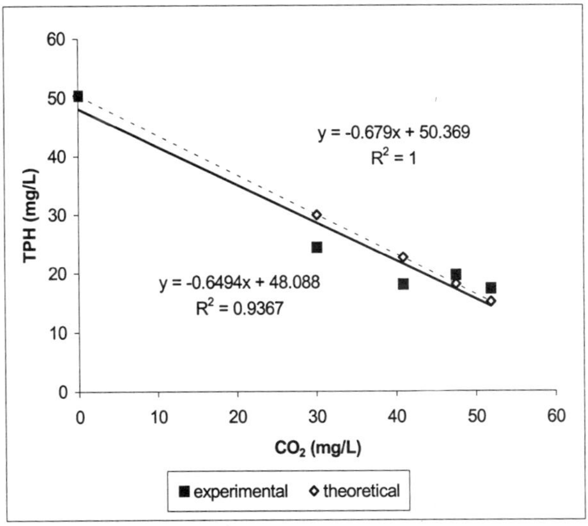

FIGURE 1. Comparison of experimental TPH/CO_2 ratio to the stoichiometrically derived TPH/CO_2 ratio.

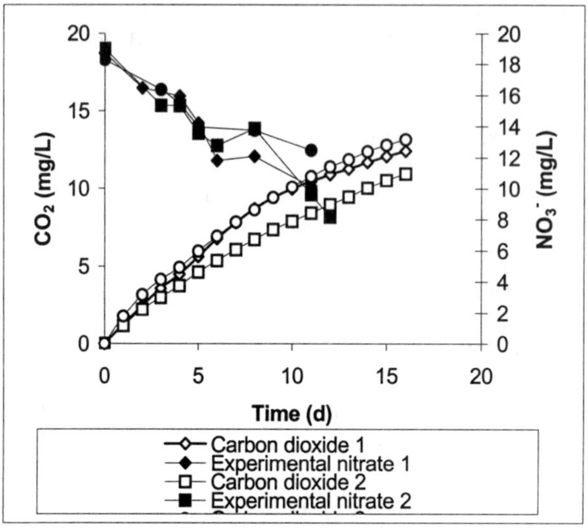

FIGURE 2. Respirometric CO_2 and measured nitrate concentrations.

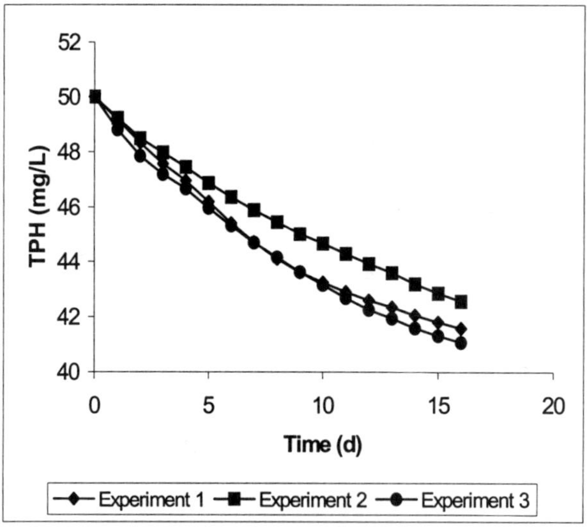

FIGURE 3. TPH reduction for nitrate-amended microcosms. First-order rate constants were 0.0117, 0.0098, and 0.012 day^{-1} for Nitrate Experiments 1, 2, and 3 respectively.

Stoichiometric relationships between nitrate and CO_2 were evaluated by comparing experimental CO_2, measured by the anaerobic respirometer, to calculated CO_2 concentrations, derived from Equation 1 and the measured nitrate concentrations. Respirometer measured CO_2 levels were higher than those predicted theoretically using stoichiometry based on bioenergetics theory (Figure 4). The theoretical was lower than the experimentally observed CO_2 because it represented the minimum amount of CO_2 produced per unit nitrate consumed assuming 100 % efficiency, from stoichiometry. This observation either means the experimental relationship between nitrate and CO_2 occurred with less than 100% efficiency and/or the stoichiometry relationship between the two is correctly balanced to actual conditions.

Groundwater Microcosms. TPH biodegradation was evaluated using fresh groundwater upstream from a TPH-contaminated aquifer. No amendments were made to the groundwater sample. TPH was added to the microcosms as the sole carbon source. Using the groundwater sample as media and the native microorganisms within the media, measured CO_2 increased from 12 to 16 mg/L, while nitrate decreased from 15 mg/L to 8 and 9 mg/L over 60 days, for duplicate runs (Figure 5). Although sulfate was present in the groundwater media, sulfate concentrations remained constant throughout the experiment, indicating nitrate was the only electron acceptor utilized during the experiments.

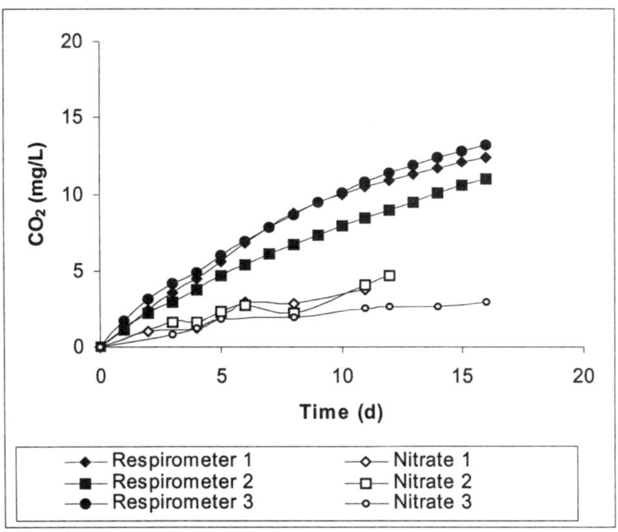

FIGURE 4. CO_2 production as measured by the respirometer and CO_2 based on nitrate measurements using stoichiometry.

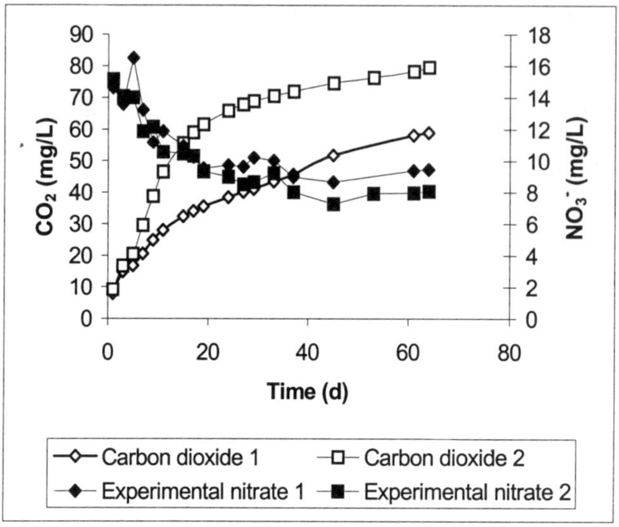

FIGURE 5. Respirometric CO_2 and measured nitrate concentrations.

Approximating the TPH concentrations using the derived stoichiometry resulted in complete exhaustion of the TPH source within 40 days for Groundwater Experiment 2 (Figure 6). Groundwater Experiment 2 had approximately 10 mg TPH/L remaining after 60 days (Figure 6). The degradation of TPH followed first-order kinetics and had degradation rates of 0.023 and 0.0959 day^{-1}. The observed rate constants for groundwater microcosms were higher than rates observed in the nitrate-amended microcosms. This is probably due to minimizing the disturbances caused by environmental factors (nitrate-amended experiment microorganisms were extracted from TPH-contaminated soil and reintroduced into a defined media). By limiting such disturbances, the microorganisms were more acclimated to the groundwater. However, nitrate-amended microcosms were inoculated with microorganisms and were not expected to be quickly accustomed to the defined media used.

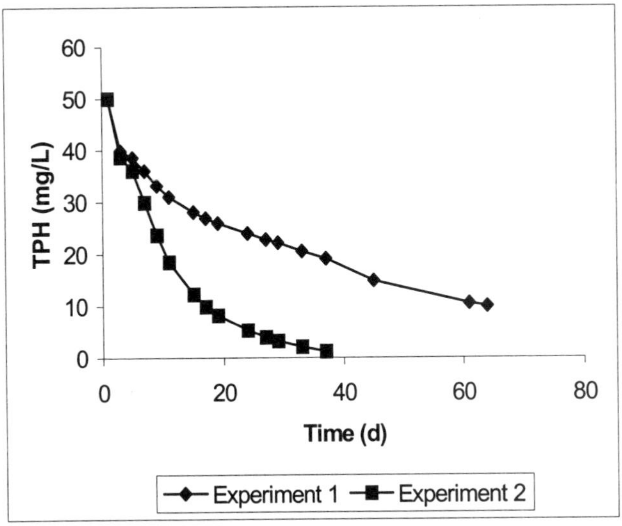

FIGURE 6. Groundwater TPH degradation based on CO_2 production. First-order rate constants were 0.023 and 0.0959 day^{-1} for Groundwater Experiments 1 and 2 respectively.

CONCLUSION

This unique and novel approach shows significant promise to indirectly measure contaminant degradation rate. The first order degradation rates obtained from this study is very similar to other published results. Experiments with other electron acceptors are currently in progress.

REFERENCES

Calabrese, E. J., and P. T. Kostecki, 1993. "Principles and Practices for Petroleum Contaminated Soils." Lewis Publishers, Chelsea, Michigan.

Kuan, R., and C. Perking. 1999. "Laboratory Estimation of Natural Attenuation via Electron Acceptor Analysis." Master's Thesis. California Polytechnic State University, San Luis Obispo.

Rittmann, B. E. and P. L. McCarty. 2001. Environmental Biotechnology. McGraw-Hill, New York.

Wiedemeier, T., H. S. Rifai, C.J. Newel, and J.T. Wilson. 1999. Natural Attenuation of Fuels and Chlorinated Solvents in the Subsurface, John Wiley & Sons, Inc., New York.

MONITORED NATURAL ATTENUATION (MNA) OF PETROLEUM HYDROCARBONS IN A HETEROGENOUS AQUIFER AFFECTED BY TRANSIENT FLOW

Christian Mossing, Lars Chr. Larsen and Hans C. L. Hansen (Hedeselskabet, Roskilde, Denmark)

Dorte Seifert and Poul L. Bjerg (Technical University of Denmark)

ABSTRACT: A 3-year study of monitored natural attenuation (MNA) was carried out at a former petrol station at Radsted, Denmark. Natural attenuation (NA) processes were described through traditional comparison of benzene, toluene, ethylbenzene and xylenes (BTEX) levels with changes in redox parameters. NA was found to occur at the site, with hydrocarbons being degraded under iron-, nitrate reducing and aerobic conditions. Average source area concentrations of Σ BTEX of 1.6 mg/L were found to attenuate to levels below 1 μg/L within 40 m downgradient of the source. The results showed strong correlation between the extent of the plume and changes in electron acceptors and metabolic byproducts in the aquifer. First-order degradation constants were calculated for BTEX under different redox conditions. $k_{\text{nitrate reducing}}$ (BTEX) = 0.3 – 6 $[10^{-3} \cdot d^{-1}]$, $k_{\text{iron reducing}}$ (T, o-X) = 3 – 16 $[10^{-3} \cdot d^{-1}]$ and $k_{\text{iron reducing}}$ (B, E, m- and p-X) = 0.3 – 3 $[10^{-3} \cdot d^{-1}]$. During the monitoring phase strong seasonal variations in the flow field were observed. It was found that the transient flow field strongly affects the shape, extent and concentration of the pollution plume. Understanding the hydrogeology and evaluating the seasonal variations in a "pre-monitoring" phase on sites with heterogeneous geology is important in order to distinguish between a temporary (seasonal) expanding plume and an overall-expanding plume.

INTRODUCTION

The use of MNA has been widely applied in the US, however the majority of sites investigated so far has been situated in sandy settings. Studies of MNA in heterogeneous aquifers such as moraine deposits are rare. In such cases the hydrogeology and flow field will be more complex and presumably affect the fate of petroleum hydrocarbons. The aim of this project was to investigate natural attenuation (NA) processes for petroleum hydrocarbons in a heterogeneous aquifer affected strongly by variations in the flow field.

The study site is a former petrol station (closed 1975) in Radsted in the southern part of Denmark. The underground storage tanks and contaminated soil in the hot spot were removed in 1996. A mass of 25-50 kg of residual contaminants (mainly BTEX compounds) in the unsaturated zone was estimated.

MNA was selected as a remedial strategy in 1997 based on:

1. Initial results indicated a significant removal of BTEX compounds within a short distance from the source area.
2. A strong correlation between disappearance of BTEX compounds and depletion of oxygen and nitrate.
3. The contamination in the source area and plume did not pose a risk to indoor climate (as no houses are found downstream the source area) or to the groundwater resource (no nearby drinking water wells).

The initial documentation of MNA on the site was based on the information obtained from groundwater analysis from a total of 9 wells, however, additional wells have been installed during the course of the study to improve the delineation of the plume. Finally, a total of 23 well-screens were installed in- and around the dissolved pollution plume and five monitoring campaigns were performed at the site in the period December 1999 to January 2001.

RESULTS

Distribution and Variation of BTEX. The results in terms of BTEX from two different sampling campaigns are shown in Figure 1 and 2.

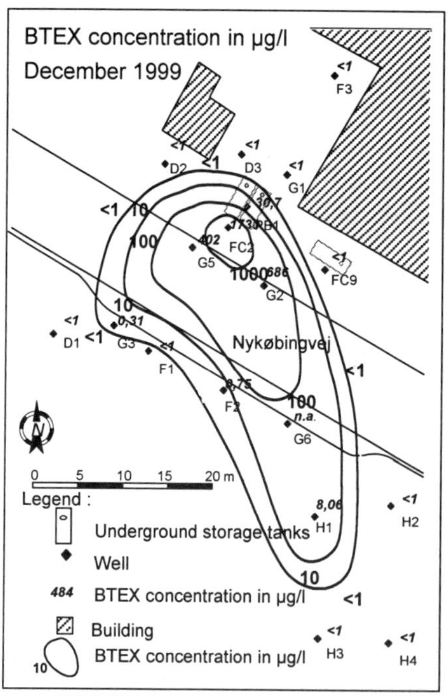

FIGURE 1. BTEX plume, December 1999.

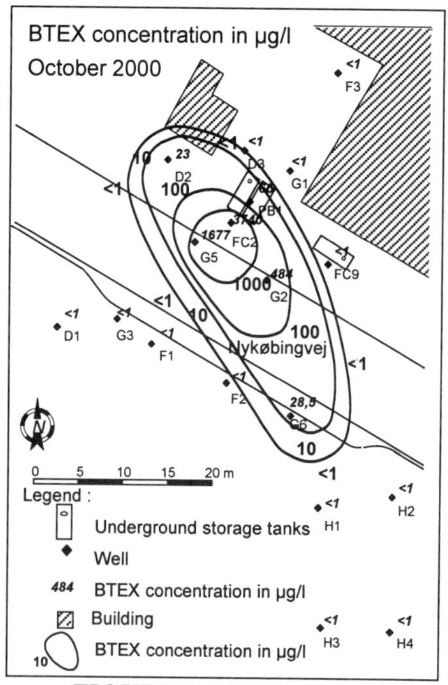

FIGURE 2. BTEX plume, October 2000.

By comparing figure 1 and 2 it seems that the concentration, shape and extent of the BTEX plume changes over time, however, concentration variations are even more evident if time series for specific screens are evaluated (Figure 3).

Variations in BTEX and Water Table. Figure 3 shows BTEX concentrations measured in screen FC2 and the groundwater table for the entire monitoring period.

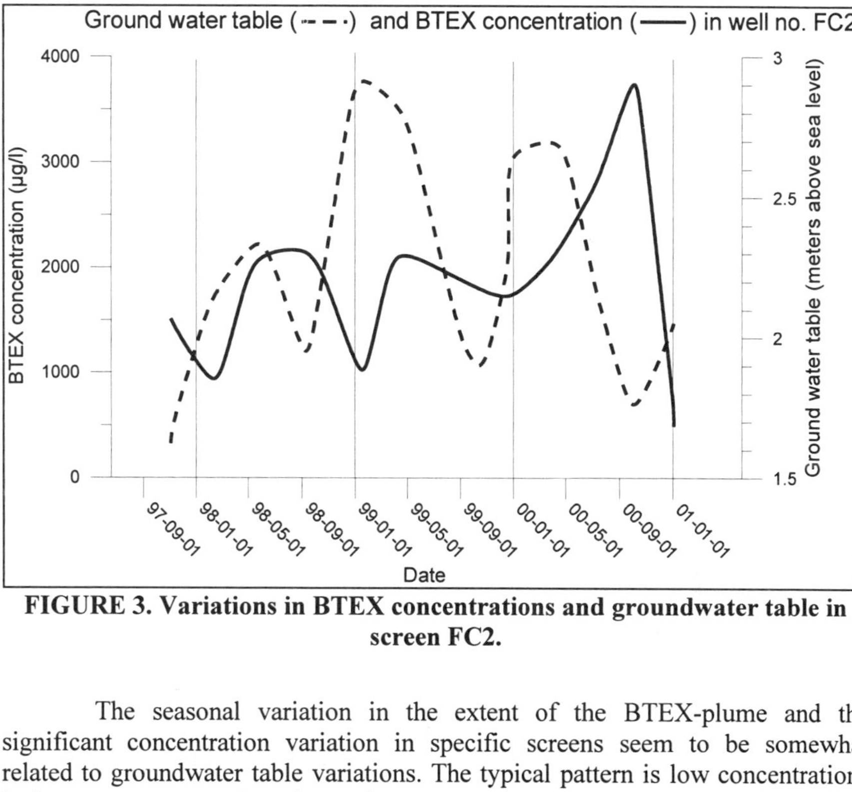

FIGURE 3. Variations in BTEX concentrations and groundwater table in screen FC2.

The seasonal variation in the extent of the BTEX-plume and the significant concentration variation in specific screens seem to be somewhat related to groundwater table variations. The typical pattern is low concentrations in the source area at wintertime, whereas concentrations rise in the summer. A fair correlation between elevated groundwater table and low BTEX concentrations and vice versa is seen.

An average hydraulic gradient of 0.0023 m/m, with a standard deviation of 0.0011 m/m, was found based on the measurements up till May 2000. The hydraulic gradient increased during winter and decreased during summer. The combination of high groundwater level and an increased hydraulic gradient leads to higher groundwater flux in the aquifer during winter. With a fairly constant outwashing of contaminants in the source area, this could lead to lower BTEX concentrations in the winter season, as a result of dilution.

Groundwater Table Variations. The fluctuations in the ground water level were measured by continuous data logging in three wells (G1, D1 and G6) (Figure 4).

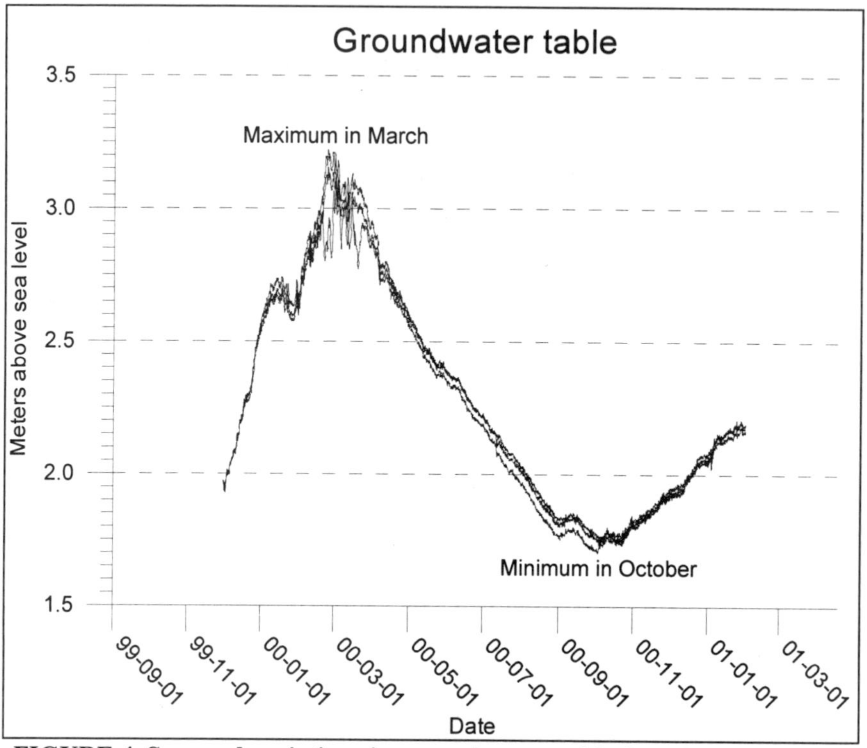

FIGURE 4. Seasonal variations in ground water table, based on continuous data logging.

Over a period of one year the groundwater table fluctuated up to 1.5 meters, with maximum level around March and minimum around October. The groundwater level in the three wells showed an almost synchronous pattern, but changes in the distance between the lines in Figure 4 indicate that the groundwater flow direction changes somewhat. It was found that the groundwater flow direction changed about 30° over a period of one month, based on data logging in the three wells.

A rise in the groundwater level of up to 1.5 meter over a 4-month period cannot be explained by a net infiltration of 75 mm/year in the area. Furthermore there are no known man-controlled activities, such as groundwater abstraction, which can account for the observed changes.

Variation in BTEX Compared to the Geology on the Site. This lead to a more detailed investigation of the overall geology in and around the site, based on information from old wells installed in the area. It was found that a wetland area about 120 meters downstream of the location probably controls the overall groundwater movement in the area. Figure 5 shows a conceptual cross section of the geology, from the source area to the wetland area.

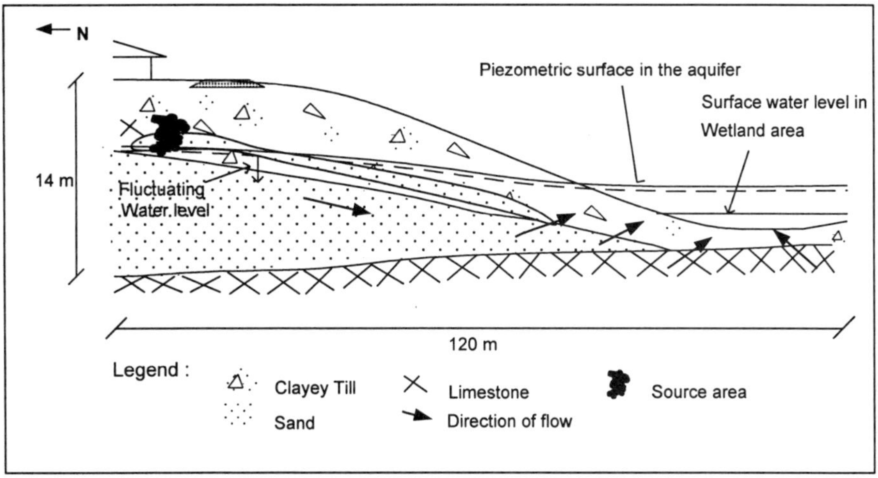

FIGURE 5. Conceptual model of the geology.

The conceptual model shows that the aquifer down-gradient of the contaminated site pinches out. Aquifer conditions change from semi-confined in the source area to confined in the area down-gradient of the source. A semi permeable layer of clay between the aquifer and the wetland down-gradient of the source causes a seasonal build up of the water level in the aquifer. It is suggested that the low-permeability layer of clay limits the hydraulic contact between the aquifer and the water in the wetland area, which causes a significant rise in the groundwater level in periods with high infiltration. This changes substantially the hydraulic gradient, the concentration of the contaminants and the extent and shape of the pollution plume.

Redox Conditions and First-Order Degradation Constants. The aquifer is initially aerobic, and the nitrate concentrations are around 10-15 mg/l. A correlation between high BTEX concentrations and depletion of electron acceptors (oxygen and nitrate)/ the formation of metabolic byproducts (dissolved iron) is observed (Figure 6.a – 6.c). This indicates degradation of the petroleum hydrocarbons under aerobic, nitrate and iron reducing conditions. Sulphate reduction and methanogenesis are presumably not taking place as no depletion of sulphate or formation of methane was found at the site. The depletion of oxygen and nitrate down gradient of the pollution plume is probably a result of the ”smoking gun” effect.

Figure 6.a – 6.c show the concentrations of oxygen, nitrate and dissolved iron, based on the monitoring campaign in December 1999.

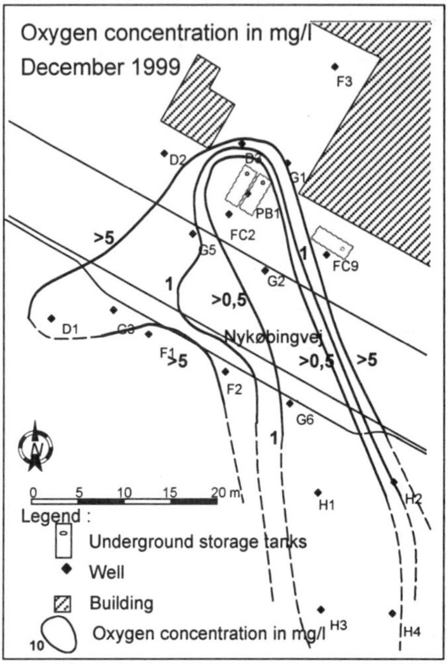

FIGURE 6.a. Oxygen concentrations, December 1999.

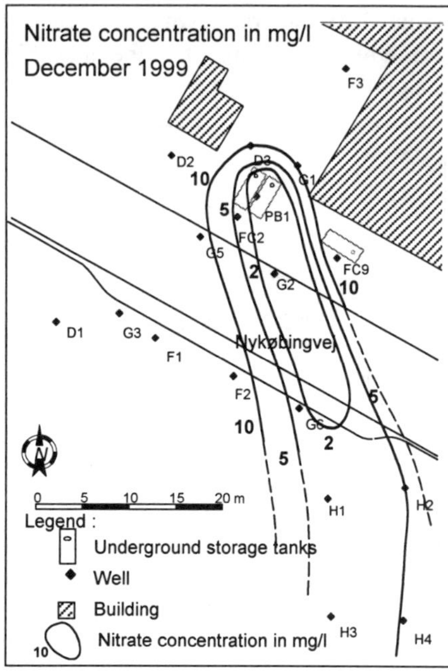

FIGURE 6.b. Nitrate concentrations, December 1999.

The wells have been divided into three groups based on an overall evaluation of the redox sensitive compound and the placement of the well. Table 1 shows average values for oxygen, nitrate and dissolved iron in three zones: upstream of the source area, in the core of the plume and in the periphery (downstream) of the plume. In the core of the plume, iron reduction seems to be the most important redox process, but small amounts of nitrate indicate a potential for nitrate reduction, as well. In the periphery, the presence of nitrate in reduced concentrations compared to upstream concentrations indicates that nitrate reduction is taking place. Dissolved iron is also observed, but is probably formed in the core of the plume. This, however, cannot be judged from dissolved redox sensitive parameters, and the prevailing redox conditions should not be seen as definitive.

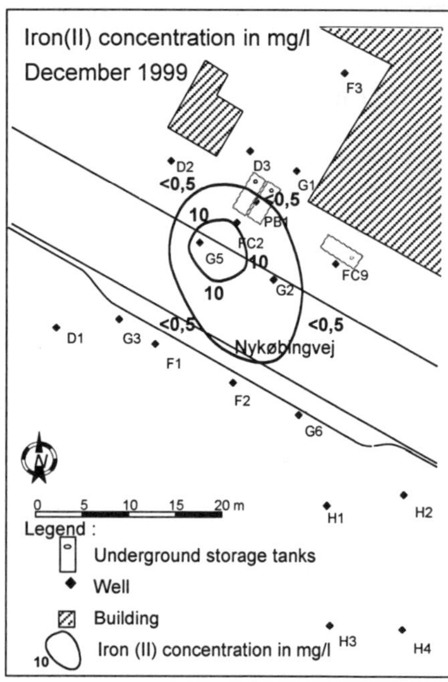

FIGURE 6.c. Dissolved iron concentration, December 1999.

A complete evaluation of the actual redox conditions will involve an array of methods as discussed in Christensen et al. (2000).

TABLE 1. Average values of redox parameters in three zones.

Parameters	Upstream (mg/L)	Core (mg/L)	Periphery (mg/L)
Oxygen	7.0	0.9	0.2
Nitrate	12.6	2.0	6.3
Iron (II)	0.08	4.6	3.9

A two dimensional geochemical solute transport model was applied to estimate first-order degradation rate constants. The model allowed for quantification of dilution, so the estimated values in Table 2 represents true first-order degradation rate constants for the core and the periphery of the plume.

The use of trimethylbenzenes (TMB´s) as conservative tracers was not possible because the TMB´s seemed to undergo degradation as well, and they are therefore not considered appropriate as tracers in this case.

TABLE 2. Estimated first-order degradation rate constant in the core and the periphery of the plume.

Placement of wells Redox conditions k	Periphery Nitrate reducing (d^{-1})	Core Iron reducing (d^{-1})
BTEX	0.3 – 6 [10^{-3}]	-
(T, o-X)	-	3 – 16 [10^{-3}]
(B, E, m- and p-X)	-	0.3 – 3 [10^{-3}]

Under nitrate reducing conditions it was found that the BTEX compounds followed the same degradation pattern, whereas degradation rate constants for toluene and o-xylene were generally higher than the other compounds under iron(III) reducing conditions (Hansen and Seifert, 2000). On the basis of simulations carried out with the model, degradation rate constants and intervals for each of the compounds were found (Table 2). Variations in the flow field were not directly incorporated into the modeling work, but an average hydraulic gradient of 0.0023 and average contaminant concentrations were used. Variability in the ground water flow direction was not incorporated either.

The first-order degradation rate constants of BTEX found under nitrate reducing conditions correspond well to degradation rate constants found under similar conditions (field observations, low temperature etc.). Equally there is a fair accordance between the first-order degradation rate constants for toluene and o-xylene found in this study and values found in the literature (Suarez and Rifai, 1999).

DISCUSSION AND CONCLUSIONS

Heterogeneous geology seems to complicate the use of MNA because the effects of seasonal variations in e.g. precipitation are amplified or altered. This can lead to drastic changes in the hydraulic gradient and thereby in ground water velocity. The changes in the flow field affect the shape, extent and concentration of/in the pollution plume.

The documentation of NA is often based on a single round of analysis – a "snapshot" in the annual cycle. If the documentation of NA is based on a situation where the pollution situation is positive (small plume, low concentrations etc.), seasonal changes in the plume extension can easily be mistaken with an overall expanding plume or vice versa.

In cases with heterogeneous geology it is therefore suggested to perform a "pre-monitoring phase" where seasonal changes are described, before setting up a detailed monitoring program. Furthermore, it may be useful to set up a proper hydrogeological model at the site.

ACKNOWLEDGMENTS

The project has been managed and funded by the Danish Petroleum Industry's Association for Remediation of Retail Sites (OM). Additional economical support has been achieved from the Danish EPA under the Program for Development of Technology (TUP).

REFERENCES

Christensen, T.H., P.L. Bjerg, S. Banwart, G. Heron, and H-J Albrechtsen. 2000: "Characterization of Redox Conditions in Groundwater Contaminant Plumes". *J. Contam. Hydrol.* 45: 165-241.

Hansen, H.C.L., and D. Seifert. 2000: *Geochemical Transport Modeling of Natural Attenuation – Case Study (in Danish)*. Master thesis from IMT/ISVA, Technical University of Denmark

Suarez, M.P., and H.S. Rifai. 1999: "Biodegradation rates for fuel hydrocarbons and chlorinated solvents in groundwater". *Biorem. J. 3*, 337-362.

COMETABOLIC AND INHIBITORY INTERACTIONS DURING ANAEROBIC BIODEGRADATION OF ALKYLBENZENE MIXTURES

Neal D. Durant (GeoTrans, Inc., Sterling, Virginia)
Edward J. Bouwer (The Johns Hopkins University, Baltimore, Maryland)

ABSTRACT: Mixed and pure culture experiments were conducted with aquifer bacteria to elucidate the effect of substrate interactions on the biodegradation of alkylbenzenes and polycyclic aromatic hydrocarbons (PAHs) under strict anaerobic denitrifying conditions. In the first experiment, a column consisting of aquifer sands collected from the site of a former manufactured gas plant (MGP) was fed a synthetic groundwater mixture containing benzene, toluene, ethylbenzene, and xylene (BTEX), naphthalene, and (1- and 2-) methylnaphthalene. Complete removal of toluene, *m*-xylene, and partial removal of *o*-xylene was evident after several months of operation. Degradation of *o*- and *p*-xylene occurred only in the presence of toluene, indicating cometabolism with toluene as the primary substrate. Benzene, ethylbenzene, and the PAHs resisted biodegradation in this 2-year study, providing further evidence of the difficulty in achieving biodegradation of these compounds under anaerobic conditions. A separate experiment evaluated the simultaneous biodegradation of toluene and *m*-xylene by denitrifying *Azoarcus* sp. strain T. In batch reactors receiving equimolar concentrations of *m*-xylene and toluene, strain T consistently degraded *m*-xylene first. Increasing the ratio of *m*-xylene to toluene caused an increase in the lag time for toluene degradation, suggesting that *m*-xylene competitively inhibits biodegradation of toluene by strain T. These experiments demonstrate that both synergistic and antagonistic interactions are possible during biodegradation of BTEX mixtures in the absence of oxygen. In addition, these data suggest that substrate interactions can significantly affect the biodegradation kinetics for individual alkylbenzenes in anaerobic groundwater.

INTRODUCTION

Researchers have known since the late 1960s that substrate interactions can affect aerobic biodegradation rates for individual polycyclic aromatic hydrocarbons (PAHs) and alkylbenzenes in the environment (Perry, 1979). A variety of studies have subsequently shown that both antagonistic (e.g., competitive inhibition) and synergistic (e.g., cometabolic) interactions can occur for biodegradation of certain aqueous mixtures of benzene, toluene, ethylbenzene, and xylene (BTEX) and PAHs (Bauer and Capone, 1988; Chang et al., 1993). Competitive inhibition occurs when two or more growth substrates are biodegraded by a common enzyme. Cometabolism is the fortuitous degradation of a non-growth substrate during microbial growth on another substrate. The occurrence of these interactions varies among bacterial species and substrates. Recent research has quantified aerobic substrate interactions in multi-substrate fuel hydrocarbon systems (Bielefeldt and Stensel, 1999; Deeb et al., 2001).

Although the number of studies evaluating aerobic biodegradation interactions for PAHs and alkylbenzenes continues to grow, few studies are available regarding interactions for these compounds under anaerobic conditions. Since most PAH- and BTEX-contaminated aquifers are oxygen-deficient, anaerobic substrate interactions are especially relevant to the problem of predicting intrinsic biodegradation rates. Moreover, findings of aerobic substrate interaction studies, in many cases, cannot be extended to anaerobic conditions because the enzymes that mediate anaerobic interactions can differ significantly from the oxygenase enzymes that mediate aerobic interactions. For example, aerobic biooxidation of methylbenzenes is initiated by oxygenases acting on the aromatic ring, while anaerobic transformation typically occurs by reaction with the methyl substituent. Because oxygenase enzymes have a low substrate-specificity, there may be a greater tendency for competitive inhibition for an aromatic substrate pair under aerobic conditions than under anaerobic conditions.

In this paper we present two experiments investigating the role of substrate interactions during biodegradation of PAHs and alkylbenzenes in the absence of oxygen. The first experiment used a laboratory column to evaluate anaerobic substrate interactions mediated by a natural, aquifer consortium. The second experiment employed a pure culture to examine the simultaneous biodegradation of *m*-xylene and toluene under nitrate-reducing conditions.

Objective. The objective of the first experiment was to investigate substrate interactions occurring in a nitrate-reducing aquifer column receiving a mixture of aromatic hydrocarbons. Toluene and *m*-xylene concentrations were varied to test whether these substrates could stimulate the cometabolism of methylnaphthalene and methylbenzenes. The column was constructed using aquifer sediments from a former manufactured gas plant (MGP) site. Nitrate-reducing conditions were chosen because biodegradation of the test compounds was thought to be feasible under these conditions, and nitrate-based bioremediation was being evaluated as a remedial option at the MGP site. The objective of the second experiment was to measure the simultaneous biodegradation of *m*-xylene and toluene by *Azoarcus* sp. strain T over a range of substrate concentrations. Strain T was chosen because it anaerobically degrades both toluene and *m*-xylene, and is representative of subsurface bacteria (the strain was originally isolated from an aquifer column).

MATERIALS AND METHODS

Aquifer Column. Aquifer sands were collected from a former MGP site as described by Durant et al. (1995). In an anaerobic glove box containing 98% N_2 and 2% H_2, an aquifer sand composite was packed into a glass column (25 cm long, 2.8 cm i.d.) equipped with septum-fitted side ports and PTFE end-fittings. The column was fed an anaerobic denitrifying basal medium (Zeyer et al., 1986) spiked with benzene, ethylbenzene, toluene, (*m*- and *o*-) xylene, naphthalene, and (1- and 2-) methylnaphthalene, each at concentrations of ~3 mg/L during column start-up. Amorphous FeS was added (10 μM final concentration) to scavenge any residual O_2. Pd catalysts and an O_2 detector maintained and verified, respectively, the absence of O_2 in the anaerobic glove box.

The growth medium and substrate feed were delivered from a gas tight syringe by means of a syringe pump (see Figure 1). Luer-lock fittings were used for the influent and effluent sample ports, and brass stopcocks at the column inlet and outlet enabled flow control. Aqueous samples (2 mL) were collected once every 2 or 3 days in a 5 mL glass barrel syringe at the influent and effluent ports. Samples were also collected periodically through the septum side ports by using a syringe. Nitrogen oxides were measured by ion chromatography (IC), and aromatic hydrocarbons were measured by gas chromatography (GC). The hydrocarbons were extracted into pentane prior to injection on the GC.

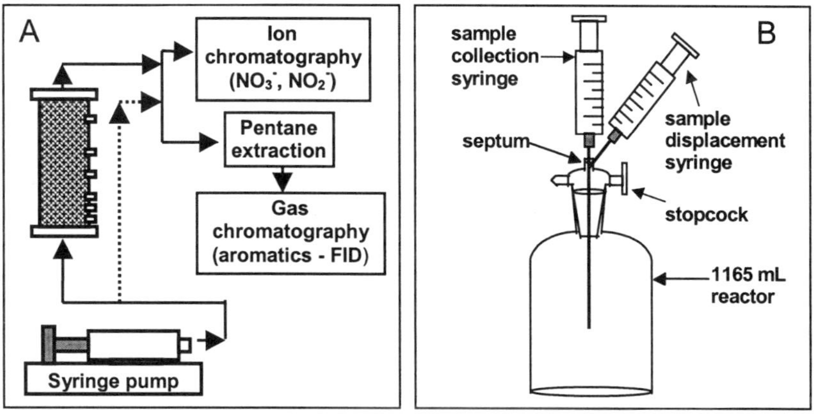

FIGURE 1. A. Column apparatus. B. Reactor for strain T experiment.

Strain T. Strain T was obtained from Deutsche Sammlung von Mickroorganismen (Braunschweig, Germany; DSM 9506), and cultivated anaerobically in a nitrate-reducing growth medium of the following composition (in mM): 8.15 Na_2HPO_4-$7H_2O$, 5.61 NH_4Cl, 5.0 $NaNO_3$, 2.03 $MgSO_4$-$7H_2O$, 1.84 KH_2PO_4, 0.86 NaCl, 0.09 $CaCl_2$-$2H_2O$, 0.07 $MnSO_4$ and 1 mL/L trace metals solution (Zeyer et al., 1986). Cells were grown in 650 mL screw-top flasks receiving equimolar concentrations of toluene and *m*-xylene (ca. 0.2 mM each). Long-term stocks were stored cryogenically and short-term stocks were maintained on agar slants. Aqueous and short-term stocks were exposed to equal parts toluene and *m*-xylene with the goal of maintaining the strain's intrinsic preferences for toluene and *m*-xylene.

The dual substrate experiment was conducted in 1165 mL batch bottles equipped with ground glass stopcocks and butyl rubber septa (see Figure 1). Biodegradation rates were measured for an array of initial toluene and *m*-xylene concentrations (Table 1). The effect of toluene on *m*-xylene biodegradation kinetics and vice versa was investigated by running a series of trials in which the initial concentration of one substrate was increased incrementally while the initial concentration of the other substrate was unchanged. Reactors were prepared in an anaerobic glove box (98% N_2 and 2% H_2). After the addition of 1165 mL inoculated medium, only a negligible (20 mL) headspace remained in the reactors.

To eliminate biomass density as a variable, all reactors received the same initial biomass concentration of ca. 3 mg/L. Prior to the initiation of the experiment, biomass (dry weight) was quantified gravimetrically by preparing a series of toluene-grown cultures, measuring optical density (A_{546}), and then weighing the contents after vacuum filtration and oven-drying. Optical measurement provided an excellent correlation with dry weight. The regression equation was: dry biomass (mg/L) = ((A_{546} – 0.0031)/0.003), with $R^2 = 0.993$.

Samples (5mL) were collected by inserting a syringe through the sampling port septum, and displacing the sample with an equivalent volume of fresh basal medium (see Figure 1). Portions of the 5 mL sample were analyzed for substrate loss by GC, nitrate consumption and nitrite production by IC, and biomass growth by absorbance. The substrate utilization and biomass growth curves were corrected for loss due to dilution prior to graphing and quantitative analysis. Although the data analysis corrected for dilution, the dilution resulting from each sampling event (0.4%) was assumed not to affect reaction kinetics.

TABLE 1. Array of *m*-xylene and toluene combinations examined in dual-substrate experiment.

m-Xylene (mM)	Toluene (mM)			
	0.02	0.04	0.07	0.1
0.02	X	X	X	X
0.04	X	X	X	X
0.1	X	X	X	X

RESULTS AND DISCUSSION

Aquifer Column. The onset of biodegradation commenced 4 months after column start-up. Of the 8 initial substrates, *m*-xylene was the only substrate other than toluene to degrade completely. The PAHs did not biodegrade significantly during this 2-year study (data not shown). Toluene was the first substrate completely degraded (120 days), followed by *m*-xylene degradation at 340 days. Partial degradation (~40%) of the *o*-xylene followed after the toluene degradation commenced and before significant *m*-xylene degradation began. *p*-Xylene, which was added at 482 days, also degraded partially. The extent of *o*- and *p*-xylene removal increased with time, but neither degraded completely during this study.

As summarized in Table 2, increases in the influent toluene concentration consistently increased the rate and extent of *o*-xylene degradation. The extent of *p*-xylene degradation also appeared to increase in response to increases in the influent toluene concentration. A 4-fold increase in the influent toluene caused *o*-xylene removal to increase from ~40 to ~55%. When the influent toluene concentration was returned to its original influent concentration at 400 days, the extent of *o*-xylene degradation was reduced back to ~40 %. There was no evident correlation between toluene concentration and *m*-xylene biodegradation (data not shown). Temporary removal of toluene from the influent confirmed that degradation of *o*- and *p*-xylene occurred cometabolically with toluene serving as the primary substrate (Figure 2). Removal of the toluene also confirmed that the biodegradation of these alkylbenzenes was coupled to nitrate-reduction. Once toluene was removed from the system, degradation of *o*- and *p*-xylene ceased, and the nitrate in the effluent returned to near its influent concentration (5 mM).

TABLE 2. Methylbenzene biodegradation rates and removal efficiencies in nitrate-reducing aquifer column.

Influent Toluene (mM)	*o*-Xylene Percent Removal	*p*-Xylene Percent Removal	BIODEGRADATION RATES Toluene (mM/hr)	*m*-Xylene (mM/hr)	*o*-Xylene (mM/hr)	*p*-Xylene (mM/hr)
0.025	40	25	0.026	0.029	0.008	0.005
0.097	54	--	0.076	0.021	0.008	--
0.097	56	--	0.083	0.022	0.008	--
0.275	86	50	0.294	0.035	0.022	0.011

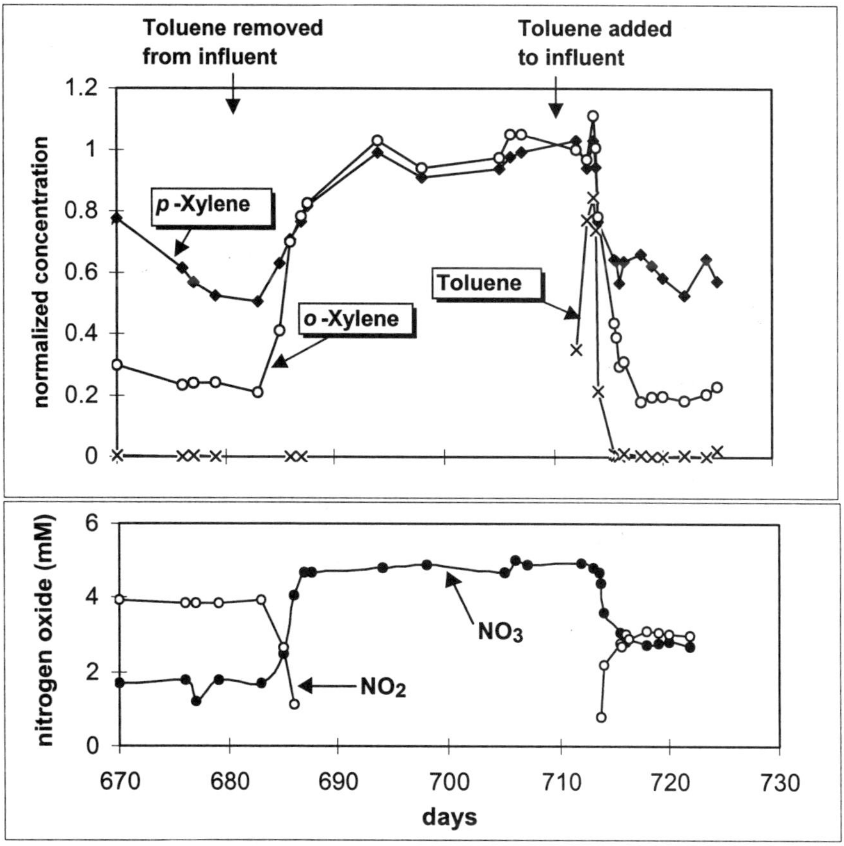

FIGURE 2. Effect of toluene on the biodegradation of *o*- and *p*-xylene in nitrate-reducing aquifer column.

Sampling along the length of the column indicated that biodegradation was rapid and limited to a zone within a few centimeters from the inlet. After 1 year of operation, approximately 75% of the toluene and *m*-xylene degradation occurred within 1 cm of the inlet, corresponding to biodegradation rates of 0.076 mM/hr (7.05 mg/L/hr) and 0.021 mM/hr (2.28 mg/L/hr), respectively. *o*-Xylene degraded at a rate of 0.008 mM/hr (0.85 mg/L/hr), but only in the presence of toluene and *m*-xylene.

This study is not the first to observe that *o*- and *p*-xylene can be biodegraded cometabolically under nitrate-reducing conditions (Evans et al., 1991; Rabus and Widdel, 1996). Nevertheless, our study is noteworthy in that we observed the cometabolic biodegradation of both *o*-xylene and *p*-xylene during biodegradation of toluene by an aquifer consortium under nitrate-reducing conditions. Our work extends previous research by demonstrating that the same cometabolic interactions that have been observed for pure denitrifying cultures are also possible in an aquifer biofilm.

Since some of the aquifer sediments from the MGP site were known to be capable of degrading naphthalene in the presence of nitrate (Durant et al., 1995), it was expected that some PAH biodegradation would have been observed during the column study. There are at least three possible explanations as to why the PAHs did not biodegrade significantly: (1) the aquifer material did not contain bacteria capable of oxidizing PAHs under nitrate-reducing conditions; (2) the substrate feed favored proliferation of alkylbenzene degraders over PAH-degraders; or (3) the alkylbenzenes competitively inhibited the biodegradation of PAHs. In any case, the persistence of PAHs in this study suggests that biodegradation of PAHs under denitrifying conditions is not as pervasive as has been suggested by others, particularly when more easily degraded compounds are present. Future research should investigate the role of competitive inhibition in limiting PAH degradation under anaerobic conditions.

Strain T. In reactors containing initially equimolar concentrations of *m*-xylene and toluene, strain T consistently degraded *m*-xylene more rapidly than toluene. Degradation of *m*-xylene characteristically proceeded first, with toluene exhibiting a longer lag period (Figure 3). Initial toluene degradation was slow or inhibited, then proceeded at a rate similar to the *m*-xylene once the *m*-xylene had dropped significantly below its initial concentration. After this point, the substrates biodegraded simultaneously. Simultaneous biodegradation, combined with the observation that biomass growth was log-linear over the entire course of substrate degradation, demonstrates that growth was not diauxic.

These data suggest that *m*-xylene competitively inhibits toluene degradation by strain T (Figure 4). As the initial concentration of *m*-xylene was increased relative to toluene, the inhibition or lag period for toluene degradation also increased. This pattern was observed repeatedly in experiments run at four initial toluene concentrations (0.02, 0.04, 0.07, and 0.1mM). As the concentration of toluene was increased relative to *m*-xylene, the inhibitory effect of *m*-xylene became less significant. Except for high initial toluene concentrations (0.1 mM), toluene did not appear to inhibit *m*-xylene degradation (data not shown).

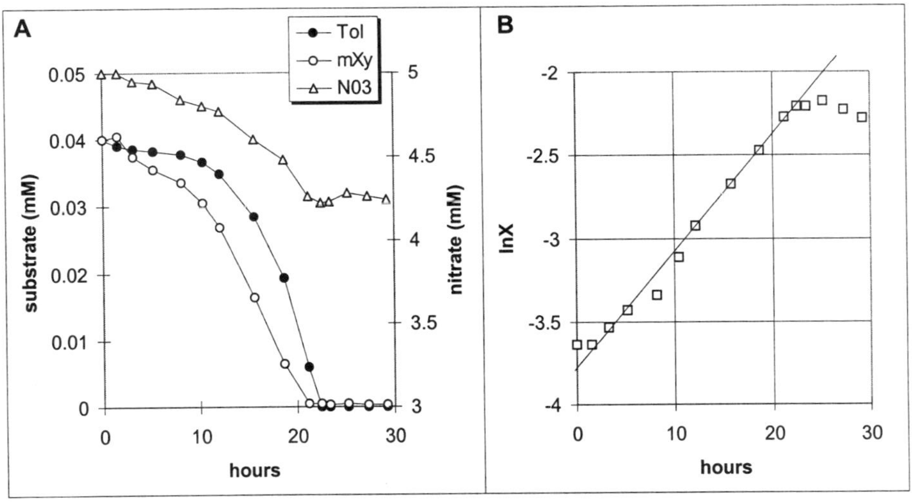

FIGURE 3. (A) Simultaneous biodegradation of toluene and *m*-xylene and (B) associated growth of strain T in nitrate-reducing reactors.

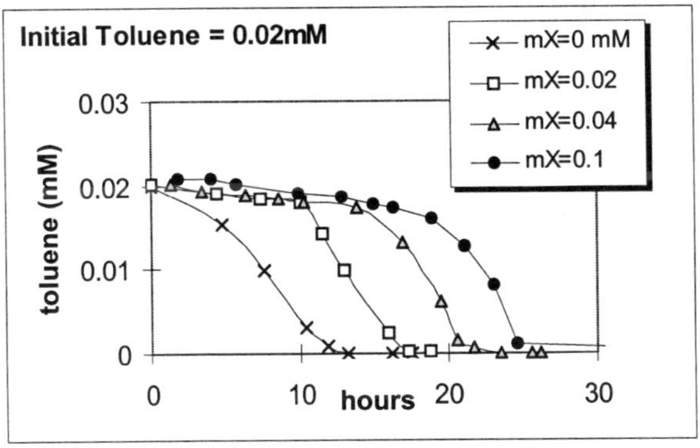

FIGURE 4. Effect of *m*-xylene on toluene biodegradation by strain T.

The finding that *m*-xylene competitively inhibits toluene biodegradation by denitrifying strain T is consistent with studies by Kreiger et al. (1999). Those researchers concluded that the same enzyme might catalyze the initial reaction for toluene and *m*-xylene biodegradation by strain T. For both substrates, the initial reaction likely involves the addition of fumarate to the methyl group. If the same enzyme were responsible for the initial reaction for both substrates, it would follow that *m*-xylene biodegradation by strain T in the presence of toluene occurs more rapidly because *m*-xylene has twice as many reactive sites as toluene. Future research should use biokinetic modeling to test whether competitive inhibition models can be used to describe the simultaneous biodegradation of toluene and *m*-xylene by strain T.

ACKNOWLEDGMENT

This work was made possible by the generous support of the Baltimore Gas Electric Company.

REFERENCES

Bauer, J.E., and D.G. Capone. 1988. "Effects of Co-occurring Aromatic Hydrocarbons on Degradation of Individual Polycyclic Aromatic Hydrocarbons in Marine Sediment Slurries." *Appl. Environ. Microbiol. 54*(7):1649-1655.

Bielefeldt, A.R., and H.D. Stensel. 1999. "Modeling Competitive Inhibition Effects During Biodegradation of BTEX Mixtures." *Wat. Res. 33*(3):707-719.

Chang, M-K., T.C. Voice, and C.S. Criddle. 1993. "Kinetics of Competitive Inhibition and Cometabolism in the Biodegradation of Benzene, Toluene, and *p*-Xylene by Two *Pseudomonas* Isolates." *Biotechnol. Bioeng. 41*(11):1057-1065.

Deeb, R.A., H-Y Hu, J.R. Hanson, K.M. Scow, and L. Alvarez-Cohen. 2001. "Substrate Interactions in BTEX and MTBE Mixtures by an MTBE-Degrading Isolate." *Environ. Sci. Technol. 35*(2):312-317.

Durant, N.D., L.P. Wilson, E.J. Bouwer. 1995. "Microcosm Studies of Subsurface PAH-Degrading Bacteria From a Former Manufactured Gas Plant." *J. Cont. Hydrol. 17:*213-237.

Evans, P.J., D.T. Mang, K.S. Kim, and L.Y. Young. 1991. "Anaerobic Biodegradation of Toluene by Denitrifying Strain T1." *Appl. Environ. Microbiol. 57*(4):1139-1145.

Kreiger, C.J., H.R. Beller, M. Reinhard, and A.M. Spormann. 1999. "Initial Reactions in Anaerobic Oxidation of *m*-Xylene by the Denitrifying Bacterium *Azoarcus* sp. Strain T." *J. Bacteriol. 181*(20):6403-6410.

Perry, J.J. 1979. "Microbial Cooxidations Involving Hydrocarbons." *Microbiol. Rev. 43*(1):59-72.

Rabus, R., and F. Widdel. 1996. "Utilization of Alkylbenzenes During Anaerobic Growth of Pure Cultures of Denitrifying Bacteria on Crude Oil." *Appl. Environ. Microbiol. 62*(4):1238-1241.

Zeyer, J., E. P. Kuhn, and R.P. Schwarzenbach. 1986. "Rapid Mineralization of Toluene and 1,3-Dimethylbenzene in the Absence of Molecular Oxygen." *Appl. Environ. Microbiol. 52*(4):490-496.

NEW MARKERS FOR MONITORING THE BIODEGRADATION OF PETROLEUM FUELS IN SOIL

Peter Child, Louise Vinnai (Investigative Science Inc., Burlington, Ontario, Canada), Wendy Mortimer (Bell Canada, Toronto, Ontario, Canada),
James Ford (Investigative Science Incorporated).

ABSTRACT: We have investigated the use of per-deuterated and phenyl-labelled alkanes as indicators of the biological degradation of petroleum fuels. In laboratory scale reactors containing soil, peat moss and the target compounds, both compounds were completely degraded at a similar rate to diesel fuel within 16 days at room temperature. Degradation slowed markedly in a nitrogen purged reactor and was virtually stopped by incubation at 4 °C and by the biocide Kathon®.

Corresponding with the disappearance of the starting materials from the reactors was the appearance of fatty acids possessing the labels present in the original compounds. ^{2}H-Tetradecanoic-d_{27} acid arising from ^{2}H-tetradecane-d_{30} and 12-Phenyldodecanoic acid formed from phenyldodecane were identified. The fatty acid metabolites were found to be incorporated into complex polar lipids and were not found as free acids.

The results of this study demonstrate that ^{2}H-tetradecane and phenyldodecane in soil are both biologically degraded, converted into the corresponding fatty acids and incorporated into higher lipids. The labels in both target molecules survived the metabolic process allowing the metabolites to be traced directly to the parent molecules. The findings show that these labelled compounds can be used as sensitive monitors for the biodegradation of petroleum fuels.

INTRODUCTION

It is generally recognized that bioremediation, the biological degradation of organic pollutants, is a powerful tool for the cleanup of petroleum contaminated water and soil. Without proper verification, however, losses attributed to biodegradation may actually be due to failure to account for sampling variability, analytical variability or abiotic losses such as sorption, leaching or volatilization (Madison, 1991).

One of the most reliable ways of verifying that biodegradation has occurred is to demonstrate the appearance of metabolites that can be unequivocally traced to the pollutant (Madison 1991, Shannon and Unterman 1993, Beller *et al* 1995) . For most petroleum fuels, however, this has proven to be difficult. This is in part because the main expected products; carbon dioxide, water, acetate and fatty acids (Atlas 1981) are common in the environment and are difficult to trace to the fuel.

Aggarwal and Hinchee (1991) have explored the use of stable isotopes to link the production of carbon dioxide to the biodegradation of petroleum fuel in contaminated soil. The approach was based on the premise that the ratio of the stable isotopes ^{13}C and ^{12}C in fuels is different than that of carbon dioxide present in the air today. Others (Beller *et al* 1995) have identified products of the anaerobic biodegradation of alkyl benzenes in groundwater contaminated with gasoline.

A different way to link metabolic products to specific fuel components is to spike the fuel with parent molecules possessing recognizable labels that can be traced to the daughter products. Several reports, reviewed by Atlas (1981), describe the addition of radioactively-labelled precursors, such as ^{14}C-labelled hexadecane, to spilled fuels followed by monitoring of the ^{14}C-labelled carbon dioxide produced. This approach has also been widely used to investigate the degradation of herbicides and pesticides in soil (Fuhr 1985). These experiments provide firm evidence of biological degradation of the substrate but possess obvious containment and licensing problems when used in the field.

In the present work, we have investigated the biodegradation of phenyldodecane and fully-deuterated tetradecane, analogues of n-dodecane and n-tetradecane, typical of the straight chain alkanes present in diesel fuel and home heating oil. These analogues were chosen because they are chemically stable, biodegradable and possess readily identifiable mass spectral characteristics. As well they are expected to survive the metabolic process with labels intact and are closely similar to normal components of diesel fuel. These features make the analogues ideal for extending the technology to the field.

EXPERIMENTAL SECTION

Reactor Design. Laboratory scale reactors were comprised of 0.5 litre Mason jars containing 100-200g Guelph sandy loam soil, 1 g commercially available peat moss and the target compounds. The target compounds ^{2}H-tetradecane and phenyldodecane were added to the soil dissolved in 1-2 ml of acetone giving final concentrations of 200-1300 μg/g. Where diesel fuel was employed, it was added directly to the soil and mixed well. Except in the case of nitrogen-purged controls, the Mason jar lid inserts were replaced with circles of filter paper to allow the passage of air into the jar. Nitrogen- control reactors were purged with nitrogen and sealed with a standard Mason jar lid. Kathon® controlled reactors were treated with 46 μg/g biocide. The reactors were kept hydrated and incubated at either room temperature (25 °C) or refrigerated (4 °C) depending on the regimen required. Soil samples (2 grams) removed from the reactors were flushed with nitrogen and stored at -20 °C prior to extraction and analysis.

Extraction and Analysis of Soil. Extraction of the non-polar target compounds and diesel fuel from soil was done by sonication using dichloromethane as a solvent (USEPA SW-846). Polar lipids were extracted from the soil using a Bligh & Dyer (chloroform/methanol/water) extraction (Kates 1972).

The target compounds in dichloromethane extracts of soil were determined by combined gas chromatography-mass spectrometry (GCMS). Quantification was done using ^{2}H-dodecane-d_{26} as an internal standard. Under these conditions fatty alcohols and aldehydes are also extracted and are detectable if present. Fatty acids present in chloroform/methanol/water extracts of soil were analyzed as the methyl esters by GCMS employing the same chromatographic conditions as above.

Identification of Neutral and Polar Lipids. Complex lipids present in chloroform/methanol/water extracts of soil were fractionated by thin layer chromatography using silica gel plates eluted with polar lipid and neutral lipid solvent systems as described (Kates 1972). The compounds were visualized by ultraviolet light or by charring with sulphuric acid. For analysis, the thin layer bands were scraped and the lipid material converted into fatty acid methyl esters.

RESULTS AND DISCUSSION

Evidence of Target Compound Depletion. In initial experiments, reactors were spiked with 400 µg/g each ^{2}H-tetradecane and phenyldodecane. The soil had a moisture content of about 20%. The reactors were incubated under aerobic conditions at 4 °C and 25 °C. One set of reactors was flushed with nitrogen, sealed and incubated at 25 °C. The soil was periodically sampled over a three week period, extracted and analyzed by GCMS.

The results, summarized in Figure 1, show that under aerobic conditions at room temperature, both target compounds are completely degraded within about 16 days. The initial rate of degradation of deuterated tetradecane displayed apparent first order kinetics with a half life of about 87 hours. There was a reproducible time lag of about 48 hours in the degradation of phenyldodecane during which time the rate was negligible. The period of maximum degradation for this compound occurred during 92 and 192 hours.

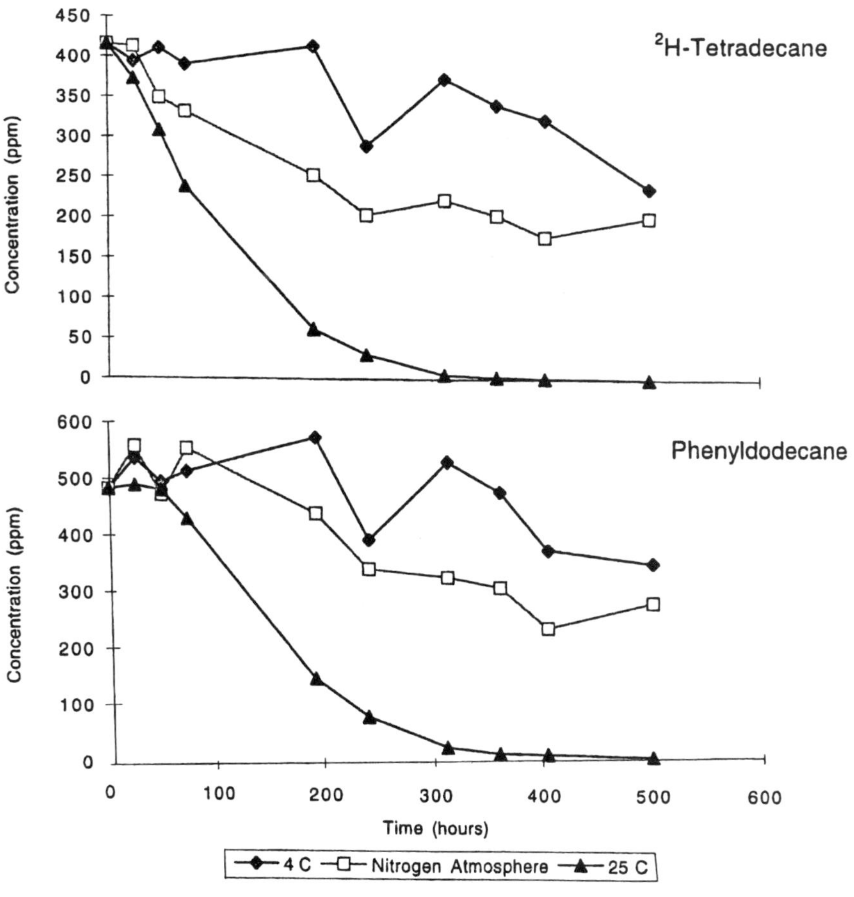

FIGURE 1: Degradation of ^{2}H-Tetradecane and Phenyldodecane in Laboratory Scale Reactors

Evidence for Biological Degradation. Evidence that the loss of target molecules from soil containing reactors was the result of biological action was collected in two ways. First, experiments were conducted to examine the effect of metabolic inhibitors on the reactor. Second, the reactors were examined for metabolic products.

As illustrated in Figure 1, the degradation of ^{2}H-tetradecane and phenyldodecane was markedly slower in soil reactors under a nitrogen atmosphere. About 50% of the material was degraded during the three week test period. No loss of either target compound was detected in the 4°C reactors before 200 hours of incubation. During the subsequent 300 hour period, the concentrations of both compounds decreased to 60-70% of their starting values.

The results of incubation of ^{2}H-tetradecane with soil in the presence of the biocide, Kathon®, is shown in Figure 2. The mode of action of Kathon® is to inhibit respiration causing bacterial, fungal, and yeast cell death (Supelco, 1997). The biocide essentially stops the reaction confirming that the observed loss of the target compounds is the result of biological activity.

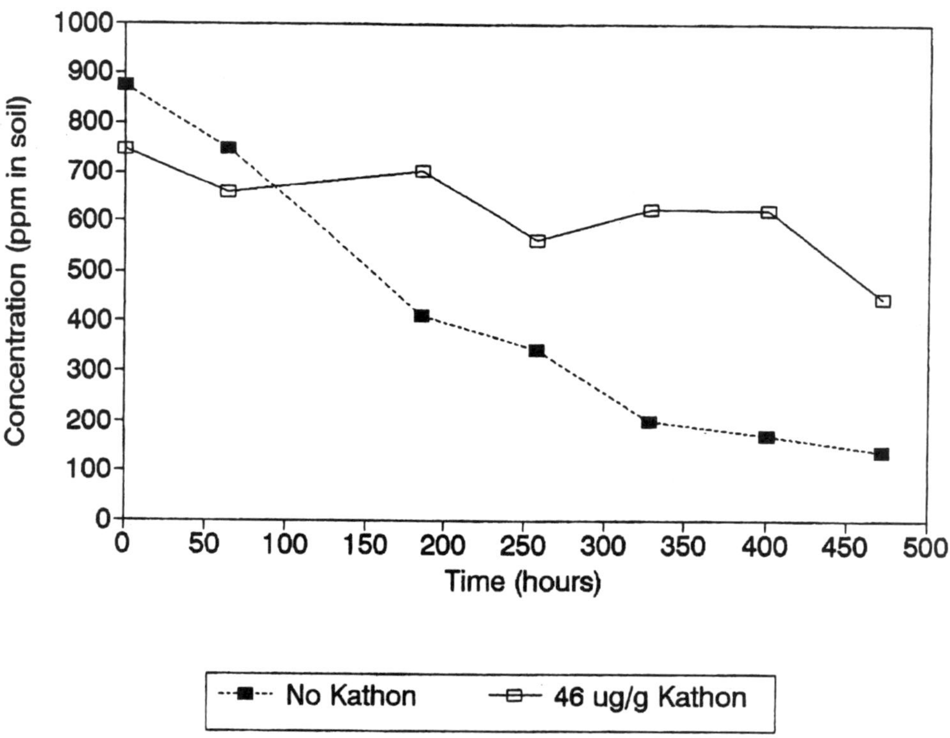

FIGURE 2. Degradation of ^{2}H-tetradecane in the presence of local garden soil and the biocide Kathon®.

Investigation of Metabolic Products. To look for metabolic products, soil samples subjected to 16 days of incubation were extracted, resolved into polar lipid fractions by thin layer chromatography and reacted to form methyl esters of any fatty acids present. The methylation reaction used converts both free fatty acids and those found in complex lipids, such as glycerides or waxes, into their methyl esters.

Analysis of the polar lipid fractions showed a range of normal fatty acid methyl esters from about 12 to 20 carbons in length in samples for both control and test reactors. These most likely represent the fatty acid component of the polar lipids arising from the microbial population.

In three of the polar lipid fractions from the reactors spiked with the target compounds and exposed to air at room temperature, compounds identified as ^{2}H-tetradecanoic acid-d_{27} methyl ester and phenyldodecanoic acid methyl ester were found. The mass spectrum and gas chromatographic retention time of the deuterated metabolite matched that of authentic ^{2}H-tetradecanoic acid-d_{27} methyl ester available commercially. The spectra of the metabolite and the authentic material are shown in Figure 3.

Authentic phenyl dodecanoic acid was not available commercially, but the mass spectrum of the phenyl-labelled metabolite possessed all of the expected elements including ions at 91 and 105 mass units characteristic of alkyl benzenes, m/z 74 typical of methyl esters, m/z 290 corresponding to the molecular weight of the compound and m/z 258 corresponding to the loss of methanol form the molecular ion. The mass spectrum was also closely similar to authentic purchased samples of the related homologues phenyldecanoic acid methyl ester and phenyl hexanoic methyl ester.

These metabolites were not detected in the control reactor which contained the same soil and peat mix without the target compounds. No fatty alcohols, fatty aldehydes or water soluble metabolites that could be traced to the starting material were detected in any of the reactors.

The nature of the polar lipids in which the metabolites were found was not firmly established, but the mobility of the material on the thin layer plates would be consistent with phosphatidyl ethanolamine or phosphatidyl glycerol. Fatty acids produced from n-alkanes by bacteria in culture have been shown to be incorporated into cellular glycerolipids including phosphatidyl ethanolamine and phosphatidyl glycerol (Makula and Finnerty 1972).

The detection of metabolic products possessing the labels present in the starting material, and their incorporation into higher glyceride structures confirms that the removal of the starting material from the reactors is the result of biological action. This is also supported by our ability to stop the reaction using a commercially available biocide.

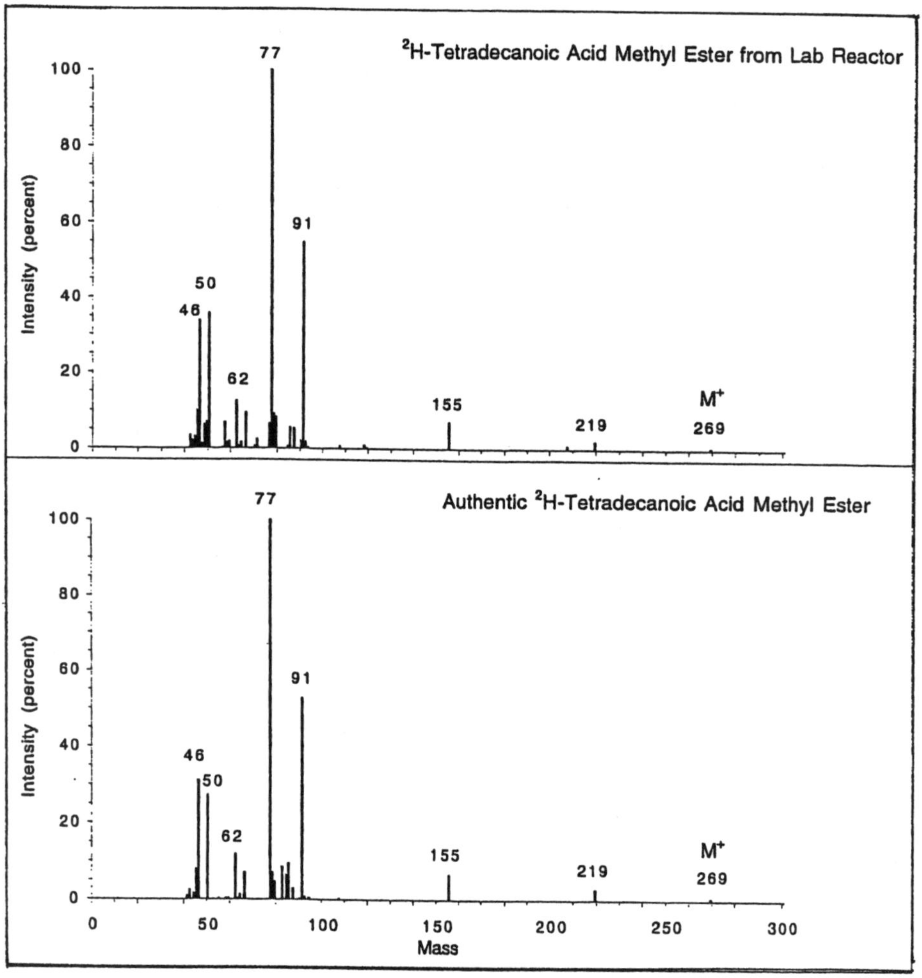

FIGURE 3. Mass spectra of ^{2}H-tetradecanoic acid-d$_{27}$ methyl ester derived from ^{2}H-tetradecane in the lab scale reactors and authentic ^{2}H-tetradecanoic acid-d$_{27}$ methyl ester.

Relationship Between Compounds and Metabolites. The results of the initial reactor experiments establish that both phenyl-labelled and fully deuterated alkane analogues are oxidized to the corresponding fatty acids and incorporated into higher lipids by micro-organisms in soil. In the following experiments, the kinetics of the appearance of these metabolites were investigated.

To track the appearance of the metabolites over time, samples of soil were taken over a three week period from the aerated reactors incubated at 4 °C and 25 °C and nitrogen-flushed reactor incubated at 25 °C. The total lipid extract of each was methylated and analyzed by GCMS. The results, summarized in Figure 4, show that under aerobic conditions at room temperature, the metabolite ^{2}H-tetradecanoic acid-d_{27}, formed from the degradation of ^{2}H-tetradecane-d_{30}, increases in concentration to a maximum of about 3 ppm after 150 hours then decreases. This trend has been observed in other studies of the rate of appearance of metabolites (Beller *et al* 1995) and may indicate further conversion of the metabolite. It is commonly accepted that fatty acids undergo beta-oxidation in bacteria, yeasts and fungi expected to be present in soil (Atlas 1981). We found no trace of chain-shortened fatty acids possessing deuterium labels in any of the samples from this study.

Much lower amounts of ^{2}H-tetradecanoic acid-d_{27} were formed in the nitrogen-flushed reactor in the first 400 hours. This is consistent with the assumption that oxygen is required for the conversion of alkanes to fatty acids by microorganisms in soil (Bartha 1986). There are reports of the biological degradation of petroleum hydrocarbons occurring under anaerobic conditions (Caldwell et al, 1999, Fuhr, 1985) but, in this case we cannot rule out that t6he metabolism that occurred under a nitrogen atmosphere may also have been due to incomplete purging or to trace amounts of oxygen leaking into the reactor.

Neither metabolite was detected in the soil from the 4 °C reactor. This is consistent with reports that metabolism of alkanes is greatly slowed at low temperatures (Bartha 1986, Pollard *et al* 1995). This result is also of interest because it implies that preservation of field samples at refrigerator temperatures may prevent biological degradation for three weeks or more, assuming that the target molecules behave in a similar manner to diesel fuel.

The overall yield of the metabolites, taken at the time point at which the maximum concentrations occurred, did not exceed 1% of the total amount of starting material on a molar basis. The total accumulation of the deuterated fatty acid reached about 2.5% of the total fatty acid content of the samples at this point. These levels of incorporation are low compared to reports of the behaviour of bacteria in culture. It has been reported (Makula and Finnerty 1968) that *Micrococcus cerificans* grown on n-tetradecane in culture produced tetradecanoic acid that accumulated in the biomass until it represented the main fatty acid. These cultures also produced the corresponding alcohol. This was not observed in our experiments. The rates of fatty acid production found in our experiments may underestimate the true rate if subsequent metabolism is occurring, however, we take the generally low yield to indicate that the bulk of the starting material is mineralized to carbon dioxide.

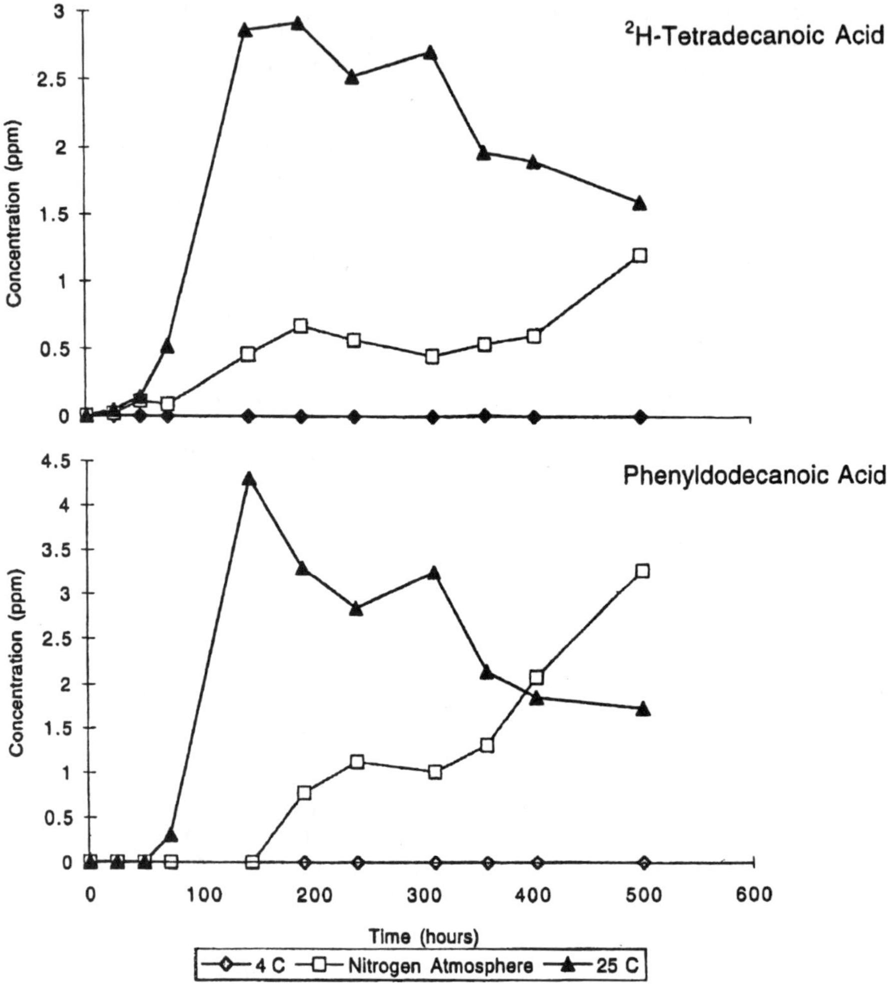

FIGURE 4: The Rates of Formation of 2H-Tetradecanoic and Phenyldodecanoic Acids at 4 °C, 25 °C, and Under Nitrogen

Degradation in the Presence of Diesel Fuel. To assess the suitability of ^{2}H-tetradecane and phenyldodecane as markers for the degradation of diesel fuel, reactors containing soil (200g) were spiked with the two target compounds at concentrations of 1300 μg/g and with diesel fuel at a concentration of 13, 000 μg/g. This level of fuel was chosen as it is typical of moderately contaminated sites and allows adequate distribution of the fuel in the soil reactor without separation of free product. The reactors (in triplicate) were incubated at 25°C and lost moisture replenished as before. Duplicate samples (2 grams each) were collected daily for 4 weeks and stored frozen until analysis.

The results, illustrated in Figure 5, show that the degradation of the diesel fuel occurred at a uniform rate for the first 2 weeks, then stopped after about three quarters of the fuel had been consumed. The time lag noted in earlier experiments during the initial 24-48 hours was also seen in these experiments. The target compounds, ^{2}H-tetradecane and phenyldodecane degraded at a uniform rate in the presence of diesel fuel in our reactors.

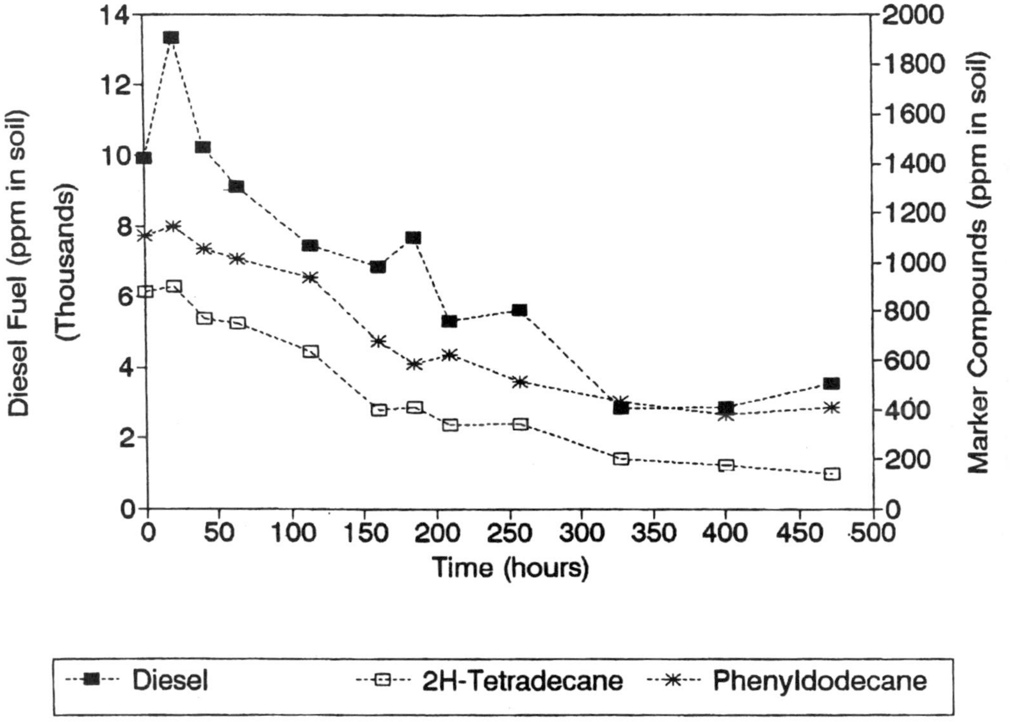

FIGURE 5: The Rates of Degradation for Diesel Fuel, 2H-Tetradecanoic and Phenyldodecanoic Acids at room Temperature

CONCLUSIONS

The results of this study demonstrate that in lab reactors containing soil, the target compounds ^{2}H-tetradecane and phenyldodecane are biologically degraded, converted into their corresponding fatty acids and incorporated into higher lipids. The labels in both target compounds survived the metabolic process and gave identifiable daughter compounds traceable directly to the parent molecules. To our knowledge this is the first demonstration of the formation of metabolic products of stable isotope and phenyl labelled alkanes during biological degradation in soils

The findings show that both phenyl and deuterium labelled alkanes can be used as sensitive tools for the confirmation of biodegradation. The rates of degradation of ^{2}H-tetradecane and phenyldodecane are similar to that of diesel fuel and their unlabelled analogues. We predict that longer or shorter chain length labelled compounds offer the opportunity for easy manipulation of the rate of degradation to suit needs in the field. Further, the metabolic products are easy to detect in soil because they appear against a blank background, unlike native metabolites. Under optimal conditions of temperature and nutrients, the metabolites can be seen within a few days. The results also indicate that the metabolites are themselves further degraded. This is an important prerequisite for the use of these labelled-compounds as markers for biodegradation at contaminated sites.

Potential application of these markers include assessment of biodegradation potential of soils in lab scale experiments or under actual field conditions and for verification of biodegradation during the bioremediation of contaminated sites. We are currently investigating these applications.

ACKNOWLEDGEMENT

The project was funded jointly by National Science and Engineering Research Council of Canada, Bell Canada and Investigative Science Incorporated.

REFERENCES:

Aggarwal, P.K. and Hinchee, R.E. 1991. "Monitoring in Situ Biodegradation of Hydrocarbons by Using Stable Carbon Isotopes." *Environ. Sci. Technol. 25*(6): 1178-1180.

Atlas, R.M. 1981. "Microbial Degradation of Petroleum Hydrocarbons: an Environmental Perspective." *Microbiological Reviews. 45*(1): 180-209.

Bartha, R. 1986. "Biotechnology of Petroleum Pollutant Biodegradation." *Microb. Ecol. 12*: 155-172.

Beller, H.R., Ding, W-H., and Reinhard, M. 1995. "Byproducts of Anaerobic Alkylbenzene Metabolism Useful as Indicators of in Situ Bioremediation." *Environ. Sci Technol. 29*(11): 2864-2870.

Caldwell, M.E., Garrett, R.M., Prince, R.C. and Suflita, J.M. 1998. "Anaerobic Biodegradation of Long-Chain n-Alkanes under Sulfate-Reducing Conditions." *Environmental Science & Technology. 32*(14): 2191-2195.

Fuhr, F. 1985. "Application of 14C-Labeled Herbicides in Lysimeter Studies1." *Weed Science. 33*(2): 11-17

Kates, M. 1972. *Techniques of Lipidology.*

Madson, E.L. 1991. "Determining in situ biodegradation: Facts and Challenges." *Environ. Sci. Technol.* 25(10): 1663-1673

Makula, R.A. and Finnerty, W.R. 1968. "Microbial Assimilation of Hydrocarbons." *Journal of Bacteriology. 95*(6): 2102-2107.

Makula, R.A. and Finnerty W.R. 1972. "Microbial Assimilation of Hydrocarbons: Cellular Distribution of Fatty Acids." *Journal of Bacteriology. 112*(1): 398-407

Pollard, S.J.T., Hrudey, S.E. and Fedorak, P.M. 1994. "Bioremediation of Petroleum- and Creosote-Contaminated Soils: A Review of Constraints." *Waste Management & Research 12*: 173-194.

Shannon, M.J.R. and Unterman, R. 1993. "Evaluating Bioremediation: Distinguishing Fact from Fiction." *Annu. Rev. Microbiol. 47*: 715-738.

Supelco Canada. 1997. "Kathon® CG/ICP Preservative for Media Biocide Applications." *Bulletin 912*: 5-6

U.S. Environmental Protection Agency. 1990. *Test Methods for Evaluating Solid Waste, Physical/Chemical Methods.* SW846, 3rd Edition, Method 3550, Rev. 1: 3550A-1-10.

INFLUENCE OF METHYLENE CHLORIDE ON BIODEGRADATION OF OTHER CHLORINATED COMPOUNDS

Benny DeHghi (Honeywell International Inc., Torrance, California)
Vanessa Riva, Terry H. Feng (Parsons Engineering Science, Inc., Oakland, California)

ABSTRACT: Methylene chloride (MC) appears to have promoted biodegradation of other chlorinated volatile organic compounds (CVOCs) in groundwater at an industrial site. The other major contaminants included tetrachloroethene (PCE), trichloroethene (TCE), trichloroethane (TCA) and their daughter products, dichloroethane (DCA) and dichloroethene (DCE). Seasonal changes in groundwater recharge cycles and intermittent operation of a dual phase extraction (DPE) system created sequential aerobic and anaerobic conditions, as measured by seasonal changes in dissolved oxygen (DO) and oxidation/reduction potential. Based on spatial correlation of MC, organic acids, and DO, MC is acting as an electron donor and enhancing the biodegradation of other CVOCs under seasonally alternating aerobic/anaerobic conditions at the site. Cometabolic processes may be the primary mechanism driving biodegradation as a result of MC acetogenesis, formate production, and the presence of methane. The presence and degradation of MC in the source zone appears to contribute to the overall stability of CVOC plume.

INTRODUCTION

Historical industrial operations at the project site resulted in migration of volatile organic compounds (CVOCs) into vadose zone soil and shallow groundwater in a 3-meter thick silty sand zone encountered between 2.5 to 3 meters below ground surface (bgs). The static water table fluctuates between 1 to 2 meters bgs, with high water levels in winter/spring seasons due to recharge from winter precipitation and low water levels in summer/fall seasons as a result of dry weather and reduced recharge. The hydrological and weather cycle has resulted in seasonally alternating anaerobic/aerobic conditions in the shallow aquifer.

Methylene chloride (MC) was detected at concentrations of up to 26,000 micrograms per liter (µg/L) and constituted more than 20 percent of the total CVOCs in some source zone groundwater samples, but has not been detected in wells located just a short distance downgradient of the source zone (Figure 1). Data from this study show that MC is readily degradable under both aerobic and anaerobic conditions and acts as a carbon source to drive the degradation of other CVOCs at the site. Based on a better understanding of the role played by MC in the biodegradation of the other CVOCs and demonstrated plume stability, monitored natural attenuation (MNA) has been implemented as the interim remedial action for groundwater and previously planned active remediation of the groundwater CVOC plume is being removed from consideration.

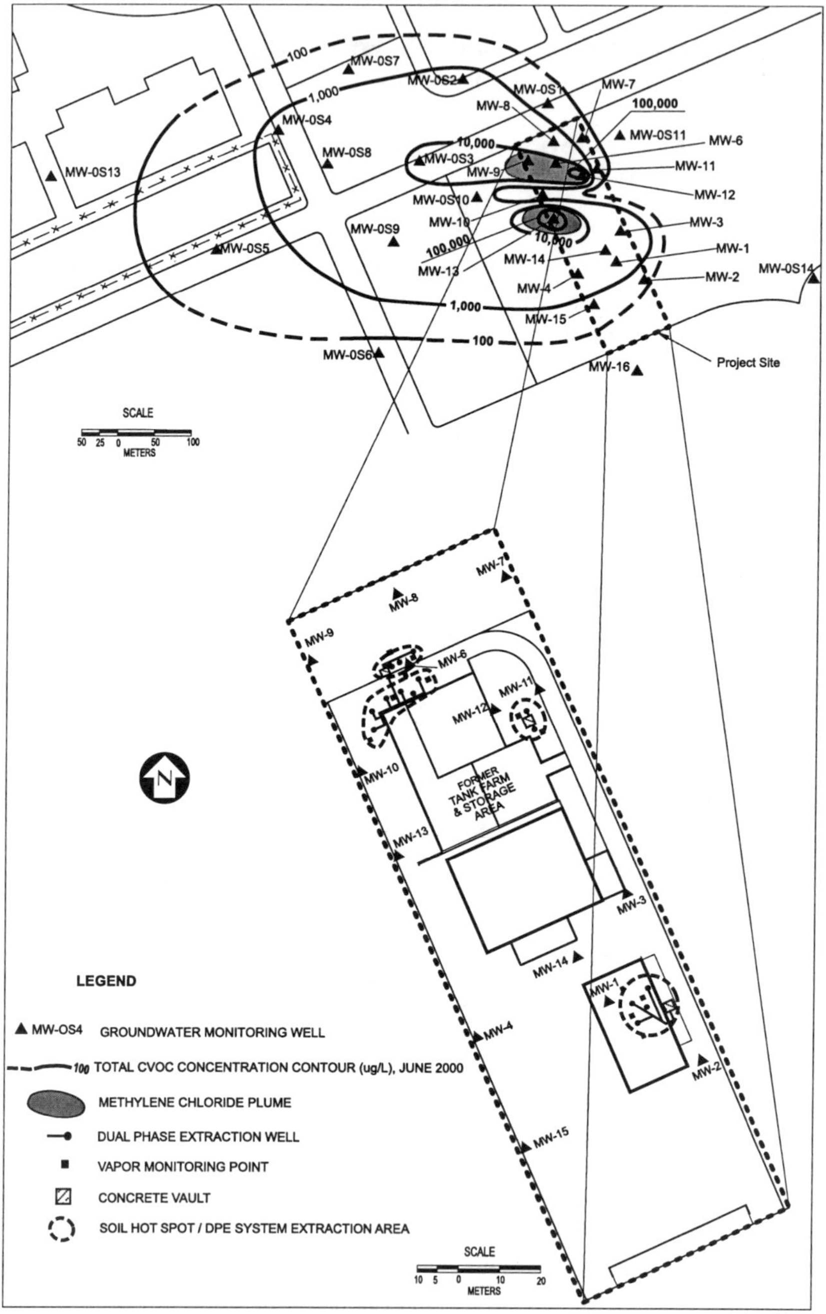

FIGURE 1. Site layout and CVOC and MC plumes in groundwater.

MATERIALS AND METHODS

This study included literature research, sampling and analysis of site soil and groundwater, and focused data evaluation. Site-specific data collected and used in this study consisted of:

- Water level and CVOC concentration data from routine quarterly groundwater monitoring conducted since 1994;
- MNA indicator parameters from three complete rounds of sampling and analysis conducted between 1998 and 2000; and
- Organic acid concentrations from a one-time sampling conducted in source zone wells where MC was detected at elevated concentrations.

A series of hydrographs were plotted to study seasonal water level fluctuations. Concentration-time plots of geochemical parameters and CVOCs were prepared to observe the patterns of temporal variations. Snap shot isoconcentration contour maps of MC and other CVOCs [including tetrachloroethene (PCE), trichloroethene (TCE), trichloroethane (TCA), dichloroethane (DCA), and dichloroethene (DCE)], and selected geochemical indicator parameters [including dissolved oxygen (DO), oxidation/reduction potential, carbon dioxide (CO_2), and methane] were prepared to examine spatial distribution patterns. Concentration ratios of parent/daughter pairs including TCE/PCE, 1,1-DCE/1,1,1-TCA, and cis-1,2-DCE (c-DCE)/TCE were calculated and plotted to evaluate their spatial variation patterns. A series of concentration-time plots were also prepared for MC, CVOCs, organic acids, and other geochemical parameters from different sampling events to compare changes among parameters and establish potential biodegradation mechanisms occurring at the site.

RESULTS AND DISCUSSION

MC is Readily Degradable under Both Aerobic and Anaerobic Conditions. MC is one of the most widely used solvents in industry and also one of the most frequently detected contaminants at hazardous waste sites. However, it is usually only present in the immediate source area coexistent with other CVOCs, such as TCA, PCE, TCE, DCA, and DCE, while the other CVOCs usually migrate much farther downgradient of the source zone. The primary factor that prohibits the widespread presence of MC in aquifers is due to its readily degradable characteristics under both aerobic and anaerobic environmental conditions (Leisinger and Braus-Stromeyer, 1995). MC degradation products generated through aerobic and anaerobic metabolisms are quickly utilized as electron donors in other microbially mediated reactions. MC degradation by aerobic methylotrophic bacteria results in the formation of formaldehyde (Leisinger and Braus-Stromeyer, 1995), which can be further converted to formate (Ma, 2000).

$$\text{MC} \rightarrow \text{formaldehyde} \rightarrow \text{formate} \qquad \text{(aerobic)} \qquad (1)$$

Under reducing conditions, MC is metabolized to acetate and formate (organic acids) (Magli et al., 1997). Acetate can be utilized by anaerobic organisms as a substrate in the degradation of highly chlorinated CVOCs such as PCE and TCE (Loffler et al., 2000). Acetate can also be utilized by methanogens to produce methane under weak reducing conditions.

$$\text{MC}\begin{array}{l}\nearrow \text{ formate}\\ \\ \searrow \text{ acetate} \rightarrow \text{methane}\end{array}\qquad \text{(anaerobic)} \tag{2}$$

Formate and methane have been shown to be preferred substrates for the cometabolic degradation of CVOCs by bacterial cultures that produce the enzyme methane monooxygenase (MMO) (Anderson, and McCarty, 1996; Oldenhuis et al., 1989). In both aerobic and anaerobic environments, MC not only quickly degrades, but also serves as an effective carbon source to enhance the degradation of other coexistent CVOCs. These findings are supported by the data collected at the project site.

MC Drives Degradation of Other CVOCs in the Source Zone. Increases in concentration ratios of c-DCE/TCE and 1,1-DCE/1,1,1-TCA pairs as a function of time and distance from the source are indicative of anaerobic degradation of highly chlorinated CVOCs. In the source zone, significant increases in these concentration ratios occur only in the two wells (MW-12 and MW-13) in which the highest concentrations of MC (26,000 μg/L and 16,000 μg/L) were also detected. MW-12 is located at one of the soil hot spot areas (Figure 1). Comparing the data from MW-12 between June 1999 and June 2000 (Figure 2A), concentrations of all parent compounds (PCE, TCE, and 1,1,1-TCA) significantly decreased while all detected intermediate daughter products (c-DCE, 1,1-DCE, and 11-DCA) increased, or remained at similar concentrations, indicating occurrence of reductive dechlorination. Although vinyl chloride (VC) and 1,2-DCA were not detected, the sharp increase in CO_2 concentration during the same period suggest that further cometabolic degradation and/or abiotic mineralization of CVOC daughter products may have occurred under the alternating aerobic and anaerobic conditions. The close association of the sharp drop in MC concentration (from 26,000 μg/L to 11,000 μg/L), moderate drop in total CVOC concentration (from 199,500 μg/L to 150,700 μg/L), and sharp increase in CO_2 concentration (from 50,000 μg/L to 85,000 μg/L) during the one year period, suggest that MC may have played a critical role in the degradation of other coexistent CVOCs.

$$\begin{array}{ccccccccl} & \nearrow Cl^- & & \nearrow Cl^- & & \nearrow Cl^- & & & \\ \text{PCE} & \rightarrow & \text{TCE} & \rightarrow & \text{DCE} & \rightarrow & \text{VC} & & \text{(anaerobic)}\\ & & \searrow & & \searrow & & \searrow & & \\ & & CO_2+3HCl & & CO_2+2HCl & & CO_2+HCl & & \text{(aerobic)}\end{array} \tag{3}$$

$$\text{1,1,1-TCA}\begin{array}{ll}\nearrow \text{ acetate} & \text{(abiotic degradation)}\\ \rightarrow \text{ 1,1-DCA} & \text{(biodegradation)}\\ \searrow \text{ 1,1-DCE} & \text{(abiotic degradation)}\end{array} \tag{4}$$

Similar patterns are also observed at MW-13 (Figure 2B). The only difference between MW-12 and MW-13 is that all analyte concentrations increased in MW-13 between June 1999 and June 2000. MW-13 is located just downgradient of a new source from which elevated concentrations of CVOCs are still actively migrating. The critical similarity between MW-12 and MW-13 is the relative increase among related constituent ratios. While the concentration of total CVOCs increased more than 50 percent, all parent CVOCs, including PCE, TCE, 1,1,1-TCA, and MC, were either unchanged or increased by a negligible amount. Concurrently, concentrations of two daughter products, c-DCE and 1,1-DCE, increased 2 and 16 times, respectively. The data indicate occurrence of anaerobic reductive degradation of PCE and TCE, and strong abiotic degradation of 1,1,1-TCA. In addition, a sharp increase in CO_2 concentration during the same period suggests either complete cometabolic mineralization or direct oxidation of CVOC daughter products under the alternating aerobic/anaerobic conditions.

Indicator Parameters Support Occurrence of Alternating Dechlorination and Cometabolization Processes. Selected indicator parameters including DO, CO_2, and methane along with acetate, formate, and MC are plotted in concentration-time diagrams for representative MC source area wells (MW-12 and MW-13) (Figures 3A and 3B). Low DO concentrations, indicative of reducing conditions, were observed during the dryer months of the year (December 1998) with corresponding higher methane and lower CO_2 concentrations. Acetate and formate (organic acids) are expected to be produced during this period due to metabolization of MC under reducing conditions and abiotic degradation of 1,1,1-TCA. Although organic acids were not analyzed in December 1998, the high methane concentration indicates that acetate was degraded via methanogenesis. The low CO_2 concentration, on the other hand, indicates that complete mineralization of CVOCs and their daughter products was limited during this season. Sequential dechlorination process may not complete under the presence of other competing substrates such as sulfate and toxic conditions that may develop during cometabolism due to the creation of certain intermediate degradation products (van Hylckama Vlieg et al., 1997).

During the predominantly aerobic seasons (June 1999 and June 2000), as represented by substantially higher DO concentrations, the corresponding concentrations of methane decreased and CO_2 increased, reflecting that limited CO_2 methanogenesis was occurring, but metabolization of substrates resulted in the complete mineralization of CVOCs through cometabolic degradation or direct oxidation. The data strongly suggests that the alternating anaerobic and aerobic conditions between summer/fall and winter/spring seasons enabled degradation of highly chlorinated parent compounds, such as PCE, TCE, and 1,1,1-TCA, to intermediate daughter products during the predominantly reducing seasons, and further cometabolic degradation or direct oxidation of CVOC daughter products during the predominantly aerobic seasons. Since MC is degradable under both aerobic and anaerobic conditions, MC is able to act as a

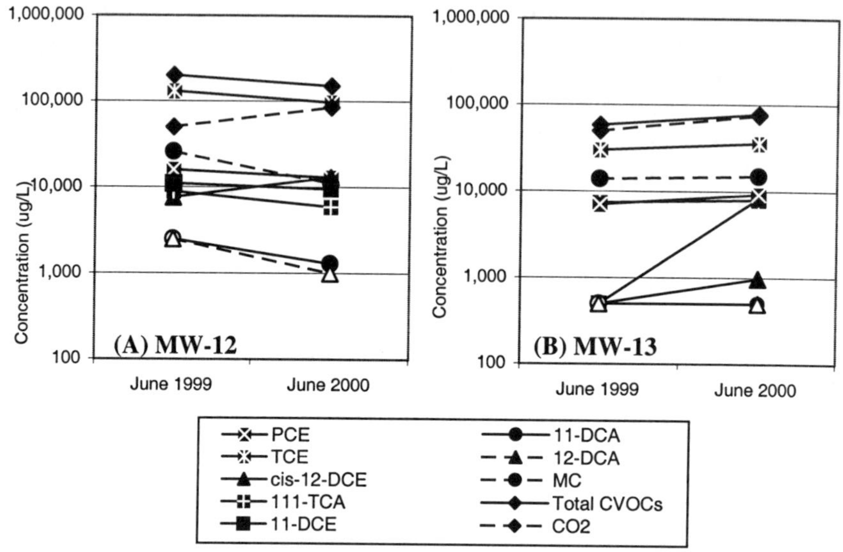

FIGURE 2. CVOC and CO2 concentrations between June 1999 and June 2000. Data suggest strong direct anaerobic degradation of 111-TCA, PCE, and TCE (decrease) to DCE and DCA (increase), and cometabolic degradation of CVOCs daughter products to CO2 (open symbols indicate concentration below detection limit).

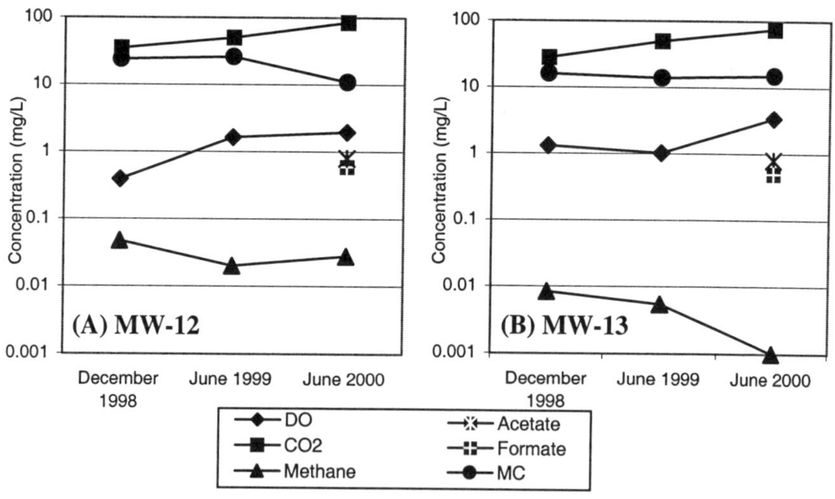

FIGURE 3. Degradation inidicators and organic acids parameters. December 1998 data are representative of weak anaerobic conditions, June 1999 and June 2000 data are representative of aerobic conditions (open symbols indicate concentration below detection limit).

carbon source, or electron donor, regardless of the seasonal conditions thereby functioning as an effective driver for the degradation of other coexistent CVOCs.

Organic Acid Data Support Degradation of MC and Production of Organic Acids. To further assess the potential metabolic pathways at the site, selected organic acids data were collected during a one time sampling event. At MW-12 (Figure 3A), one of the locations with the highest levels of MC, formate was detected at 0.6 milligrams per liter (mg/L), while acetate was not detected above the detection limit (0.8 mg/L). Formate can be produced by the degradation of MC under both aerobic and anaerobic conditions. When MC is anaerobically degraded, acetate is also produced. This is consistent with the non-detection of acetate in June 2000 when aquifer conditions were predominantly aerobic. It has been shown that acetate can in turn be reduced to methane under slightly reducing conditions or used as a substrate in the dechlorination of PCE and TCE. Formate can also accumulate from methane degradation (Ma, 2000) and cometabolic/abiotic degradation of other CVOCs such as TCE. The detection of formate but not acetate at MW-12 during the predominantly aerobic June 2000 sampling event is consistent with the sequential aerobic/anaerobic degradation processes. Formate can act as a carbon source in the cometabolic degradation of CVOCs mediated by methanotrophs expressing both soluble and particulate MMO. Organic acid data for MW-13 are very similar to those for MW-12.

Organic acid data strongly suggest the occurrence of MC degradation. The organic acids produced act as a carbon source for the degradation of other CVOCs. Although MC is only detected at elevated concentrations in the source zone, the organic acids produced during MC degradation would enhance the degradation of the CVOC plume downgradient from the source.

Biodegradation Contributes to Overall Plume Stability. Evaluation of historical CVOC data on concentration-time plots indicates that the CVOC plume has been stable since 1995 as evidenced by negligible spatial expansion and stable or declining CVOC concentrations in downgradient wells. Natural attenuation processes including dilution, adsorption, volatilization, and biodegradation are considered the primary factor for the observed plume stability. The occurrence of biodegradation is evident from observed steady increases in concentration ratios of TCE/PCE and 1,1-DCE/1,1,1-TCA pairs as a function of distance downgradient from the source area. Between the source area well MW-6 and downgradient well MW-OS4 (Figure 1), the concentration ratios increased approximately 30 and 7 times, respectively, for the TCE/PCE and 1,1-DCE/1,1,1-TCA pairs. While these ratio increases indicate occurrence of sequential anaerobic dechlorination of highly chlorinated CVOCs, the absence of common daughter products including c-DCE and VC in the plume area suggests that the sequential anaerobic dechlorination process is either incomplete or the less chlorinated daughter products are destroyed by other mechanisms.

Since the total CVOC plume has been stable for a long time already under the potential presence of residual dense non-aqueous phase liquids (DNAPL) in the source zone, CVOC mass must have been destroyed. Because the site undergoes sequential alternations between aerobic and anaerobic conditions, it is likely that the less chlorinated daughter products, such as VC and c-DCE, may have been destroyed throughout the plume via direct oxidation or cometabolic aerobic degradation. Cometabolic aerobic degradation of VC and c-DCE have been shown to be at least an order of magnitude greater than for TCE and 1,1-DCE when mediated by methanotrophs expressing particulate MMO (Anderson, and McCarty, 1996). It is also possible that TCE, the primary CVOC in the plume, may have been destroyed through cometabolic degradation to TCE-epoxide, which in turn, through abiotic processes, would lead to the production of formate, carbon monoxide (CO), glyoxylate, and chlorinated acids (Whittaker et al., 2000; van Hylckama Vlieg *et al.*, 1997).

$$\text{TCE} \rightarrow \text{TCE-epoxide} \rightarrow \text{formate} + \text{glyoxylate} + \text{CO} + \text{chlorinated acids} \quad (5)$$

As discussed in preceding sections, MC degradation in the source area reduces CVOC mass loading to the plume and produces alternative carbon sources (i.e. organic acids) which contribute directly to the degradation of and stability of the CVOC plume.

CONCLUSION

Analysis performed using site-specific MC and other CVOC data, organic acid data, and biodegradation indicator parameters have shown that MC readily degrades under both aerobic and anaerobic conditions resulting in the production of electron donors, such as formate, acetate, and methane, which drive the degradation of other coexistent CVOCs such as PCE, TCE, and TCA. Under alternating aerobic/anaerobic conditions, highly chlorinated CVOCs are rapidly degraded during anaerobic periods while low to moderately chlorinated CVOCs are destroyed by direct oxidation and/or cometabolic degradation.

MC has only been detected at and just downgradient of the source area, while the more extensive downgradient plume defined by the other CVOCs has been shown to be stable over a five-year period. VC has not been found at the site possibly due to incomplete dechlorination, fast cometabolic degradation in the presence of formate and methane, or direct oxidation. Based on a better understanding of the role played by MC in the biodegradation of the other CVOCs and demonstrated plume stability, active remediation of the groundwater CVOC plume is being removed as a planned remedial action at the site.

REFERENCES

Anderson, J. E., and P. L. McCarty. 1996. "Transformation Yields of Chlorinated Ethenes by a Methanotrophic Mixed Culture Expressing Particulate Methane Monooxygenase." *Applied and Environmental Microbiology. 63(2): 687-693.*

Leisinger, T., and S. A. Braus-Stromeyer. 1995. "Bacterial Growth with Chlorinated Methanes." *Environmental Health Perspectives. 103(Suppl) 5.*

Loffler, F. E., Q. Sun, J. Li, and J. M. Tiedje. 2000. "16S rRNA Gene-Based Detection of Tetrachloroethene-Dechlorinating *Desulfuromonas* and *Dehalococcoides* Species."

Applied and Environmental Microbiology. 66(4): 1369-1374.

Ma, J. 2000. "C1 Metabolic Cycle Pathway Map." *http://www.labmed.umn.edu/umbbd/C1cyc/C1cyc_map.html.* University of Minnesota, Twin Cities, MN.

Magli, A., M. Messmer, and T. Leisinger. 1997. "Metabolism of Dichloromethane by the Strict Anaerobe Dehalobacterium formicoaceticum." *Applied and Environmental Microbiology. 64(2): 646-650.*

Oldenhuis, R., R. L. Vink, D. B. Janssen and B. Witholt. 1989. "Degradation of Chlorinated Aliphatic Hydrocarbons by Methylosinus trichosporium OB3b Expressing Soluble Methane Monooxygenase." *Applied and Environmental Microbiology. 55(11): 2819-2826.*

van Hylckama Vlieg, J. E. T., W. de Koning and D. B. Janssen. 1997. "Effect of Chlorinated Ethene Conversion on Viability and Activity of Methylosinus trichosporium OB3b." *Applied and Environmental Microbiology. 63(12): 4961-4964.*

Whittaker, M., D. Monroe, and D. J. Oh. 2000. "Trichloroethylene Pathway Map." *http://www.labmed.umn.edu/umbbd/tce/tce_map.html.* University of Minnesota, Twin Cities, MN.

NATURAL ATTENUATION OF A TRICHLOROTRIFLUOROETHANE PLUME IN GROUNDWATER SUPPORTED BY CO-MINGLED ACETONE

Timothy R. Marshall (URS Corporation, Santa Ana, California)

ABSTRACT: Subsurface investigations conducted at an integrated circuits research and manufacturing facility in Southern California detected a variety of chlorinated solvents and other compounds in soil and groundwater. Releases of process wastewater in a former metals coating application area and releases from a neutralization clarifier resulted in impacted soil and groundwater. Additional releases had occurred from an exterior drum storage area. The predominant dissolved contaminants were identified to be 1,1,2-trichlorotrifluoroethane (Freon-113) and acetone. Other contaminants identified in soil and groundwater at the facility included 1,2-dichlorobenzene, trichloroethene, toluene, and xylenes. Sampling data indicated that the dissolved plume of Freon-113 was attenuating downgradient from the suspected identified sources. In addition, acetone was not detected in significant concentrations in groundwater as compared to the concentrations remaining in soil, indicating that potential degradation/sorption was occurring. A conceptual model of natural attenuation at the facility was developed based on the observed data that included: shallow groundwater at the facility was impacted by volatile organic compounds; attenuation of Freon-113 in shallow groundwater; probable biodegradation of Freon-113 driven/supported by an exogenous carbon source, acetone; deeper groundwater to approximately 75 feet below ground surface was not impacted and was most likely protected by discontinuous clay/silt layers; source removal was conducted to the extent possible; and results of a bench-scale study indicated that indigenous microbial populations existed in soil and groundwater at the facility. The information presented herein was conducted as part of a feasibility analysis for the dissolved constituents in groundwater at the facility.

INTRODUCTION

1,1,2-Trichlorotrifluoroethane (Freon-113), a chlorofluorocarbon (CFC), is a highly volatile, nearly odorless liquid, and fairly recalcitrant compound that can be released to the environment in many ways. Its industrial uses as a solvent include degreasing electrical equipment, photographic films, magnetic tapes, precision instruments, plastics, glass, elastomers, or metal components. The United States annual production capacity for Freon-113 is approximately 300 million pounds. Of that amount, an estimated 24.6 million pounds may be released into the environment. An estimated 2,000 pounds and 9,000 pounds were released to surface and ground waters combined and to land, respectively (www.epa.gov/opptintr/chemfact). If released to soil, Freon-113 would rapidly volatilize from soil surfaces due to its high vapor pressure, or migrate through the soil into groundwater.

CFCs have not been demonstrated to degrade aerobically to an appreciable extent in the environment (Lovley and Woodward, 1992). However, CFCs have been shown to degrade under anaerobic conditions (Denovan and Strand, 1992). Biochemical oxygen tests indicate that Freon-113 degrades by an average of about 2% (www.citi.or.jp). Table 1 lists some chemical and physical properties of Freon-113.

TABLE 1. Chemical/physical properties of Freon-113.

Characteristic/Property	Data
Molecular Formula	$C_2Cl_3F_3$
Chemical Structure	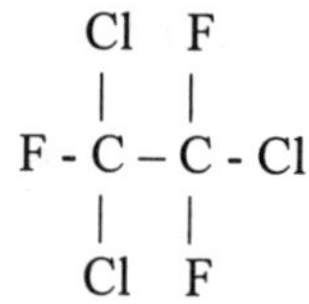
Physical State	colorless liquid
Molecular Weight	187.38
Melting Point	-35°C
Boiling Point	48C
Water Solubility	170mg/L
Density	1.5635 @ 25°C
Vapor Density (air = 1)	6.5
Vapor Pressure	284 mm Hg at 20°C
Flash Point	nonflammable
Henry's Law Constant	5.3×10^{-1} atm m^3/mol
Odor Threshold	135 ppm (in air)

Acetone is primarily used as a chemical intermediate and solvent. Acetone is readily biodegradable in the presence of oxygen. Significant volatilization of acetone also will occur. Table 2 lists some physical and chemical properties of acetone.

TABLE 2. Chemical/physical properties of acetone.

Characteristic/Property	Data
Molecular Formula	C_3H_6O
Chemical Structure	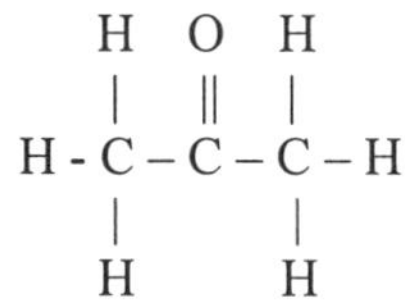
Physical State	colorless liquid
Molecular Weight	58
Melting Point	-95.35°C
Boiling Point	56.2°C
Water Solubility	Miscible
Density	0.7899 @ 20°C
Vapor Density (air = 1)	2
Vapor Pressure	181
Flash Point	-20C
Odor Threshold	33 ppm (in air)

Remedial actions conducted at the facility included excavation and on-site treatment of acetone-impacted soil at the location of the former metals room. Other actions completed at the facility included biotreatment of excavated soil impacted by 1,2-dichlorobenzene and other organic compounds from the former drum storage facility. A cone penetrometer test survey was conducted to evaluate the subsurface lithology.

Remaining acetone concentrations in soil above the shallow groundwater table have the potential to result in continuous groundwater impacts. It appears that the acetone, upon dissolving in the shallow groundwater, is readily biodegraded. The Freon-113 plume attenuates by an order of magnitude before reaching the downgradient property boundary, approximately 150 to 200 feet southwest of the probable release areas. The attenuation mechanism appears to be aerobic decomposition of the co-mingled acetone which provides a readily usable source of carbon and energy. Biodegradation of the dissolved acetone reduces the supply of oxygen, allowing partial dehalogenation of Freon-113 to occur.

FIELD INVESTIGATIONS

Figure 1 shows the distribution of Freon-113 and Figure 2 shows the distribution of acetone in shallow groundwater at the facility. Table 3 is a summary of selected dissolved concentrations of Freon–113 and acetone. Other compounds, including 1,2-dichlorobenzene, were also detected in minor concentrations in groundwater. Dissolved oxygen concentrations in upgradient groundwater were on the order of 1 to 1.5 mg/L. Dissolved oxygen concentrations within the groundwater plume were on the order of 0.2 mg/L, indicating microbial consumption of oxygen in biodegradation of dissolved acetone.

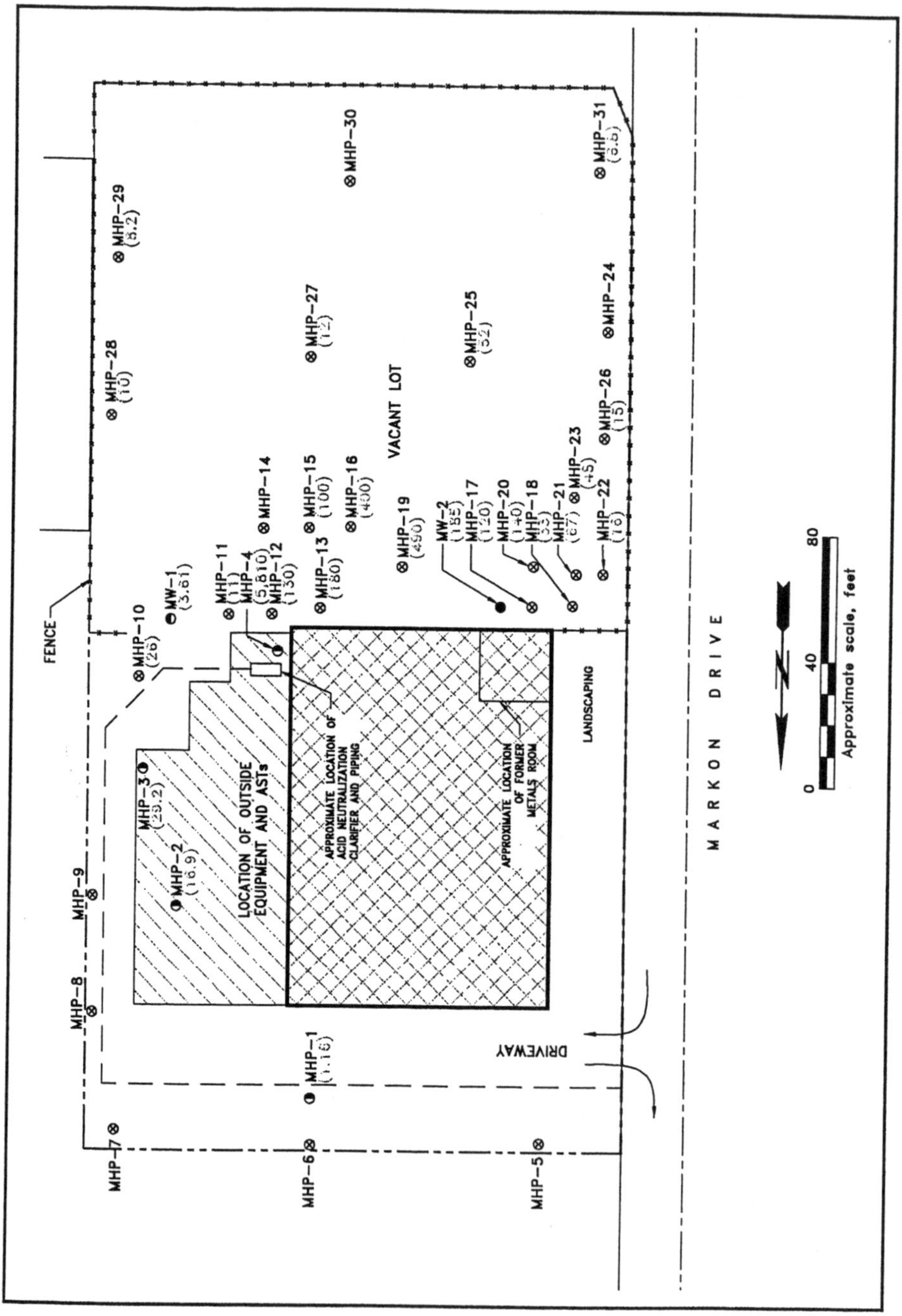

FIGURE 1. Freon 113 concentrations ($\mu g/L$) detected in groundwater samples.

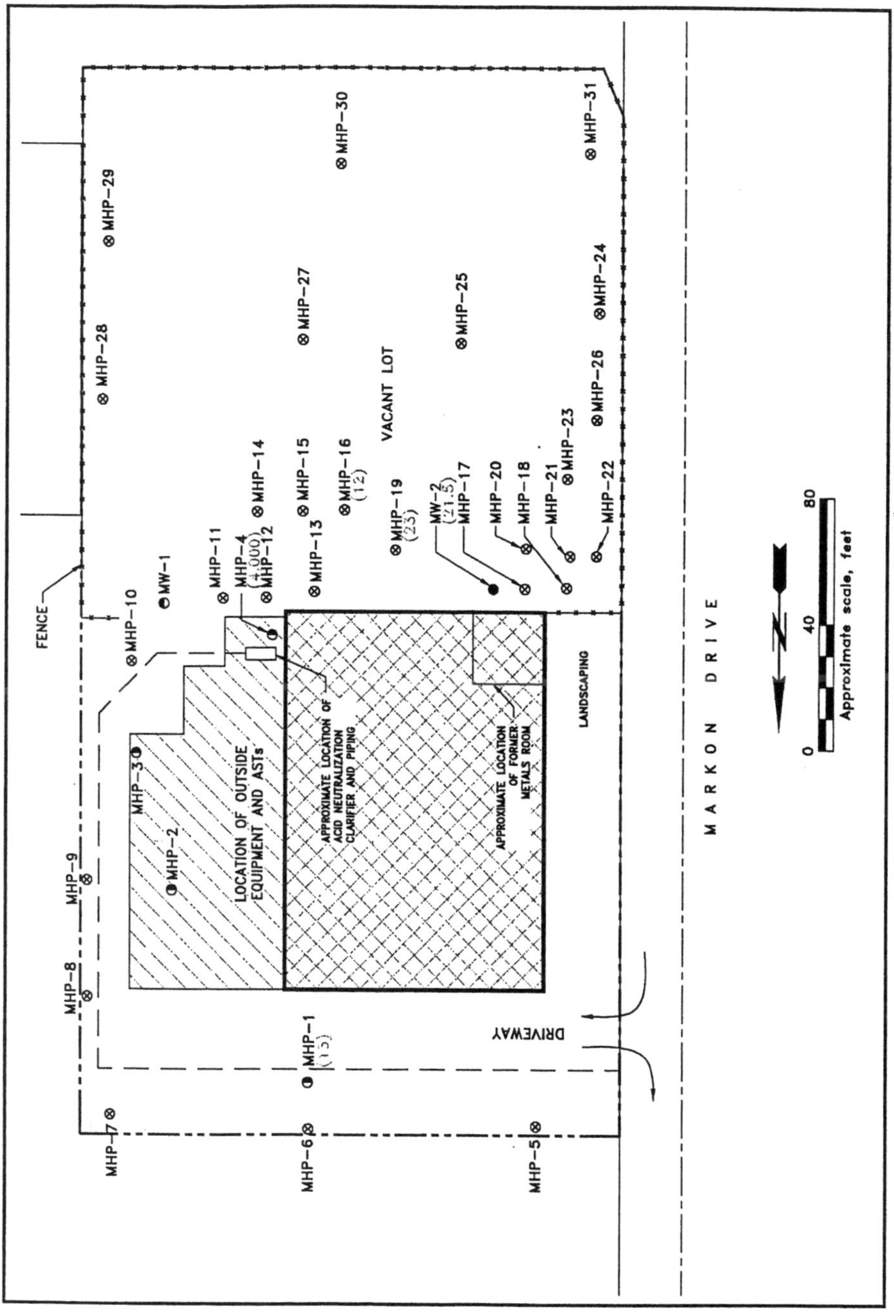

FIGURE 2. Acetone concentrations (μg/L) detected in groundwater samples.

TABLE 3. Selected dissolved concentrations of Freon-113 and acetone listed in order of increasing distance from source area.

Sample Point	Freon-113 Concentrations (μg/L)	Acetone Concentrations (μg/L)
MHP-4	5,810	4,000
MHP-19	490	23
MHP16	400	12
MHP-25	52	<5
MHP-31	6.8	<5

Seven Cone Penetrometer Test (CPT) borings were completed to evaluate subsurface lithology to support the conceptual model of natural attenuation and to evaluate the presence of deeper aquifers. The CPT borings were advanced to between 72 feet (21.9 m) and 75 feet (22.8 m) below ground surface (ft-bgs). Pore pressure dissipation tests were conducted at various depths to evaluate probable water-bearing zones.

The results of the CPT borings indicate that there are generally five continuous stratigraphic units across the site: Unit I (shallowest) through Unit V (deepest). Unit I is encountered between 0 to 20 feet (6.1 m) bgs across the site, and generally consists of an interbedded unit of silty sand to sandy silt, with occasional layers of silt. Unit I also contains an approximate 5-foot (1.5 m) thick silty sand/sand layer between 8 feet (2.4 m) to 13 feet (3.9 m) bgs at CPT-1 and CPT-2, located in the northern parking lot.

Unit II is encountered at the site between approximately 20 feet (6.1 m) to 40 feet (12.2 m) bgs, and consists primarily of clay, with lesser amounts of silty clay, clayey silt, and silt. Additionally, an approximate 4-foot (1.2 m) thick unit of sand, silty sand, and silt interbeds is present between approximately 28 feet (8.5 m) to 32 feet (9.7 m) bgs. Unit III is an approximately 10-foot (3.1 m) thick sequence of primarily silt and sandy silt, with lesser amounts of sand, clay, silty clay, and is generally present between 40 feet (12.2 m) and 50 feet (15.2 m) bgs across the site.

A bench-scale treatability study was conducted to evaluate the feasibility of bioremediation in the removal of 1,2-dichlorobenzene (1,2-DCB) in soil by land treatment. The impacted soil resulted from historical practices at the former drum storage area. This activity was conducted to support remedial decisions for soil primarily impacted by 1,2-DCB and secondarily ethylbenzene and toluene. A mineralization study was conducted to assess the optimum concentrations of nutrients required for land treatment and a microcosm study was conducted to monitor the decline of 1,2-DCB over a 42-day period. Some of this data was used in developing the conceptual model of attenuation of the dissolved Freon-113/ acetone plume.

The physical and chemical properties of the soil were characterized in terms of pH, inorganic nitrogen (NH_4-N and NO_3-N), inorganic phosphorus (orthophosphate), moisture content, soluble salts and particle size analysis. Methodologies can be found in the bench-scale report. The following is a summary of the analyses; pH – 8.28, NH_4-N – 4.6 mg/kg, NO_3-N – 5.2 mg/kg, orthophosphate – 20 mg/kg, moisture content – 3% dry weight, soluble salts – 0.95 dS/m, and particle size analysis – 80% sand, 12.5% silt, 7.5% clay. The data suggests that nitrogen species are relatively low in soil at the property, and may limit natural attenuation. Other parameters do not appear to be limiting for natural attenuation, although soil moisture content is low.

The total heterotrophic microbial count on nutrient agar was found to be 31 x 10^7 colony-forming units (cfu) per gram dry soil. Attempts were made to enumerate the soil microorganisms active in the degradation of 1,2-DCB as a sole carbon and energy source by using the most probable number technique, but the attempts were unsuccessful. Thus, cometabolism may be involved in the breakdown of chlorinated compounds in impacted soils at the facility.

CONCEPTUAL MODEL

The conceptual model of natural attenuation of the Freon-113/acetone plume includes the following:

- Shallow aquifer impacted by releases of solvents, deeper aquifers not impacted.

- Facility upgrades have mitigated additional releases.

- Resident microbial populations exist in adequate densities, but may not be able to grow on chlorinated compounds as sole carbon source. This suggests that the populations may be adapted for cometabolic functions. The fact that the degradation of Freon-113 is incomplete suggests that either cometabolism may be occurring or that the environment cannot sustain complete dehalogenation.

- The presence of co-released acetone provides an available carbon source to drive the system energetics. It is possible that the slow release of acetone from shallow soil at the facility to groundwater creates a low yet steady supply of hydrogen from metabolic processes. This low-concentration yet steady supply of hydrogen may allow dehalogenating microorganisms to outcompete methanogens or sulfate-reducers.

- The predominant mechanism of Freon-113 degradation is not yet known; cometabolic degradation of Freon-113 may occur during acetone utilization or anaerobic dehalogenation may proceed to some extent when oxygen levels are reduced from aerobic degradation of acetone. It is likely that the mechanism is acetone oxidation coupled with Freon–113 dehalogenation.

- Additional electron acceptor/donor and mass balance studies are necessary at the facility to elucidate the attenuation mechanisms. It is necessary to determine whether dehalogenation is occurring or if the degradation mechanism is cometabolic. Attenuation mechanisms may need to be enhanced to accelerate natural processes to prevent further off-site migration and to promote a reasonable Freon-113 attenuation half-life.

REFERENCES

Denovan, D.A. and S.E. Strand. 1992. Biodegradation of Chlorofluorocarbons in Anaerobic Environments. *Chemosphere. 24:* 935-940.

Lovley, D.R. and J.C. Woodward. 1992. Consumption of Freon CFC-11 and CFC-12 by Anaerobic Sediments and Soils. *Environmental Science and Technology. 26:* 925-929.

BIOREMEDIATION OF DIESEL OIL CONTAMINATED SUBANTARCTIC SOILS (CROZET ISLAND)

Daniel Delille (CNRS, Banyuls-sur-mer, France)
Emilien Pelletier (ISMER, Rimouski, Canada)
Charles Gerday (Université de Liège, Sart Tilman, Belgium)

ABSTRACT : An accidental contamination occurred in the subAntarctic Crozet Island between July and November 1997 near the "Alfred Faure" scientific station (51° 51'E - 46° 25'S). More than 20 000 L of diesel fuel were spilled in the soil in the direct vicinity of the power station. A field study was initiated in 1999 with the objective of determining the long-term effects of some bioremediation agents on the biodegradation rate and the toxicity of oil residues under severe subAntarctic conditions. In order to evaluate bioattenuation efficiency, the contaminated area was sampled in December 1999 and bacteria were enumerated. All samples were analysed for total bacteria, heterotrophic viable assemblages and hydrocarbon utilising microflora. The results of this first analytical survey clearly show a significant response of subAntarctic microbial soil communities to hydrocarbon contamination. Significant increases of total, hetrotrophic and hydrocarbon utilizing bacteria occurred in the more contaminated zone (from less than 5 x 10^4 MPN ("Most Probable Number") g^{-1} to more than 10^8 MPN g^{-1} for hydrocarbon degrading bacteria). The very high numbers of hydrocarbon-degrading bacteria present in the more contaminated zone are probably an indication of bioattenuation activities. However, biostimulation treatments may accelerate the restoration of the contaminated soils. A bioremediation treatement of the contaminated area by addition of a slow release fertilizer (Inipol EAP-22, Elf Atochem) and temperature elevation (agricultural plastic sheets) will begin in January 2001.

INTRODUCTION

The patterns of the hydrocarbons found in Antarctica indicated that their origin was biogenic and there was no strong evidence for anthropogenic hydrocarbons in the Antarctic ecosystem (Cripps, 1990). However, accidental contamination of Antarctic soils with fuel oil occurs frequently. (Bej *et al.*, 2000). There is a continuing need to understand the fate of hydrocarbon contaminants in these extreme systems. However, relatively few data are available for Antarctic soils (Delille, 2000). Little is known about hydrocarbon degradation processes in cold environments. While it is recognized that microorganisms play a critical role in the breakdown of hydrocarbons, the impact of petroleum pollutants on the metabolism and abundance of natural microbial communities is poorly understood. Crude oil, for example, has been shown to enhance (Delille and Delille, 2000), reduce (Griffiths *et al.*, 1981), or have no effect (Bauer and Capone, 1985) on total abundance of bacteria. Biostimulation treatments have been shown to stimulate the biodegradation of oil on a number of contaminated shorelines (Bragg *et al.*, 1994; Delille and Delille, 2000). However, questions raised about the applicability of bioremediation in cold regions. Some bioremediation experiments have been conducted on Antarctic soils (Delille, 2000; Kerry, 1993), but, to our knowledge, never in subantarctic soils.

Objective. The purpose of this present study is to provide a synthetic overview of the bacterial data collected during a first analytical survey conducted in subAntarctic soils, 2 years after a relatively minor, but significant, diesel oil spill which occurred in 1997 in the Crozet Island

MATERIAL AND METHODS

Study Area. An accidental contamination occurred in the subAntarctic Crozet Island between July and November 1997 near the "Alfred Faure" scientific station (51° 51'E - 46° 25'S) in the direct vicinity of the power station. The spill began in July 1997, but become apparent only in November 1997 at a point located approximately 200 m downstream of the origin of the spill (Fig.1). The leak was then rapidly filled in, and a precise analysis of consumption data of the power station demonstrated that the leak had a flow of approximately 1700 l/week. Thus, more than 20 000 l of diesel fuel were spilled in the soil between the original leak and the appearance of a visible contaminated area. On October 7th 1998, the contaminated area in the surface soils was estimated to measure approximately 1200 m^2. 4 months later, a 13 % increase of the surface area of the contaminated zone had been observed. Presently, the contaminated area has not visibly incresased. The contaminated area is located in a transition zone between an arid fell-field (upstream) and a wet vegetated area (downstream). Surface soils are thermally unstable, experiencing large temperature fluctuations with temperatures frequently dropping near 0°C and reaching much more than 20°C during sunny afternoons A field study of the surface contaminated area was initiated in 1999. In order to evaluate bioattenuation efficiency, the area was sampled on December 7th 1999 for bacterial enumerations. Surface soil samples were collected within a 30 m grid.

Bacterial Enumerations. Total bacteria were determined by acridine orange direct count (AODC) on black nuclepore filters (0.2 μm) using an Olympus BHA epifluorescence microscope.A minimum of 500 fluorescing cells with a clear outline and definite cell shape cells were counted under oil immersion (x 1000) in a minimum of 10 randomly chosen fields. Hydrocarbon-degrading bacteria were counted using the M.P.N. method using a basal mineral medium without carbon, supplemented with "Arabian Light" crude oil. Rezasurin was used as a growth indicator. After inoculation (3 tubes per dilution) the tubes were incubated at 12°C for 3Ø days (.).

RESULTS

All of the results of this first analytical survey clearly show a positive quantitative response of subAntarctic microbial soil communities to diesel oil contamination. Total bacterial counts were ten times higher in the center of the

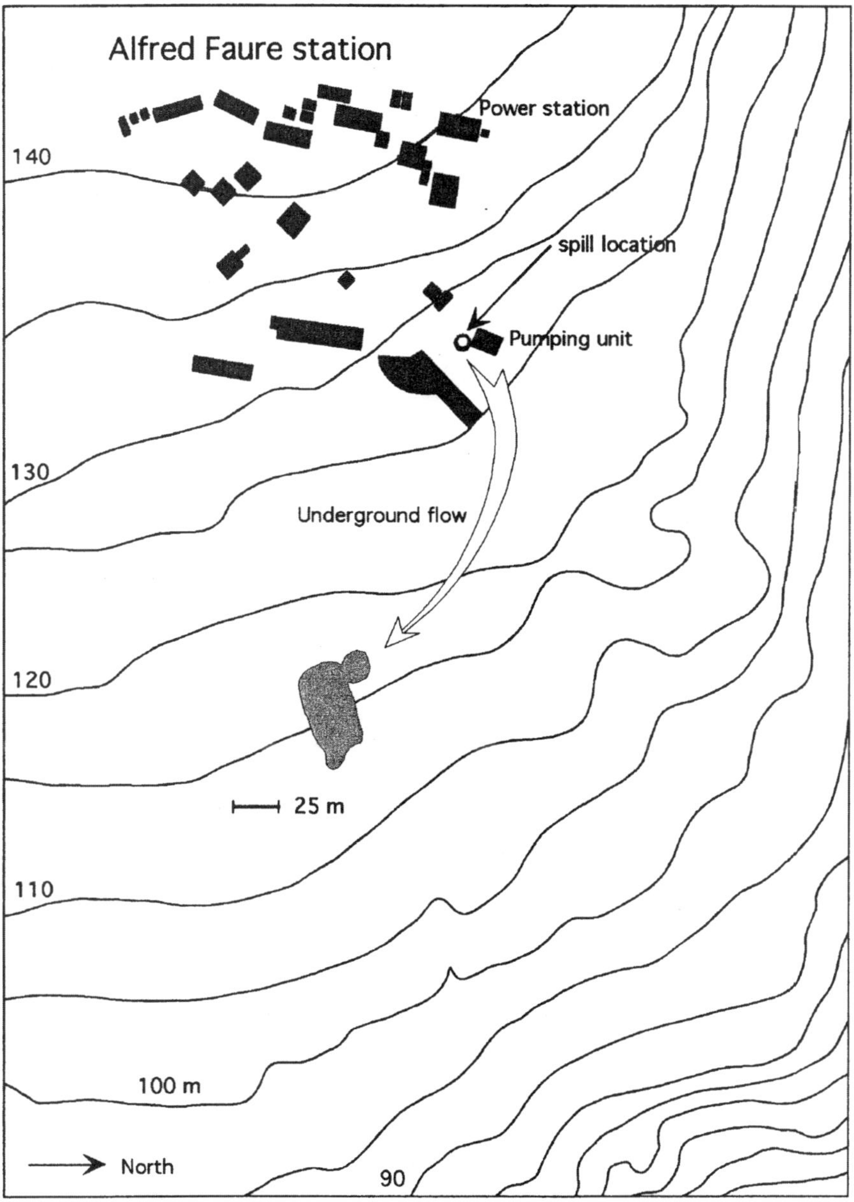

FIGURE 1. Surface occurrence (dark grey) of diesel oil in the vicinity of the Alfred Faure station (Possession Island, Crozet Archipelago) two years after an accidental contamination by diesel fuel.

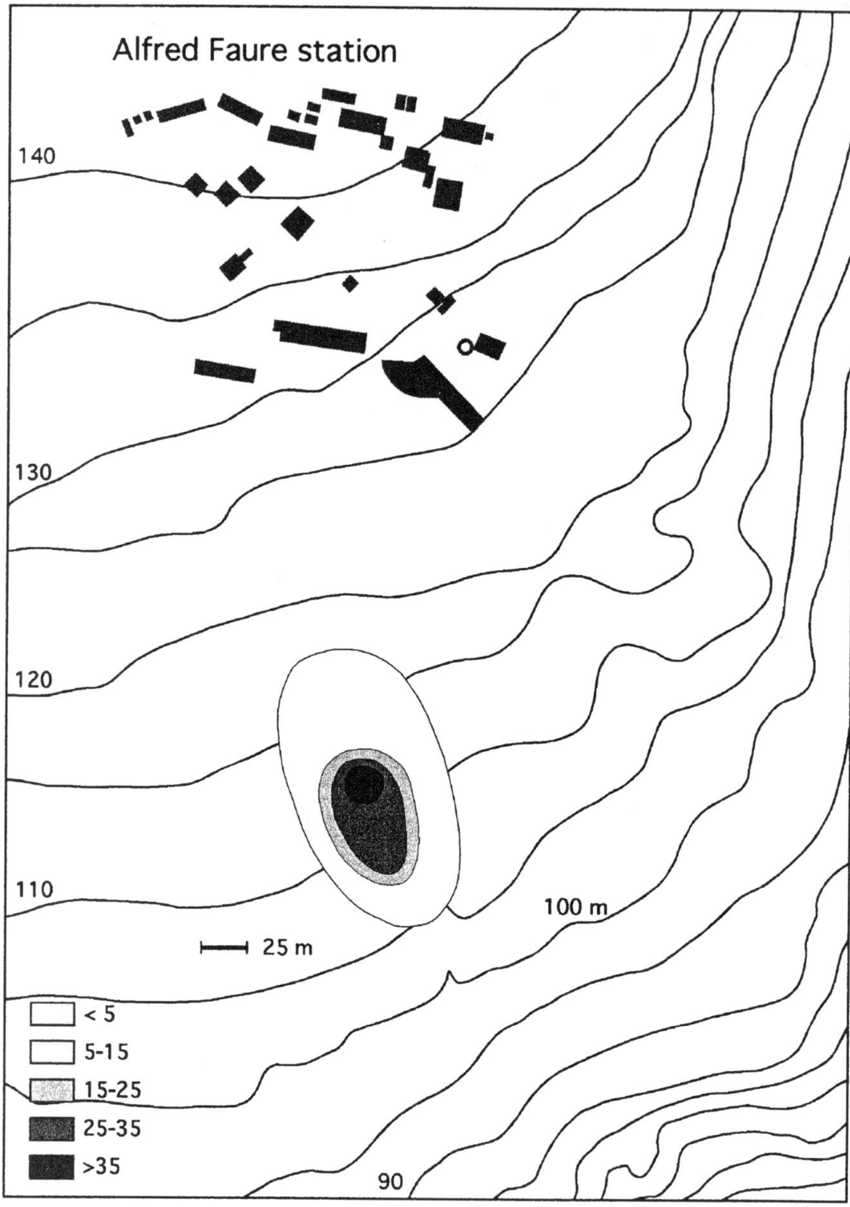

FIGURE. 2. Spatial distribution of total bacterial abundance in the vicinity of the contaminated area (10^{10} cells g-1)

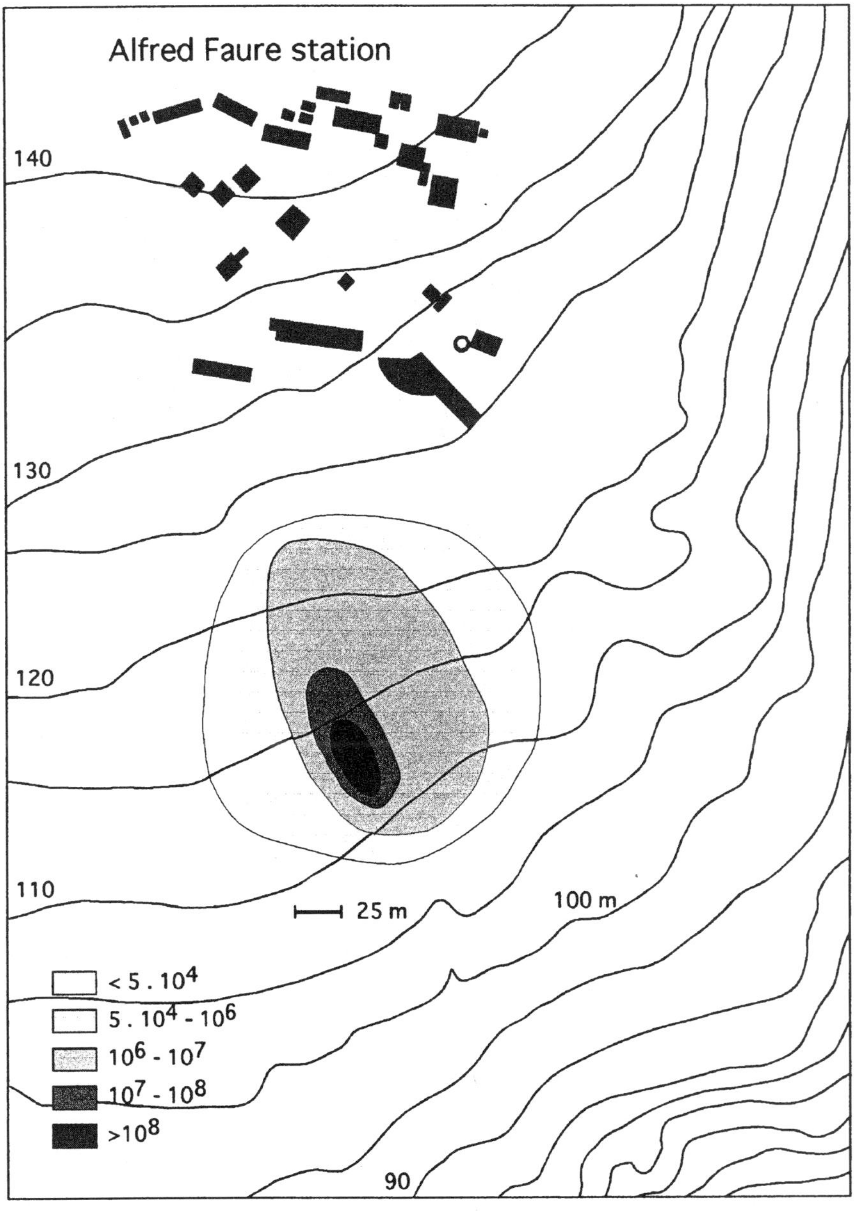

FIGURE 3. Spatial distribution of hydrocarbon-degrading bacterial abundance in the vicinity of the contaminated area (MPN g-1)

contaminated area than in the non-contaminated zone (Fig. 2). Moreover, the spatial distribution of the total bacterial abundance corresponded exactly to the occurrence of diesel oil in the surface soils.

The data concerning hydrocarbon-degrading bacteria were even more convincing (Fig. 3). Numbers of specific bacteria observed from contaminated areas increased 4 orders of magnitude from pristine habitat (from 1.5×10^4 to 4.8×10^8 MPN g^{-1}). The spatial distribution paralleled those of other bacterial parameters but the observed zone of enrichment was larger than the corresponding zone for the total and heterotrophic assemblages. This is related to the heightened sensitivity of hydrocarbon-degrading bacteria to diesel oil contamination.

DISCUSSION

Given that an increase of total bacterial counts by a factor of 2 will be significant (Delille and Bouvy, 1989), the ten-time increase of total bacteria observed after contamination suggests that a positive quantitative response of microbial soil assemblages occurred after the spill. Biological degradation (Bioattenuation) represents one of the main route through which organic chemicals are removed from contaminated environments (Bragg *et al.*, 1994). While numerous factors control enzyme-mediated degradation mechanisms, it is still difficult to predict biodegradation rates under various environmental conditions encountered in nature. However, measurements of hydrocarbon-degrading microorganisms are good indicators of the level of exposure of natural environments to hydrocarbons (Braddock *et al.*, 1995) and their increase may indicate a stimulation of the degradation potential (Kämpfer *et al.*, 1993). In this study hydrocarbon-degrading bacteria were always present in pristine samples. The ubiquitous distribution of hydrocarbonoclastic bacteria has already been reported in a wide variety of niches (Leahy and Colwell, 1990). Microscopic direct counts overestimate the number of true active bacteria. Total number of living bacteria will roughly range between 10^9 and 10^{10} bacteria ml^{-1}. Thus, hydrocarbon-degrading bacteria ranged from less than 0.01% of the total bacterial assemblage in pristine soils to more than 10% in the more contaminated area. Such increase of the hydrocarbon-degrading assemblage is a clear indication of bioattenuation activities Thus, biological degradation by naturally occurring populations of micro-organisms will be a significant mechanism for the removal of petroleum from the contaminated area. However, under the severe Antarctic conditions bioattenuation processes may be insufficient

The rate of microbial degradation of hydrocarbons in soils is affected by several physicochemical and biological parameters including the numbers and species of microorganisms present; the conditions for microbial activity (e.g., concentration of nutrients, pH, oxygen and temperature); the quality, quantity and bioavailability of the contaminants and the soil characteristics (Margesin and Schinner, 1997). It is well established that nutrients are one of the major limiting factors of hydrocarbon biodegradation at sea. In sub-Antarctic marine environments, total bacterial abundance increased by two orders of magnitude following the addition of fertilizer to hydrocarbon contaminated sites (Delille and

Delille, 2000, Delille and Vaillant, 1990). The contaminated soils of Crozet Island are relatively poor in mineral nutrients, and thus a biostimulation study, using a slow release fertilizer was begun in January 2001.

Microbial metabolism increases as temperature increases (Leahy and Colwell, 1990), usually doubling for each 10°C increase in temperature from 10 to 40°C. However, the metabolism of psychrophilic and psychrotrophic bacteria is adapted to work and function at low temperature. Antarctic soils are thermally unstable, experiencing large temperature fluctuations with temperatures dropping well below 0°C at night and reaching much more than 20°C during sunny afternoons. These extreme fluctuations may seriously affect bacterial activity. Bacteria must acclimate continuously and be able to rapidly switch activity on and off. Stability of temperature (like in sea-ice or seawater) may render a habitat more favorable for bacterial growth than a relatively high but fluctuating temperatures, such as that existing in Antarctic soils. In order to evaluate the possible effect of increasing temperature, experiments using agricultural sheets was begun in January 2001.

The main objective of the future studies will be to evaluate the effectiveness of biostimulation and "biopiles" treatments.

ACKNOWLEDGEMENTS.

This research was supported by the "Institut Français pour la Recherche et la Technologie Polaires" .

REFERENCES

Bauer, J.E. and Capone D.G. 1985. "Effects of four aromatic organic pollutants on microbial glucose metabolism and thymidine incorporation in marine sediments". *Appl. Environ. Microbiol. 49*: 828-835.

Bej, A.K., Saul, D. and Aislabie, J. 2000. "Cold-tolerant alkane degrading Rhodococcus species from Antarctica". *Polar Biol. 23* : 100-105.

Braddock, J.F., Lindstrom, J.E.and Brown, E.J. 1995. "Distribution of hydrocarbon-degrading microorganisms in sediments from Prince William Sound, Alaska, following the Exxon Valdez oil spill". *Mar. Pollut. Bull. 30*: 125-132.

Bragg, J.R., Prince, R.C., Harner, E.J. and Atlas, R.M. 1994. "Effectiveness of bioremediation for the Exxon Valdez oil spill". *Nature 368*: 413-418.

Cripps, G.C. 1990. "Hydrocarbons in the seawater and pelagic organisms of the southern ocean". *Polar Biol. 10*: 393-402

Delille, D. 2000. "Response of Antarctic soil bacterial assemblages to contamination by diesel fuel and crude oil". *Microbial Ecol. 40*:159-168

Delille, D. and Bouvy, M. 1989. "Bacterial responses to natural organic inputs in marine sub-Antarctic area". *Hydrobiologia 182*: 225-238.

Delille, D. and Delille, B. 2000. "Field observations on the variability of crude oil impact on indigenous hydrocarbon-degrading bacteria from subAntarctic intertidal sediments". *Mar. Environ. Res. 49*:403-417.

Delille, D. and Vaillant, N. 1990. "The influence of crude oil on the growth of subantarctic marine bacteria". *Antarctic Sci. 2*:655-662.

Griffiths, R.P., MacNamara, T.M., Cadwell, B.A.and Morita R.Y. 1981. "Field observations on the acute effect of crude oil on glucose and glutamate uptake in samples collected from Arctic and subArctic waters". *Appl. Environ. Microbiol. 41*: 1400-1406.

Kämpfer, P., Steiof, M., Becker, P.M. and Dott, W. 1993. "Characterization of chemoheterotrophic bacteria associated with the in situ bioremediation of a waste-oil contaminated site". *Microbial Ecol. 26*:161-188.

Kerry, E. 1993. "Bioremediation of experimental petroleum spills on mineral soils in the Vesfold Hills, Antarctica". *Polar Biol. 13*: 163-170.

Leahy, J.G. and Colwell, R.R. 1990. "Microbial degradation of hydrocarbons in the environment". *Microbiol. Rev. 54*: 305-315.

Margesin, R. and Schinner, F. 1997. "Efficiency of indigenous and inoculated cold-adapted soil microorganisms for biodegradation of diesel oil in alpine soils". *Appl. Environ. Microbiol. 63*: 2660-2664.

EFFECT OF CARBON SOURCE ON BIODEGRADATION OF CHLORINATED ETHENES

Laurie T. LaPat-Polasko (Parsons Engineering Science, Phoenix, Arizona)
Brian S. Aiken (Parsons Engineering Science, Phoenix, Arizona)
Karthik Narayanaswamy (Parsons Engineering Science, Phoenix, Arizona)

ABSTRACT: Natural attenuation of chlorinated ethenes in groundwater has been evaluated at numerous sites within the United States. Because terminal electron accepting processes have a significant effect on the biodegradation rate of various chlorinated organics, scientists have spent a significant amount of time collecting data related to various terminal electron acceptors, such as dissolved oxygen, nitrate, iron, sulfate and methane. However, it is not only important to understand which terminal electron accepting process is predominating at a site to establish biodegradation rates, but it is also essential to obtain an understanding of the total organic carbon (TOC) source and how this geochemical indicator is affecting biodegradation rates. This paper discusses how TOC concentration and type affects the degradation rates of trichloroethylene (TCE), *cis*-1,2-dichloroethylene (DCE) and vinyl chloride (VC) in three different plumes in the Phoenix area. One of the plumes was the result of chlorinated solvents leaching from a landfill material. The other two plumes were due to contamination from a variety of sources, such as leaking underground storage tanks, dry wells, leachfields and other surface disposal incidences. All three plumes are typically anaerobic. It appears that in some locations where reductive dechlorination has occurred, current TOC concentrations are low. This may be the result of the indigenous microbial population utilizing the available biodegradable carbon sources in the groundwater. Based on evaluating the data from 3 chlorinated solvents sites in the Phoenix area, it appears that the type and concentration of TOC is important to understanding and estimating biodegradation rates of chlorinated solvents in groundwater. Likewise, monitoring and understanding the geochemical parameter TOC is of primary importance in evaluating natural attenuation of chlorinated organics in groundwater.

INTRODUCTION

Natural attenuation of chlorinated solvents has been evaluated for several sites around the United States. Terminal electron acceptor considerations are critical in reductive dechlorination of chlorinated ethenes and have been studied extensively. The collection of terminal electron acceptor data, such as dissolved oxygen, nitrate, iron, sulfate, carbon dioxide and acetate, is prevalent in studies related to chlorinated solvent contamination. However, the importance of a carbon source is not often recognized.

This paper discusses how the amount and type of total organic carbon (TOC) affects the degradation rates of trichloroethylene (TCE), *cis*-1,2-dichloroethylene (DCE) and vinyl chloride (VC) in three independent field sites in Phoenix, Arizona. One plume is the result of chlorinated solvents leaching from a landfill, while the other two are from industrial facilities.

METABOLIC FUNDAMENTALS

Energy producing reactions require an electron acceptor, an electron donor and a carbon source. The electron donor is usually the organic substrate, while the electron acceptor could be oxygen, nitrate, sulfate, carbon dioxide, acetate, etc. The electron acceptor which yields the highest energy is preferred, so in the presence of oxygen, nitrate, for example, will not be used as the electron acceptor. Certain electron acceptors may be detrimental to microbial growth. For example, tetrachloroethene (PCE) has not been shown to undergo biodegradation under aerobic or nitrate-rich environments.

In addition, there are anaerobic bacteria that reduce chlorinated solvents directly and derive energy from the reductive dechlorination process (McCarty, 1997). These bacteria use the chlorinated solvents as an electron acceptor, and reduce the parent compound completely to ethylene. Some dehalorespirers can tolerate very high chlorinated solvent concentrations, and may thrive near the source zones of such contamination. However, such bacteria are not always naturally present in the subsurface, and hence dehalorespiration is not applicable for every site.

METHODS AND MATERIALS

Groundwater quality data from various sites were collected and used to analyze degradation of chlorinated solvents. For each site, a pair of closely located monitoring wells along the gradient were used to calculate the degradation rates of the significant parent compound (e.g. TCE) based on correlating the measured concentrations at the paired wells.

This analysis assumes the aquifer behaves as a plug-flow reactor. Mass conservation is a reasonable assumption if the wells are close together because losses of total VOC due to dilution are minimal. If TCE was the parent compound (i.e. no PCE present) then the total amount of TCE initially present that contributed to the current molar concentrations of TCE, DCE and VC is represented by:

$$[TCE]_o = [TCE] + [DCE] + [VC] \tag{1}$$

If the degradation rate of each component is denoted by k_i, where i is the component name, and t is the time elapsed in transport from the first well to the second, then the molar concentrations of each compound can be written as:

$$\frac{[TCE]}{[TCE]_o} = e^{-k_{TCE}t} \tag{2}$$

$$\frac{[DCE]}{[TCE]_o} = \frac{k_{TCE}}{k_{DCE} - k_{TCE}}(e^{-k_{TCE}t} - e^{-k_{DCE}t}) \tag{3}$$

$$[VC] = [TCE]_o * a * e^{-k_{VC}t}\left(\left(\frac{e^{bt}-1}{b}\right)+\left(\frac{1-e^{ct}}{c}\right)\right) + DCE_o * d * e^{-k_{VC}t}(e^{-k_{DCE}t}-1) + [VC]_o * e^{-k_{VC}t}$$

Where: $a = k_{TCE}/(k_{DCE}-k_{TCE})$
$b = k_{VC}-k_{TCE}$
$c = k_{VC}-k_{DCE}$
$d = k_{TCE}/k_{DCE}$

(4)

The previously described methods were applied to three sites with chlorinated solvent plumes associated with them - a leaching landfill and two semiconductor facilities. Table 1 briefly describes the contaminant concentrations and TOC ranges at each site.

TABLE 1. Maximum concentrations of dissolved chlorinated solvents and daughter products at the various sites

Location	TCE (µg/L)	*Cis*-1,2-DCE (µg/L)	VC (µg/L)	TOC Range (average) (mg/L)	Type of TOC
Landfill	200	12,000	10,000	0.5 - 94 (5.6)	Mixture from various sources
Semiconductor facility No.1	4,700	330	150	2.5 - 13 (4)	Anthropogenic origin
Semiconductor facility No.2[1]	1,380,000	230	110	N/A	Anthropogenic origin

[1] Only alluvium wells

ESTES LANDFILL

From 1953 to 1972, the Estes Landfill was used by commercial trash haulers, septic tank effluent haulers and other private users. The site included a liquid waste disposal pit in the southeastern portion of the landfill, which received several hazardous chemicals including chlorinated solvents, such as TCE. The VOC plume is approximately 3,400 ft long, consisting primarily of DCE and VC. TOC levels in the vicinity of the landfill range from 0.5 - 94 mg/L, and is expected to be the result of a variety of biodegradable and non-biodegradable sources.

The subsurface geology under the landfill consists of four distinct layers - a shallow alluvial layer described as unconsolidated sand, gravel, cobbles and boulders to an average depth of 50 to 60 ft bgs (15 to 18 m) designated as unit F1, a middle alluvial layer described as a 30 to 40 ft (9 to 12 m) thick sequence of unconsolidated to semiconsolidated gravel and cobbles supported in a clayey silt matrix designated unit F2, a deep alluvial layer described as unconsolidated sand, gravel and cobbles that ranges from 15 to 60 ft (4.5 to 18 m) thick designated as unit F3, underlain by sedimentary bedrock designated as F4. The F2 layer is present under most of the site and possibly to the south and east but tends to disappear west and northwest of the site. On the basis of aquifer test data, hydraulic conductivities were calculated to be approximately 1.8×10^{-1} cm/s for unit F1 and approximately 7.2×10^{-2} cm/s for unit F3. Hydraulic conductivity for unit F2 is believed to be orders of magnitude less than the F1 and F3 units, while that for unit F4 is approximately 4×10^{-6} cm/s. The direction of groundwater flow is typically to the west except when special recharge events occur.

Anaerobic conditions were prevalent below most of the site. Subsurface nitrate levels varied from less than 1 mg/L as N to about 5 mg/L as N. The data indicate pockets of higher nitrate, whereas in other areas, it is likely that sulfate is the primary electron acceptor.

TCE degradation rate could not be computed because of the low TCE concentrations currently recorded at the site. The DCE degradation rate was computed by selecting a well pair, EW-PZ3 and EW-6, and comparing the TCE concentrations as shown in Figure 1.

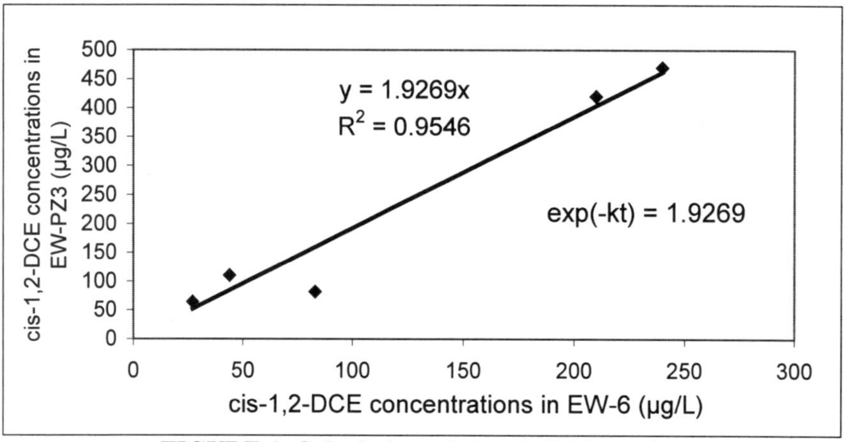

FIGURE 1. Calculation of *cis*-1,2-dichloroethene degradation rate at the Estes Landfill

SEMICONDUCTOR FACILITY NO. 1

For this site, solvents including TCE and PCE, as well as nitric and sulfuric acids were used for semiconductor processing steps, and were subsequently discharged to the subsurface environment.

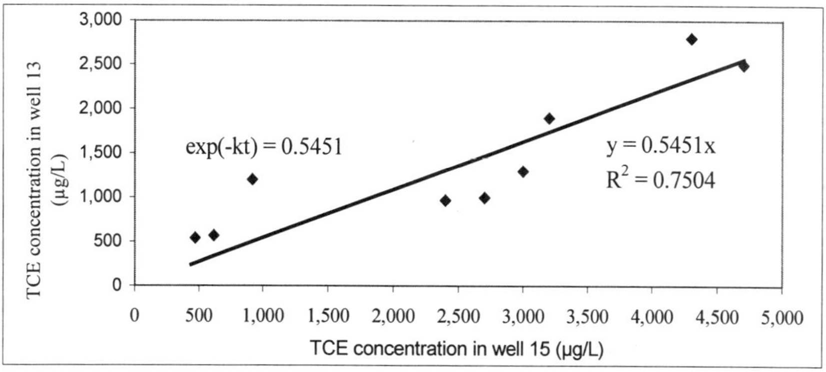

FIGURE 2. Correlation of TCE concentrations in two wells about 200 feet apart to determine TCE degradation rate.

The site is characterized by two alluvial units - a shallow alluvial unit, which ranges from 40 to 50 feet thick and is composed of gravel, sand and silt, and a deep alluvial unit which ranges from 10-40 ft (3 to 12 m) thick, composed of about 10 to 75% sand and gravel. Hydraulic conductivities were estimated to be 46 ft/day (14 m/day) based on pumping test data for the aquifer (Woodward-Clyde, 1995). Bedrock underlies the alluvial units and is believed to have hydraulic conductivities ranging from 0.001 to 1 ft/day (0.0003 to 0.3 m/day). Groundwater flow is to the southwest.

In the vicinity of the site, the aquifer was found to be anaerobic. TOC levels in this area ranged from 2 - 13 mg/L and were generally less than 4 mg/L, while nitrate levels ranged from 4 - 14 mg/L. The VOC plume is about 10,500 ft long.

TCE degradation rates for this site were estimated by comparing TCE concentrations in two wells about 200 ft (60 m) apart along the flow path as shown in Figure 2. An Excel spreadsheet was then used to predict degradation rates for DCE and VC using the plug-flow equations described in the Methods section.

SEMICONDUCTOR FACILITY NO.2

On this site, solvents such as TCE and PCE were used and discharged in a manner that is believed to be responsible for a DNAPL source. This DNAPL source has resulted in dissolved chlorinated solvent concentrations of thousands to hundreds of thousands of parts per billion at the site and a VOC plume more than 18,500 feet long. Nitric and sulfuric acids were also utilized at this site.

This facility is underlain by an area of uplifted bedrock, eroded and overlain by alluvial deposits to form an irregular erosional surface with small bedrock highs extending close to the land surface. The Quaternary alluvium underlying the facility varies in thickness from 25 to 75 feet (7.5 m to 22.6 m) and consists of poorly consolidated to unconsolidated fine- to medium-grained silty sandy gravel and silty gravelly sand deposits. The bedrock units are distributed in a complex pattern that is controlled by faulting. Bedrock beneath the facility dips from east to west. Groundwater beneath the facility flows west-southwest. Groundwater is encountered beneath the facility at depths ranging from 35 feet bgs to 52 feet (10.6 to 15.8 m) bgs. The estimated horizontal hydraulic conductivity values calculated for the alluvium beneath the site is 30 to 40 ft/day (9 to 12 m/day), though some lower and higher values have been calculated. The average horizontal hydraulic conductivity for the fractured bedrock is approximately 0.02 ft/day (0.006 m/day), but local fracture zones have hydraulic conductivity values as high as 3.8 ft/day (1.2 m/day).

TCE degradation rates for this site were calculated based on a linear correlation between the TCE concentrations observed in two wells, 307 and 308 as described above. Figure 3 shows the concentration for the paired wells for semiconductor facility No.2. This figure was used to compute the TCE degradation rate as with the other sites. The DCE and VC degradation rates were then computed using a spreadsheet programmed with the plug flow reactor formulas described earlier.

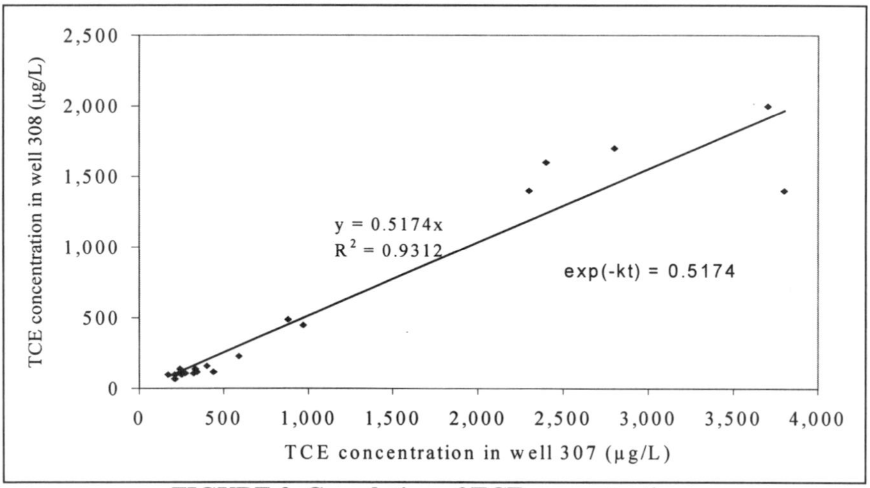

FIGURE 3. Correlation of TCE concentrations at 2 wells at semiconductor facility No.2

RESULTS AND DISCUSSION

Table 2 summarizes the computed degradation rates for TCE, DCE and VC at the three sites, as well as mean field/*in situ* degradation rates provided in Aronson and Howard (1997).

While it was not possible to compute TCE degradation rates at the Estes Landfill, the low TCE levels and the high rate of production of DCE indicates a higher degradation rate than at the two semiconductor facilities.

This is also apparent in terms of the DCE degradation rate, which is higher at the landfill. This is consistent with the amount of TOC historically present at the site as well as the nature of TOC, since it would be expected that the landfill would include highly biodegradable TOC.

TABLE 2. Degradation rates calculated at the three sites

Location	Electron Acceptor	TCE day^{-1} ($x10^{-3}$)	DCE day^{-1} ($x10^{-3}$)	VC day^{-1} ($x10^{-3}$)	TOC Range (average) mg/L	TOC Type
Landfill	Sulfate	Rapid	10	N/A	0-94 (5.6)	Mixture from various sources
Semiconductor facility No.1	Nitrate	3	3	N/A	2 - 13 (4)	Anthropogenic origin
Semiconductor facility No.2	Nitrate	1	3	9	N/A	Anthropogenic origin
Aronson and Howard mean field/*in situ* rates		2.5	N/A	7.3		

CONCLUSION

The amount and source of organic carbon is very crucial to degradation of chlorinated solvents and their daughter products. In addition to measurements of electron acceptors, efforts should be made to characterize possible sources and nature of carbon sources, such as leaching landfills and leaking underground storage tanks in a chlorinated solvent plume, as this could aid demonstrating natural attenuation of these compounds.

REFERENCES

Aronson, D., Howard, P.H. 1997. *Anaerobic Biodegradation of Organic Chemicals in Groundwater: A Summary of Field and Laboratory Studies.* Prepared for American Petroleum Institute.

Harding Lawson Associates 1997. *Estes Landfill RI/FS Remedial Investigation Draft Report.* Prepared for the City of Phoenix.

McCarty, P.L. 1997. "Breathing with Chlorinated Solvents", *Science*, Vol. 276, pp. 1521-1522

Woodward Clyde, Inc. 1995, *Motorola 48th Street Pilot-Scale Treatment Facility Baseline Report.* Prepared for Motorola, Inc.

NATURAL ATTENUATON OF TCE AT SPACE LAUNCH COMPLEX 4 (SITE 10) VANDENBERG AFB

Larry Tyner (IT Corporation, Knoxville, Tennessee)
David Grainger, Bruce Sibbett, Dorota Rek, and Tom Perina (IT Corporation, San Bernardino, California); and Amena Atta (Vandenberg AFB, Lompoc, California)

ABSTRACT: Space Launch Complex 4 (SLC-4) is an active facility located on south Vandenberg Air Force Base (AFB). SLC-4 is 3,000 feet from the Pacific Ocean, on the Lompoc Terrace, and is underlain by unconsolidated sand deposits 115 to 200 feet thick. The SLC-4 area includes installation restoration program (IRP) Sites 8, 9 and 10. Waste handling practices prior to 1986 have resulted in significant concentrations of chlorinated solvents in groundwater. Trichloroethene (TCE), used for cleaning operations at the launch towers on SLC-4 east and west (Sites 8 and 9), was released down paved drains into concrete-lined retention basins. A large volume of water was used for blast control during rocket launches and drained into the retention basins. The deluge water was discharged into Spring Canyon (Site 10), carrying with it TCE. The Site 10 contaminant plume is under Spring Canyon and flows west to the Pacific Ocean. The Spring Canyon plume has TCE concentrations of 56 µg/L and less. Based on the preliminary screening guidance from the Air Force Center for Environmental Excellence (AFCEE) technical protocol the flowpath along Spring Canyon scored a 12. This area is anoxic, which is required for reductive dechlorination. Biologic activity is occurring along this flowpath, as shown in the plate counts, biomass calculations, and the production of metabolic by-products, dichloroethene (DCE), vinyl chloride, fatty acids, and methane. The enzyme methane monooxygenase (mmox) was detected at Site 10. Mineralization of TCE could occur through this organism under aerobic conditions.

INTRODUCTION

Sites 8, 9, and 10 are located in the southwestern portion of the Lompoc Terrace, a wide erosion platform overlain with shallow marine, windblown, and alluvial sand deposits above the Santa Ynez River floodplain. The sites share common boundaries and are arranged approximately in a straight line from east to west. Site 8 is the easternmost location, Site 9 is in the middle, and Site 10 is the westernmost location (Figure 1).

Sites 8 and 9 are Titan missile launch facilities, designated Space Launch Complex 4E (SLC-4E) and SLC-4W, respectively. Each facility consists of a launch gantry, support facilities, and chemical storage areas. Launches totaled 58 at SLC-4E from 1964 through 1993 and 85 at SLC-4W from 1963 through 1994. A concrete-lined deluge water channel extends from the flame bucket to a lined retention basin. Water released from the retention basin is routed to the base of

Spring Canyon through a discharge pipe and erosion control system. The system consists of a concrete-lined channel inside the fence and culvert piping from the south fence to the bottom of Spring Canyon (Site 10). The discharge system has been in place since 1974. Prior to that, water from Site 8 discharged to the canyon slope approximately 45 feet from the southwest corner of the retention basin and water from Site 9 discharged to the canyon slope south of the retention basin. From 1974 to 1986, wastewater was diluted in the retention basin and discharged into Spring Canyon. Since 1986, wastewater that enters the retention basin has been tested to determine the method of disposal. Water that meets quality standards is discharged to grade. Water that requires treatment is transported to a treatment pond. Launches totaled 58 at SLC-4E from 1964 through 1993 and 85 at SLC-4W from 1963 through 1994.

Site 10 is a dry topographic depression at the mouth of Spring Canyon and is referred to as Spring Canyon Pond. Ponding of water in this area has been observed only very rarely for short periods. Surface water occasionally flows in Spring Canyon, south of Sites 8 and 9, during the rainy season, but generally infiltrates into the sandy soil as it flows into Site 10. Water released from the retention basins at Sites 8 and 9 could reach Site 10 along this surface pathway.

SITE SETTINGS

Regional Geologic. The southern portion of Vandenberg AFB is located within a geologically complex transition zone between the Transverse Range geomorphic province to the south and the Coast Range geomorphic province to the north (Versar, 1991). Sites 8, 9, and 10 are located on a wide paleomarine terrace commonly known as the Lompoc Terrace, which is bounded on the north by the Santa Ynez River and on the south by the Santa Ynez Mountains. The terrace consists predominantly of regressive paleomarine sediments overlain by dissected stream terrace, alluvial fan, and eolian deposits (Dibblee, 1988). The sedimentary deposits at Sites 8 and 9 are characterized as Pleistocene dune sand deposits overlying sediments of the Pleistocene Orcutt Formation and the Tertiary Careaga Formation.

Regional Hydrologic. Sites 8, 9, and 10 are located on the southern margin of the south Santa Ynez River groundwater basin/Lompoc Terrace subbasin. The Lompoc Terrace aquifer predominantly exists within the Careaga and Orcutt Formation sands and encompasses an area approximately 2 miles wide (measured north-south) by 5 miles long (measured east-west). The Lompoc Terrace subbasin is probably in hydraulic continuity with groundwater beneath the alluvial plain of the Santa Ynez Valley to the northeast. The west end likely is in hydraulic continuity with the Pacific Ocean (Evenson and Miller, 1961). Groundwater flows to the west and northwest in the SLC-4 area.

Groundwater at Sites 8, 9, and 10 occurs in the alluvial sediments in Spring Canyon and in the Careaga Formation deposits underlying the sites. The recharge to these areas is from infiltration of seasonal rainwater falling within the upper reaches of Spring Canyon and directly on site soils. No perennial surface water is found at these sites; however, seasonal surface water flows in Spring Canyon. Perennial springs along the coastline flow from the base of the Careaga sand, at the top of the diatomite bedrock. The coastal cliffs are being rapidly eroded by wave action and the location or flow volume of a spring can change from year to year.

Site Biotic. The predominant habitats surrounding Sites 8 and 9 are coastal sage scrub and valley foothill riparian. A partially stabilized dune field with coastal strand vegetation is present between Surf Road and the coast. The coast itself consists of cliffs, a small intertidal beach zone, and near-shore marine habitats. At Site 10, the closed basin referred to as Spring Canyon Pond is typically dry for many years, with standing surface water occurring only for brief periods during exceptionally heavy precipitation events. Therefore, freshwater emergent/aquatic habitat with wetland plants is not found at Site 10. Site 10 consists of valley foothill riparian habitats.

GROUNDWATER PLUME

Recharge to the aquifer occurs from Spring Canyon, via surface-water infiltration into the sediments, groundwater flow from east-southeast, and precipitation. Groundwater elevations indicate that the creek in Spring Canyon is a losing stream (i.e., contributes water to the saturated zone). Discharge occurs at the cliffs into the ocean.

Shallow groundwater (5 to 20 feet) occurs in Spring Canyon and the basin area at Site 10. The saturated thickness varies from 2 to 47 feet, with the greatest saturated thickness found in Spring Canyon in well 10MW1. The well locations with the greater saturated thickness correspond to areas of expected high recharge from Spring Canyon, and clayey, low hydraulic conductivity beds are present. The saturated thickness of the aquifer varies annually and peaks after the rainy season. The changes in groundwater gradient appear to be influenced by the relief on the underlying bedrock and variation in permeability.

Groundwater under Spring Canyon flows west and the contaminant plume consists of relatively low concentrations of TCE, as determined from the April 1999 sampling round. TCE concentrations of 46.7 and 38.0 μg/L were detected in this plume during monitoring in 1994. No contaminants were detected in the coastal springs directly west of Site 10 during the September 1997 sampling round, but a TCE concentration of 3.3 μg/L was detected in spring location 10SW11 during September 1998 sampling. To the northwest, spring locations 10SW12 and 10SW13 contained TCE concentrations of 11 and 9.4 μg/L, respectively, when sampled in September 1997, but were dry in 1998. The southern limit of the Spring Canyon plume is defined by non-detect values for TCE in wells 8MW5 and 9MW2, and coastal spring 10SW10.

RESULTS AND DISCUSSION

The *Technical Protocol for Evaluating Natural Attenuation of Chlorinated Solvents in Groundwater* was used to evaluate the potential for natural attenuation at the SLC-4 area. Based on the preliminary screening, the flowpath along Spring

Canyon scored a 12. A value of 6 to 14 is interpreted as having limited evidence for biodegradation of chlorinated organics. A score of 15 is defined as having adequate evidence for biodegradation of chlorinated organics. This area is anoxic, which is required for reductive dechlorination. Soil samples from 10MW4 were analyzed for microbial enumeration (Table 1) and biomass calculations (Table 2). Biologic activity is occurring along this flowpath, as shown in the plate counts, biomass calculations, and the production of metabolic by-products, DCE (Figure 1), and fatty acids.

The enzyme methane monooxygenase (mmox) was detected at 10MW4 (Table 3). Mmox allows TCE biodegradation to occur by serving as a co-metabolite with methanotrophs. In most environments, this occurs at the fringe of a readily degradable carbon plume where methane was produced, but aerobic conditions are maintained. When these bacteria are active, dechlorinated daughter products such as DCE and vinyl chloride are typically observed.

Sphingomonas was detected in all samples (Table 3). This genus is known to biodegrade many complex chlorinated contaminants. Mineralization of TCE could occur through this organism under aerobic conditions. The general water chemistry data gathered during the January sampling round is presented on Figure 2.

The flowpath along the Spring Canyon groundwater plume shows evidence for biodegradation of chlorinated solvents. The TCE degradation product cis-1,2-DCE occurs in concentrations proportional to TCE concentrations. Anoxic conditions suitable for reductive dechlorination exist.

Table 1. Anaerobic and Aerobic Microbial Enumerations

Sample ID	Aerobic Heterotrophs [a] (cfu/gm)	Anaerobic Heterotrophs[b] (cfu/gm)
10MW4S	$5.0x10^4$	$1.5x10^2$
10MW4S	$5.3x10^4$	$3.7x10^2$
10MW4S	$7.5x10^4$	$4.4x10^2$

[a]Two-week incubation period.
[b]Three-week incubation period.
cfu/gm – colony – forming units per gram of dry soil

Table 2. Biomass Calculations Groundwater Analysis

Sample	picomoles total PLFA/ml filtered	Cells/mL filtered[a]	picomoles prokaryote PLFA/mL filtered	picomoles eukaryote PLFA/mL filtered	ratio prokaryote/ eukaryote
9MW6W-001	0.2	3.41E+03	0.2	0.01	20
10MW4W-001	4.9	9.71E+04	4.6	0.2	23
8MW7W-001	0.4	7.74E+03	0.4	ND	NC
9MW5W-001	0.2	3.78E+03	0.2	ND	NC
8MW6W-001	0.2	3.07E+03	0.2	ND	NC
8MW5W-001	0.3	6.02E+03	0.3	ND	NC
8MW8W-001	0.5	1.01E+04	0.5	0.01	50

[a]The Cell equivalent value is calculated from experiments with typical bacteria isolated from soil and water. This value is based on $2.0X10^{12}$ cells per gram dry weight of cells and 10^8 picomoles of phospholipid/gram dry weight of cells. The number of cells/gram of dry weight may vary and is dependent on the environmental conditions from which the microorganisms were recovered.

Table 3. Results of DNA Analysis, Groundwater Analysis

Sample	Total DNA (Number of Cells/mL)	Amplifiable bacterial DNA	Presence of Sphingomonas	Presence of mmox
9MW8W-001	10^3	Yes	10^2	No
9MW6W-001	10^3	Yes	10^2	No
10MW4W-001	10^3	Yes	10^2	10^1
8MW7W-001	$<10^3$	Yes	10^2	No
9MW7W-001	$<10^3$	Yes	10^1	No
9MW5W-001	$<10^3$	Yes	10^1	No
8MW6W-001	$<10^3$	Yes	10^1	No
9MW9W-001	$<10^3$	Yes	10^2	No
8MW5W-001	10^4	Yes	10^2	No
8MW8W-001	$<10^3$	Yes	10^1	No

CONCLUSIONS

The groundwater plume under Spring Canyon contains TCE concentrations of 47 µg/L and less with low concentrations of degradation products (cis-1,2-DCE). The plume boundaries are defined with TCE values at the coastal springs less than 5 µg/L. The flowpath along Spring Canyon shows evidence for biodegradation of chlorinated solvents. Biologic activity is occurring along this flowpath, as shown in the plate counts, biomass calculations, and the production of metabolic by-products, DCE, and fatty acids. The TCE degradation product cis-1,2-DCE occurs and anoxic conditions suitable for reductive dechlorination exist.

REFERENCES

Weidermeier, T. H. et. al, 1995, Technical Protocol for Implementing Intrinsic Remediation with Long-Term Monitoring for Natural Attenuation of Fuel Contamination Dissolved in Groundwater, Vols. II Air Force Center for Environmental Excellence Technology Transfer Division, Brooks AFB, San Antonio, TX.

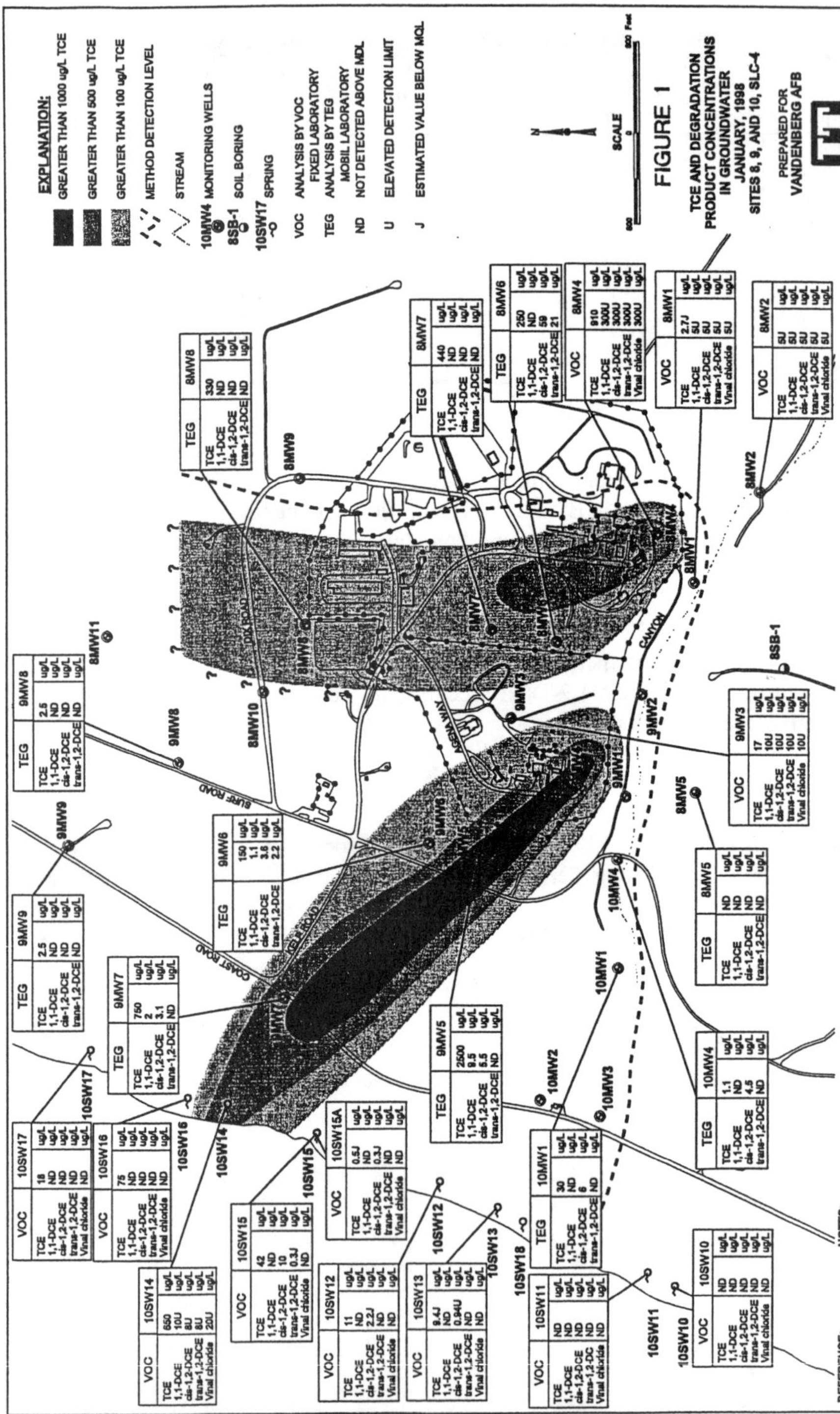

FIGURE 1
TCE AND DEGRADATION PRODUCT CONCENTRATIONS IN GROUNDWATER JANUARY, 1998 SITES 8, 9, AND 10, SLC-4

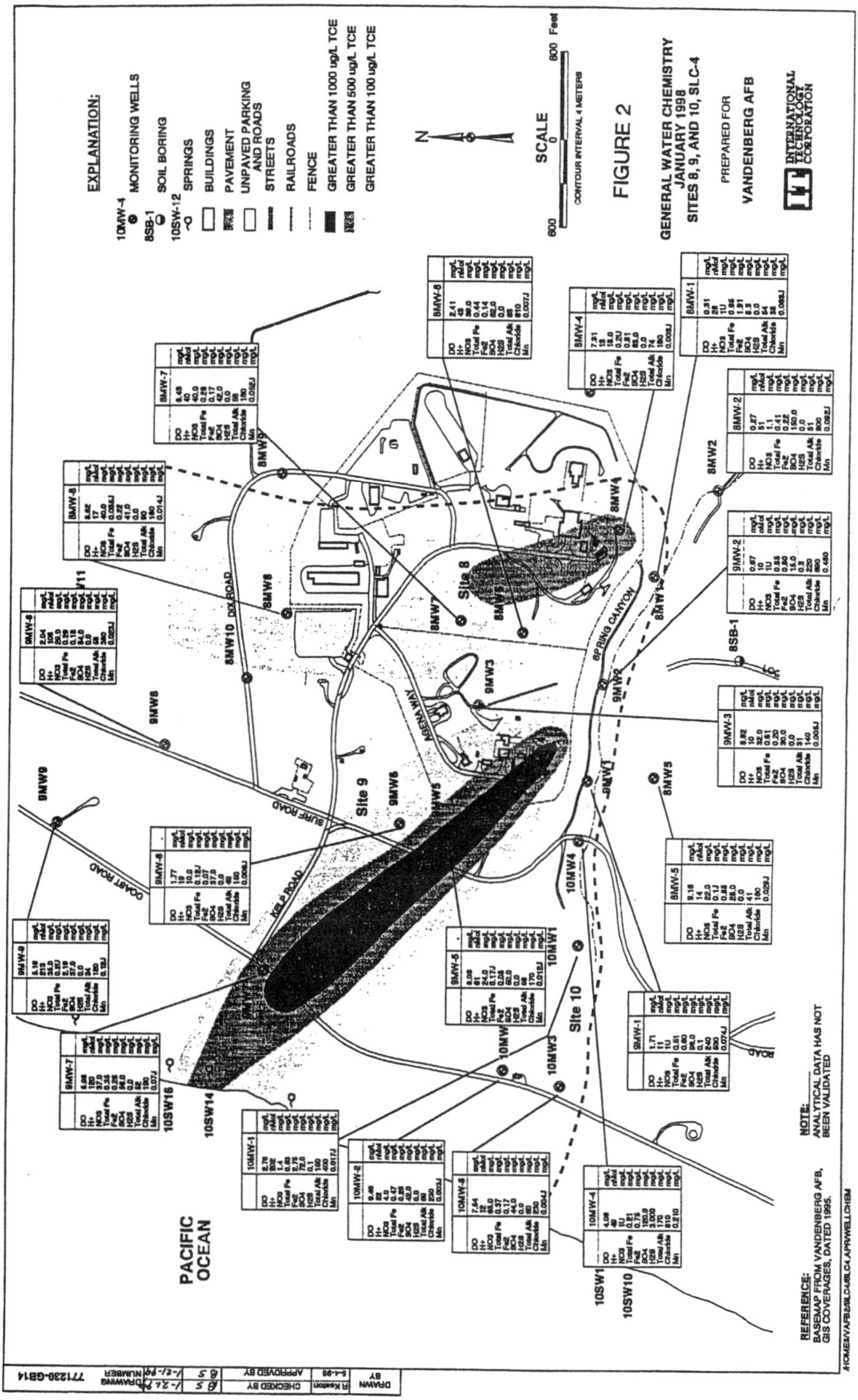
EXPLANATION:
MONITORING WELLS
SOIL BORING
SPRINGS
BUILDINGS
PAVEMENT
UNPAVED PARKING AND ROADS
STREETS
RAILROADS
FENCE
GREATER THAN 1000 ug/L TCE
GREATER THAN 500 ug/L TCE
GREATER THAN 100 ug/L TCE
SCALE
600 0 600 Feet
CONTOUR INTERVAL 4 METERS
FIGURE 2
GENERAL WATER CHEMISTRY
JANUARY 1998
SITES 8, 9, AND 10, SLC-4
PREPARED FOR
VANDENBERG AFB
INTERNATIONAL TECHNOLOGY CORPORATION
PACIFIC OCEAN
Site 8
Site 9
Site 10
SPRING CANYON
SURF ROAD
COAST ROAD
KELP ROAD
AGENA WAY
NOTE:
ANALYTICAL DATA HAS NOT BEEN VALIDATED
REFERENCE:
BASEMAP FROM VANDENBERG AFB, GIS COVERAGES, DATED 1995.
771230-GB14

IN-SITU BIODEGRADATION OF CARBON TETRACHLORIDE IN A STRATIFIED AQUIFER SYSTEM

E. Erin Mack and John E. Vidumsky (E.I. du Pont de Nemours and Company, Wilmington, DE), Michelle M. Thomson (URS Corporation, Wilmington, DE)

ABSTRACT: Carbon tetrachloride was used at a former chemical manufacturing site and is detected in groundwater underlying the site. The aquifer system consists of distinct upper and lower water bearing zones separated by a leaky clay confining unit. Carbon tetrachloride plumes are present in both the upper and lower aquifers, apparently originating from the same source. Background conditions in both aquifers are anoxic, but not strongly reducing, with minimal dissolved oxygen and total organic carbon and high sulfate concentrations. The presence of chloroform and dichloromethane indicate that carbon tetrachloride degradation is proceeding via sequential reductive dechlorination pathways in both aquifers. However, it is only in the downgradient portion of the upper aquifer that these processes result in complete plume attenuation. The downgradient portion of the upper aquifer is characterized by a significant increase in organic carbon, increases in pH and alkalinity, and a decrease in redox potential. It is concluded that the addition of lime to the surficial marsh sediments (during a former remedial action) leached soil organic material into the upper aquifer and created physicochemical conditions conducive to the complete reductive dechlorination. In contrast, carbon tetrachloride degradation in the source area region of the upper aquifer and in the entire lower aquifer appears to be limited by the low levels of total organic carbon.

INTRODUCTION

Degradation of carbon tetrachloride (CT) in anoxic environments can occur as a result of both biotic and abiotic pathways. Biologically mediated pathways of carbon tetrachloride degradation follow either sequential reductive dechlorination of CT to produce chloroform (CF), dichloromethane (DCM) and ultimately CO_2 or methane, or net oxidation via substitutive dehalogenation of CT to CO_2 with little formation of CF or DCM. The former pathway has been observed under sulfate reducing (de Best et al., 1998) and methanogenic conditions (Van Eekert et al., 1998) and in cultures of acetogenic bacteria (Egli et al., 1988), while the latter pathway has been observed in a *Pseudomonas* isolate grown under nitrate reducing conditions (Criddle et al., 1990). CT is abiotically reduced to less chlorinated daughter products by reduced iron minerals (Matheson and Tratnyek, 1994) and reduced soil organic matter (Collins and Picardal, 1999). Groundwater CT contamination at other sites has been treated in situ by biostimulation of native CT degrading activity through addition of electron donor (Devlin and Müller, 1999), bioaugmentation with cultures of *Pseudomonas stutzeri* KC (Dybas et al., 1998), and zero valent iron permeable reactive barriers (Brown, 1997).

Site Description. During previous site investigation activities, carbon tetrachloride was detected in the groundwater beneath a former manufacturing plant. The alluvial aquifer underlying the site is divided into two major local units referred to as the upper and lower aquifers. The upper aquifer extends from the ground surface to a depth of approximately 17 m, with the groundwater surface located approximately 1.5 m below grade. The upper and lower aquifers are separated by a leaky, clay confining unit that varies in thickness from 0.5 to 1.5 m. The lower aquifer extends from the bottom of this confining unit to a depth of 36.5 m where it encounters a thick, competent, clay confining layer. Both aquifers consist primarily of fine to coarse-grained sands mixed with some silt.

The extent and behavior of the CT plume differs in the two aquifers. In the upper aquifer the CT plume attenuates within 400 m of the source area. In the lower aquifer attenuation of the CT plume is not observed within 700 m of the source. The groundwater at the site is naturally high in sulfate, typically low in total organic carbon (TOC), and contains 1,2 dichloroethane (DCA) as a co-contaminant with CT. The background geochemical conditions in both aquifers are anoxic but not strongly reducing with little or no dissolved oxygen (DO).

MATERIALS AND METHODS

Groundwater samples were collected from existing monitoring wells using low flow purging techniques. Field parameters pH, Eh, specific conductance, DO, and temperature were measured using calibrated field meters and a flow through cell. Concentrations for CT, CF, and DCM were measured following Standard Method 8260B, and methane concentrations were measured using a modified form of Standard Method 8015B. Chloride concentrations were determined by ion chromatography using Standard Method 300.0.

RESULTS AND DISCUSSION

Upper Aquifer. The upper aquifer is characterized by an upgradient source area, that likely contains residual phase CT, and a down gradient dissolved plume area. These zones differ in their physicochemical parameters and apparent fate of CT. Groundwater measurements in the upgradient source area indicate that this part of the aquifer is anoxic, but not reduced (redox ~ 100 mV), and low in TOC (<1ppm). The pattern of terminal electron acceptors is consistent with an unreduced anoxic environment in that nitrate becomes depleted from background levels of 12 ppm to less than 1 ppm in this area. However, there is no evidence of reduced iron, or sulfide, and methane levels are low (less than 1 ppm). Nor is there evidence of sulfate depletion from background levels of 300 ppm. Chloride levels in this area are slightly elevated above background concentrations (375 ppm as compared to 175 ppm), and the presence of the CT degradation daughter products, CF and DCM, indicates that some CT degradation is proceeding via sequential reductive dechlorination. However, the CT degradation appears to be severely limited in the source area because overall loss of CT mass is not readily apparent (CT levels are 50-100 ppm) and daughter products are present only at low concentrations (about 1 ppm). CT degradation coupled to microbial nitrate reduction is not expected to produce the less chlorinated daughter products, therefore the significance of nitrate depletion in this area is not clear and may be associated with degradation of other chlorinated site constituents.

In contrast to the upgradient source area, the down gradient dissolved plume area is nitrate depleted, characterized by a decreased redox potential (-150 mV), and exhibits increases in pH (from background of 7.3 to 7.9), TOC (up to 50 ppm) and nutrients (total nitrogen and phosphorous). The increase in TOC and nutrients is localized to an area under/downgradient of a former evaporation basin that was excavated into the marsh sediments. Upon closure this basin was filled with a sand and lime mixture to immobilize any possible residual metal contaminants. It is hypothesized that the lime addition sufficiently raised the pH in the basin area to extract organic matter from the marsh derived soils, and that the extracted organic matter was leached into the upper aquifer (alkali extractions are routinely used for the extraction of soil organic matter in the laboratory [Anderson and Schoenau, 1993]).

The pattern of electron acceptors in the downgradient portion of the upper aquifer is difficult to interpret. The low levels of reduced iron and lack of sulfide or depletion of sulfate indicate minimal iron reduction and no sulfate reduction, yet methane levels increase from 1 to 10,000 ppb. Chloride levels are significantly elevated above background concentrations in this area (up to 4,000 ppm), and there are low (less than 1 ppm) levels of the CT degradation daughter products CF and DCM. In contrast to the upgradient source area, a significant loss of CT mass is observed in the down gradient zone resulting in the complete attenuation of the CT plume in the upper aquifer. Simultaneous attenuation of the co-contaminant DCA is also observed at this point in the plume. It is likely that the accelerated degradation of CT in this area is associated with the increased levels of organic carbon and nutrients leached from the limed surface soils. Because the daughter products are present, it appears that at least some of the CT degradation occurs via sequential reductive dechlorination with the ultimate production of methane. The low levels of daughter products suggest that the rates of CF and DCM degradation may also be accelerated in this zone. The observed methane increase is comparable on a molar basis to the amount of CT mass that is lost. This may be the result of methanogenic conversion of any possible DCM degradation byproducts (acetate, formate, CO_2) to methane. Alternatively, the methane may arise from increased microbial activity resulting from the increase in nutrients and organic carbon in this area. It is unlikely that degradation of DCA contributes to methane in this zone.

The increase in chloride in this area is suggestive of dechlorination of CT and DCA. However, the 51 ppm increase in chloride expected from the complete dechlorination of the 20 ppm DCA and 40 ppm CT present in the downgradient portion of the plume is negligible compared to the observed 3,800 ppm increase in chloride. Clearly, the chloride increase cannot be attributed solely to degradation of chlorinated solvents and is likely due to infiltration of dissolved chloride salts associated with the historic waste streams that were discharged into the evaporation basin. Figure 1 illustrates the concentration trends for CT and its related degradation products moving from upgradient background conditions, through the source area, and then downgradient of the plume.

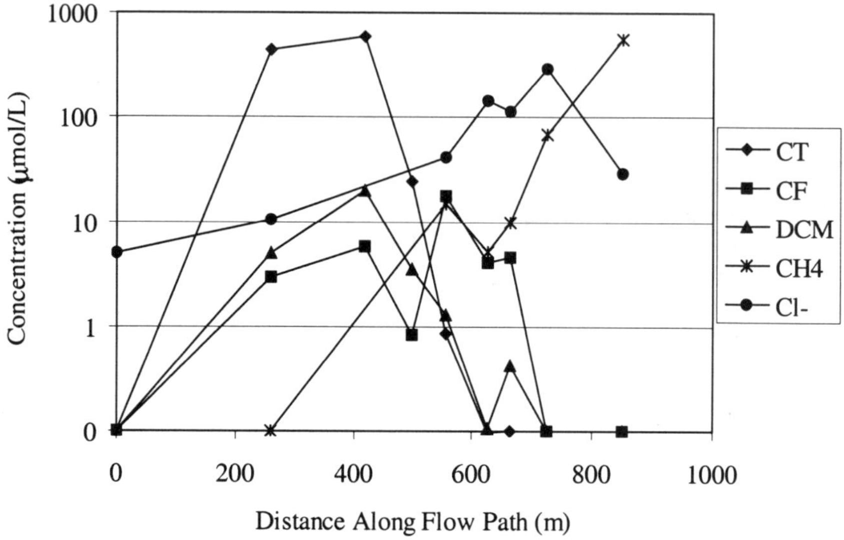

FIGURE 1. Upper aquifer transect of carbon tetrachloride related compounds – July 1999 (note: chloride concentrations are x10^{-3}).

Lower Aquifer. The lower aquifer makes an interesting comparison to the upper aquifer. Like the source area of the upper aquifer, the lower aquifer is anoxic but not reduced, and has low levels of nutrients and organic matter. Similarly, low levels of CT degradation daughter products are present in this aquifer but loss of CT mass is not observed. As is the case with the upper aquifer, the pattern of electron acceptor concentrations and chloride concentrations is difficult to interpret. Nitrate is depleted in the lower aquifer and there are low levels of reduced iron present (less than 1 ppm). Sulfate concentrations do decrease somewhat in this zone (from 400 ppm to 300 ppm) but sulfide levels are below detection. Methane increases slightly from 0 to 100 ppb. Chloride concentrations are above background but, as in the case of the upper aquifer, the degree to which this is due to CT degradation is obscured by the presence of other chlorinated compounds and the chloride salts present in the historic waste streams.

A limited amount of CT degradation via sequential reductive dechlorination does occur, as evidenced by the presence of small amounts of chlorinated daughter products and methane. The absence of nitrate, presence of minimal amounts of reduced iron, and observed decrease in sulfate concentration in this zone make it plausible that CT degradation is coupled to sulfate reduction. It seems likely that CT degradation in the lower aquifer is limited by the extremely low TOC concentrations (less than 1 ppm). Figure 2 illustrates a transect of the concentration trends for CT and its related degradation products in the lower aquifer plume area.

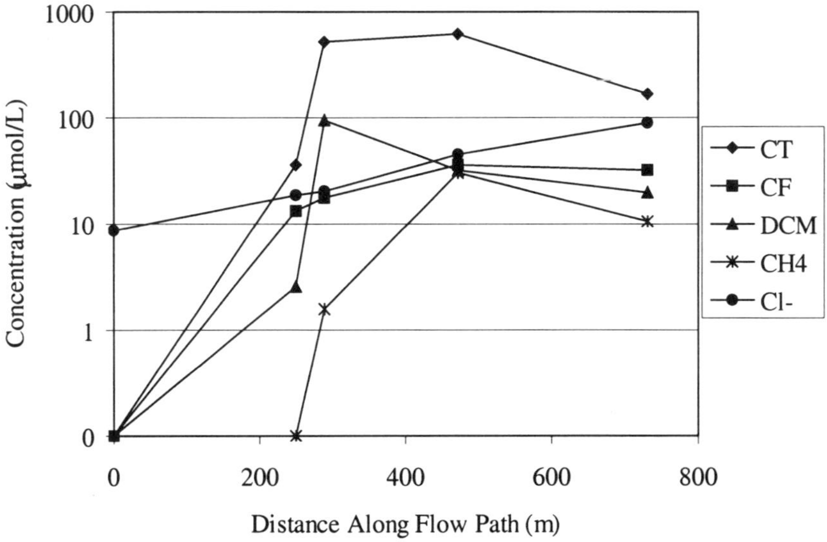

FIGURE 2. Lower aquifer transect of carbon tetrachloride related compounds – July 1999 (note: chloride concentrations are x10^{-3}).

Role of TOC in CT Degradation. The amount of TOC present in the aquifer appears to be the significant physicochemical factor determining the fate of CT this environment. There are at least two possible and compatible mechanisms for the observed acceleration of CT degradation in the presence of increased TOC. Organic matter may act as an electron donor and stimulate activity of native microbial populations competent for the degradation of CT. Alternatively, the alkali extracted soil organic matter may become reduced by microbial activities (either directly or indirectly) and act as an efficient electron donor for reductive dechlorination of CT. Degradation of CT in cultures of *Shewanella putrefaciens* 200 has been shown to be enhanced in the presence of alkali extracts of soil (Collins and Picardal, 1999). This enhancement was reduced significantly when the soil was oxidized with H_2O_2 (removing 62 % of the soil organic matter) prior to the alkali extraction. Curtis and Reinhard (1994) have shown a similar role for humic acids in mediating the reduction of CT by ferrous iron or sulfide in laboratory experiments. The results of this study, taken in context of these observations in the literature, suggest that TOC may be an important determiner of in situ rates of CT degradation.

CONCLUSIONS

The observed presence of CF, DCM, and methane indicates that CT is degrading via sequential reductive dechlorination in both aquifers. However, only in the downgradient zone of the upper aquifer are degradation rates fast enough to result in a significant loss of mass and attenuation of the CT plume. This zone is characterized by a reduction in redox potential and the presence of increased levels of TOC and methane. The extremely localized presence of TOC is likely due to the alkali extraction of soil organic matter from an area of surficial marsh sediments that were limed during an earlier remedial action. The presence of organic carbon appears to significantly enhance reductive transformation of CT and may be the key factor determining the environmental fate of CT in this aquifer system.

REFERENCES

Anderson, D. W., and J. J. Schoenau. 1993. "Soil Humus Fractions". In M. R. Carter (Ed.), *Soil Sampling and Methods of Analysis*, pp. 391-395. Lewis Publishers: Ann Arbor, MI.

Brown, M.J. 1997 "Passive and Semi-Passive Techniques for Groundwater Remediation". *In Situ and On-Site Bioremediation,* 4:181-186. Battelle Press, Columbus, OH.

Collins, R., and F. Picardal. 1999. "Enhanced Anaerobic Transformations of Carbon Tetrachloride by Soil Organic Matter". *Environ. Tox. Chem.* 18(12):2703-2710.

Criddle, C. S., J. T. DeWitt, D. Grbic-Galic, and P. L. McCarty. 1990. "Transformation of Carbon Tetrachloride by *Pseudomonas sp.* Strain KC under Denitrification Conditions". *Appl. Environ. Microbiol.* 56(11):3240-3246.

Curtis, G. P., and M. Reinhard. 1994. "Reductive Dehalogenation of Hexachloroethane, Carbon Tetrachloride, and Bromoform by Anthrahydroquinone Disulfonate and Humic Acid". *Environ. Sci. Technol.* 28(13):2393-2401.

de Best, J. H., E. Salminen, H. J. Doddema, D. B. Janssen, and W. Harder. 1998. "Transformation of Carbon Tetrachloride Under Sulfate Reducing Conditions". *Biodegradation* 8:429-436.

Devlin, J. F., and D. Müller. 1999. "Field and Laboratory Studies of Carbon Tetrachloride Transformation in a Sandy Aquifer under Sulfate Reducing Conditions". *Environ. Sci. Technol.* 33(7):1021-1027.

Dybas, M. J., M. Barcelona, S. Bezborodnikov, S. Davies, L. Forney, H. Heuer, O. Kawka, T. Mayotte, L. Sepulveda-Torres, K. Smalla, M. Sneathen, J. Tiedje,

T. Voice, D. Wiggert, M.E. Witt, and C. S. Criddle. 1998. "Pilot-Scale Evaluation of Bioaugmentation for In-Situ Remediation of a Carbon Tetrachloride-Contaminated Aquifer". *Environ. Sci. Technol.* 32:3598-3611.

Egli, C., T. Tschau, R. Scholtz, A. M. Cook, and T. Leisinger. 1988. "Transformation of Tetrachloromethane to Dichloromethane and Carbon Dioxide by *Acetobacterium woodii*". *Appl. Env. Microbiol.* 54:2819-2824.

Matheson, L. J., and P. G. Tratnyek. 1994. "Reductive Dehalogenation of Chlorinated Methanes by Iron Metal". *Environ. Sci. Technol.* 28:2045-2053.

Van Eekert, M. H., T. J. Schröder, A. J. M. Stams, G. Schraa, J. A. Field. 1998. "Degradation and Fate of Carbon Tetrachloride in Unadapted Methanogenic Granular Sludge". *Appl. Env. Microbiol.* 64:2350-2356.

RESULTS OF LABORATORY MICROCOSM STUDIES TO INVESTIGATE INTRINSIC BIODEGRADATION OF CARBON TETRACHLORIDE IN A FRACTURED BASALT AQUIFER

Michael E. Witt, Robert J. West, John W. Davis, Gary M. Klečka (The Dow Chemical Company, Midland, Michigan)

ABSTRACT: Groundwater contamination beneath a manufacturing facility has been the subject of an extensive remediation investigation over the past four years. Beneath this manufacturing facility exists a highly fractured basalt geologic formation containing groundwater contaminated with carbon tetrachloride (CT). Results of recent field investigations have indicated that dissolved-phase contamination appears to be attenuating downgradient from the manufacturing facility. Examination of historical groundwater data for footprints of natural attenuation suggests that biodegradation may play an important role in the fate and transport of CT at this site. A laboratory study was initiated to evaluate the potential for intrinsic bioremediation of CT and related chlorinated methanes. Laboratory microcosms were prepared with crushed basalt and groundwater to simulate actual field conditions, and were spiked with 5 mg/L of radiolabeled [^{14}C]CT. Results show that approximately 85% of the parent compound was degraded after 109 days. Compared to the observed 15% loss of CT in sterile (control) microcosms, this difference (70%) can be attributed to biological degradation of CT to non-chlorinated endproducts. Transient metabolites, namely chloroform and methylene chloride, were occasionally observed over the course of the study. However, complete dechlorination of CT to CO_2 was observed in most microcosms at the end of the one-year study. The results of this study indicate that intrinsic biodegradation of CT is a significant mechanism for natural attenuation of chlorinated methanes in this fractured basalt aquifer system. As a result, monitored natural attenuation may be employed as a component of the remedy at this chlorinated solvent-contaminated site.

INTRODUCTION

A groundwater sampling program was initiated to understand the type and extent of groundwater contamination in the vicinity of a manufacturing facility. The groundwater beneath the site contains elevated levels of chlorinated methanes, specifically CT, chloroform, and methylene chloride. Figure 1 details the regions where contamination has been measured in the groundwater. The chlorinated organic concentrations detected have generally been consistent over several sampling rounds, indicating no rapid advance of the groundwater plume. The overall distribution of chlorinated organic compound concentrations is consistent with the apparent groundwater flow direction, which is to the southeast of the site.

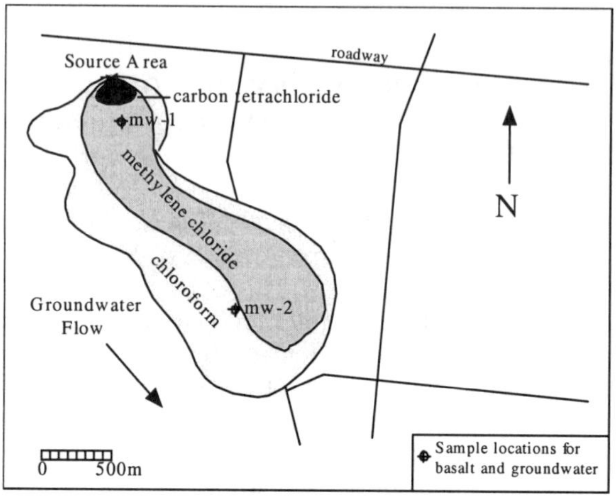

FIGURE 1. Plume map showing extent of groundwater contamination in a fractured basalt aquifer.

Examination of groundwater data provide evidence that natural attenuation of CT may be occurring in the subsurface. Monitored natural attenuation is a remedy that has come under increased awareness as a potentially important process that can lead to the destruction of many subsurface contaminants. In recent years, an increasing number of regulatory bodies view monitored natural attenuation as a viable process that may result in the destruction and/or immobilization of groundwater contaminants (USEPA, 1999). One of the most important natural attenuation processes, intrinsic biodegradation, can result in the destruction of the contaminant molecule to a chemical form that is harmless to human health and the environment. The focus of this study was to assess the importance of intrinsic biodegradation of chlorinated methanes in the aquifer beneath the manufacturing facility.

Radiolabeled carbon tetrachloride was used as a test chemical to assess the biodegradation potential of chlorinated methanes. Carbon tetrachloride has been shown to biodegrade under a variety of environmental redox conditions (Criddle et al., 1990; Egli et al., 1990). Chloroform is one of the predominant compounds found in site groundwater, and use of CT as a test compound provided an assessment of chloroform biodegradation.

The primary objective of this investigation was to determine if natural attenuation processes of contaminant destruction are occurring in the subsurface beneath this manufacturing facility. Laboratory studies were conducted to examine the biodegradation of CT in basalt/groundwater microcosms under conditions that simulate those in the aquifer.

MATERIALS AND METHODS

Groundwater and Basalt Samples. Basalt samples from the region of suspected contamination were obtained from new, temporary borings located near two existing monitoring wells. One core was collected from each sampling location using a 1.5-meter stainless steel core barrel. After each core was lifted to the surface, it was subdivided into three 50-cm sections, covered with plastic core sleeves, then placed within rigid PVC pipe to facilitate shipment. To assist in maintaining an anaerobic environment, pure nitrogen gas was used to purge the headspace inside two of the three PVC pipes prior to sealing the endcaps. The remaining section from each sampling location was sealed under ambient (aerobic) conditions. All sections were placed on ice in an insulated cooler and shipped overnight by Federal Express to the research laboratory.

Groundwater samples were also collected from existing wells adjacent to these boreholes. Groundwater was collected in one-liter polyethylene bottles and filled to zero headspace. The samples were placed on ice in an insulated cooler and shipped overnight by Federal Express to the laboratory.

Test Chemicals. The test chemicals used in the intrinsic biodegradation study were of the highest available purity. Radiolabeled (^{14}C) carbon tetrachloride was used during these experiments to facilitate determination of a material balance over the duration of the experiment, and also used to aid in identification and quantification of transformation products. All other chemicals used in preparation and analysis of the microcosms were of reagent-grade and obtained from commercial sources.

Preparation of Laboratory Batch Microcosms. Immediately prior to preparing the microcosms, solid basalt cores were crushed using a device constructed in the laboratory. This device consisted of a heavy stainless steel tube (4 inches in diameter and 12 inches long) with a stainless steel plug welded on one end of the tube. Pieces of the basalt cores were placed inside this tube and broken into smaller pieces by pounding the core with a foot-long stainless steel rod (3/4-inch diameter). Once the basalt pieces were crushed to a consistent size, the mixture was homogenized and portions were added to each microcosm as described below. This crushing procedure was performed in either an anaerobic chamber or in a fume hood (aerobic) in order to maintain the desired redox condition of the samples.

Laboratory batch microcosms (LBM) were constructed using crushed basalt and groundwater collected near the site. For this study, 5 grams of crushed basalt was placed in a sterile 20 mL glass vial (see Figure 2). Ten milliliters of authentic site groundwater were added to each bottle, resulting in ~50% headspace. Five milligrams per liter of CT was added to each LBM, and contained approximately 0.75 μCi of [^{14}C]CT activity. After addition of the test chemical, each microcosm was sealed immediately with a Teflon-lined red butyl rubber septum and an aluminum crimp seal. Sufficient quantities of LBM were prepared in order to allow for sacrificial analysis of triplicate samples at six time points during the course of the study.

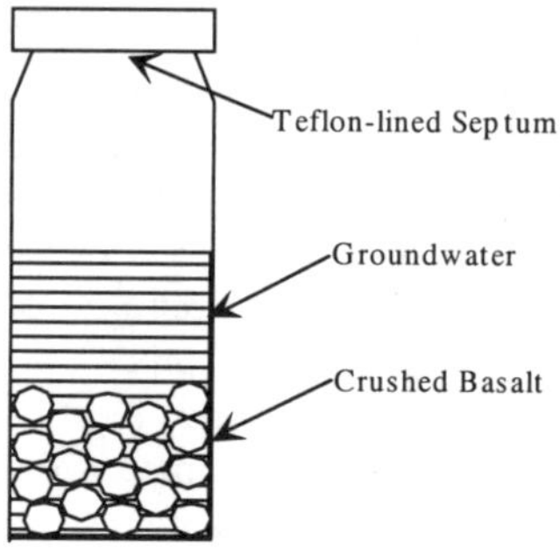

FIGURE 2. Diagram showing contents of laboratory batch microcosms (LBM) used to assess intrinsic biodegradation of CT.

Loss of the test chemicals due to irreversible sorption, permeation of the septum barrier, or other abiotic processes was measured using biologically-inhibited (killed) LBM. These microcosms were constructed in a manner similar to that of the active microcosms, with the incorporation of crushed basalt that was sterilized by autoclaving (1 hr for three consecutive days), filter sterilizing the groundwater through a 0.22 μm membrane filter, and adding a chemical sterilant (500 mg/L $HgCl_2$).

Redox Conditions. In order to evaluate the effects of redox potential on degradation, biotransformation of the test chemicals was examined under both aerobic and anaerobic conditions. Aerobic LBM were constructed as described above in laboratory fume hoods (under standard indoor atmospheric conditions). Anaerobic microcosms were prepared in a glove box under an anaerobic atmosphere equilibrated from a mixed gas containing N_2 and CO_2. A redox indicator dye, resazurin, was added to each microcosm as a concentrated solution to result in a final concentration of 2 mg/L of the total aqueous volume. All LBM were incubated at 20±2°C. Aerobic microcosms were gently agitated during incubation while anaerobic microcosms remained static.

Analytical Methods. The distribution of ^{14}C-label among parent compound, intermediate, and terminal degradation products was determined by headspace gas chromatography with a radioactivity monitor. Samples were analyzed using a system composed of a Tekmar model 7000 headspace autosampler and a Hewlett Packard model 5890 gas chromatograph equipped with an INUS Systems, Inc. model GC3 Radioactivity Detector. The headspace autosampler was operated as follows: 30-minute equilibration at 85°C, sampling valve temperature of 105°C, sample loop size of 3 mL at 105°C, transfer line temperature of 110°C with helium flow rate of 20 cm^3/min. The valve sequence used initially pressurized each vial to 10 psi for 10 seconds, vented to sample loop for 8 seconds, and then injected the sample volume for 2 minutes. The gas chromatograph was operated according to the following specifications. The injector was held at 200°C in continuous split mode with a 3 cm^3/min septum purge, 5 psi head pressure of helium, and an approximately 10:1 split ratio (35 cm^3/min total split flow). Separations were achieved using a Chrompack Poraplot Q HT column (25 m x 0.53 mm). The oven was operated at 45°C for 3.5 minutes, ramped to 180° at 30°/min, held for 8 minutes, ramped to 195° at 25°/min, held for 4 minutes, ramped to 250° at 25°/min, and held for 10 minutes.

Groundwater samples were also submitted to Midwest Laboratories, Inc., Omaha, Nebraska, for analysis of major minerals and trace metals, inorganic anions, and volatile organic compounds to confirm parameter concentrations in the groundwater used for this study.

RESULTS AND DISCUSSION

Biodegradation of CT was examined under aerobic and anaerobic conditions in microcosms containing crushed basalt and groundwater from two locations, MW-1 and MW-2 (see Figure 1). These microcosms were constructed to evaluate the potential for indigenous bacteria to degrade 5 mg/L of [^{14}C]CT in groundwater from the two indicated well locations. The primary difference between the two well locations was that groundwater from MW-1 contained approximately 1,100 mg/L total chlorinated organic compounds (TCOC), versus 2 mg/L TCOC in groundwater from MW-2. As a result, this experiment examined the potential for biodegradation of CT at one location where these contaminants persisted and bacteria had presumably adapted to the presence of high levels of dissolved CT (MW-1), and at another location where chlorinated groundwater constituents were detected at lower concentrations (MW-2) and where adaptation may not have occurred.

Analysis of the groundwater biogeochemistry suggests that anaerobic conditions are predominant in the vicinity of the manufacturing facility. Dissolved oxygen levels were low (or non-detect) in MW-1 during all sampling events in 1998 through 2000. Downgradient of MW-2 at a distance of 400 meters, dissolved oxygen levels increase to near 2 mg/L indicating that slightly aerobic conditions predominate. Between these two regions exists a transition zone where mildly reducing conditions prevail. Ferrous iron concentrations were quite high in MW-1, with levels in excess of 35 mg/L being consistently measured throughout 1999 and 2000. Nitrate levels were very low to non-detect. Sulfate levels near MW-1 were measured at 90 mg/L, with higher levels further downgradient (388 mg/L at MW-2). However, no sulfide was measured in this region, which suggests that sulfate reduction is not an active process in this portion of the aquifer. Dissolved methane levels were also very low or non-detect. As a result, the conceptual model of the biogeochemistry in the aquifer beneath the manufacturing facility shows that slightly reducing conditions predominate, with iron reduction being the dominant microbial process.

In anaerobic microcosms prepared using material from MW-1, approximately 45% of the added CT disappeared after 24 days (see Figure 3). Twenty-two percent (22%) of the initial radioactivity was recovered as chloroform and methylene chloride (11% each), and 3% was converted to [^{14}C]CO_2. After one year all of the [^{14}C]CT disappeared while 1% was recovered as [^{14}C]CF, 6% was recovered as [^{14}C]MC, and 7% was recovered as [^{14}C]CO_2. The estimated time to achieve 50% disappearance of the parent compound in these microcosms was ~35 days. In killed control microcosms, 49% of the initial radioactivity was recovered as CT, while 1% was recovered as [^{14}C]CF and 5% as [^{14}C]CO_2 after one year. No [^{14}C]MC was recovered after one year in the killed controls.

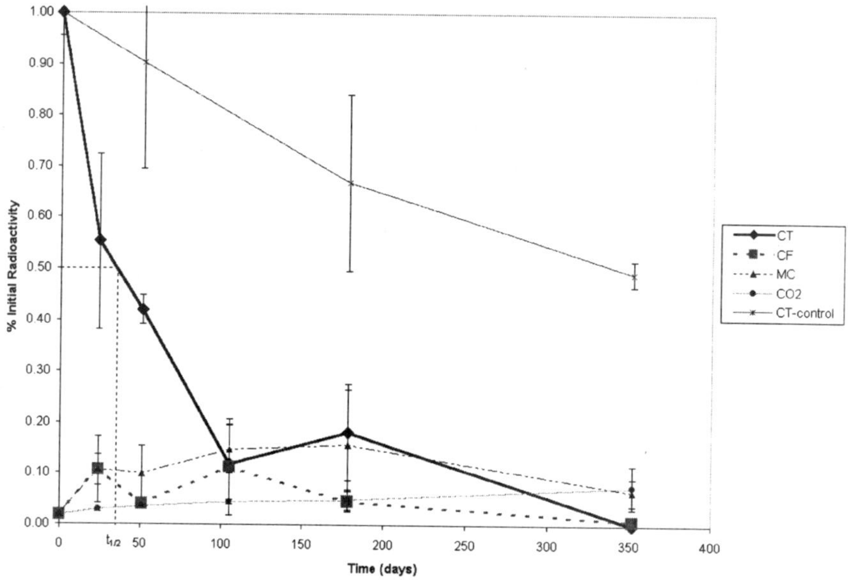

FIGURE 3. Biodegradation of 5 mg/L carbon tetrachloride in anaerobic microcosms prepared using crushed basalt and groundwater from MW-1.

In anaerobic microcosms prepared using material from MW-2, biological degradation of CT was observed after 24 days when approximately 42% of the added CT had disappeared (see Figure 4). Five percent of the initial radioactivity was recovered as chloroform, 4% as methylene chloride, and 9% as [^{14}C]CO_2, indicating that some degree of complete dechlorination had occurred. After one year, all of the [^{14}C]CT had disappeared from viable microcosms and 28% of the initial radioactivity was recovered as [^{14}C]CO_2. No chloroform or methylene chloride was measured after one year. The estimated time to achieve 50% disappearance of the parent compound in the anaerobic microcosms was ~45 days.

No significant biodegradation of [^{14}C]CT was observed in aerobic microcosms after a one-year period (as compared to the killed controls).

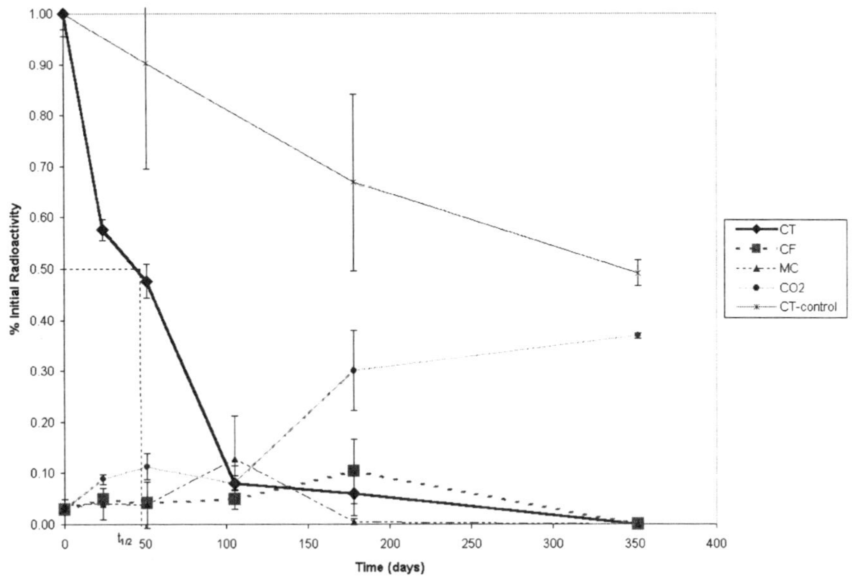

FIGURE 4. Biodegradation of 5 mg/L carbon tetrachloride in anaerobic microcosms prepared using crushed basalt and groundwater from MW-2.

As the study progressed, a decline in the material balance of [^{14}C]-labeled compounds in sterile (control) microcosms was observed. It appeared that a portion of the initial radioactivity had become incorporated into new biomass, was bound to the solid phase, or was lost from the test system. After 352 days of incubation, only 42% of the initial radioactivity was recovered as [^{14}C]CT in the aerobic (killed) controls and 1% was recovered as CF. In the anaerobic (killed) control microcosms, 49% of initial radioactivity was recovered as [^{14}C]CT, 1% as [^{14}C]CF, and 5% as [^{14}C]CO_2. These are microcosms where all biological activity was halted by adding 500 mg/L $HgCl_2$. At the completion of the study, septa were placed in scintillation cocktail and analyzed on a liquid scintillation counter for total radioactivity. Results indicated that about 30% of the total radioactivity added to aerobic (killed) control microcosms was associated with radiolabeled compounds absorbed by rubber septa. Septa from viable aerobic microcosms also showed similar results (29% of total radioactivity recovered). For the anaerobic control microcosms 32% of total added radioactivity was associated with sorbed radiolabeled compounds, and 23% radioactivity was recovered from viable anaerobic microcosms. The total radioactivity recovered from anaerobic microcosm septa was lower as a result of the degree of biodegradation of the parent compounds in those active microcosms. Given these results, it is evident that much of the radioactivity "lost" in the control microcosms was a result of ^{14}C-labeled compounds being absorbed by the Teflon-lined red butyl rubber septa.

CONCLUSIONS

Experimental results show that CT was degraded in microcosms from two different regions of the site. The time required for 50% degradation of CT ranged from 35 days in MW-1 microcosms to 45 days in MW-2 microcosms. Extensive mineralization of the test chemical CT was observed, as approximately 28% of the initial radioactivity was converted to $[^{14}C]CO_2$ after 352 days of incubation. The balance of radiolabeled material in each microcosm was likely sorbed by microcosm components (e.g. septa) and may have been incorporated into biomass during growth. These results suggest that an active iron-reducing microbial community exists in the aquifer that may have the ability to dechlorinate CT to non-hazardous endproducts. Dehalorespiring populations may also be contributing to the observed loss in CT with perhaps the ability to dechlorinate other types of chlorinated solvent contamination.

The results of field investigations, trends in contaminant levels, and laboratory microcosm experiments together provide some understanding of the importance of biological processes in the destruction of groundwater contamination at the manufacturing site. Field results suggest that biological degradation may be occurring in the aquifer and that chlorinated methanes are being degraded to less-chlorinated species. Biogeochemical parameters indicate that a reducing environment exists, which is typically needed to support microbial communities that have the ability to biodegrade highly chlorinated species. These results also indicate the potentially important role that monitored natural attenuation may have in the overall strategy for addressing groundwater contamination at this manufacturing site.

REFERENCES

Criddle, C.S., J.T. DeWitt, D. Grbic-Galic, and P.L. McCarty. 1990. "Transformation of Carbon Tetrachloride by *Pseudomonas* sp. Strain KC under Denitrification Conditions." *Appl. Environ. Microbiol.* *56*(11): 3240-3246.

Egli, C., S. Stromeyer, A.M. Cook, and T. Leisinger. 1990. "Transformation of Tetra- and Trichloromethane to CO_2 by Anaerobic Bacteria is a Non-enzymatic Process." *FEMS Microbiol. Lett.* *68*: 207-212.

USEPA. 1999. Use of Monitored Natural Attenuation at Superfund, RCRA Corrective Action, and Underground Storage Tank Sites. Office of Solid Waste and Emergency Response Directive 9200.4-17P.

NATURAL DEGRADATION OF A TCE PLUME IN PEAT

James A. Buss and Bernd W. Rehm. RMT, Inc., Madison, Wisconsin, USA
Gregory A. Davis, Microbial Insights, Inc., Rockford, Tennessee, USA

ABSTRACT: Chlorinated degreasing solvents released from a former wastewater lagoon form a groundwater plume that migrates through permeable outwash, ablation till, and peat to a wetland 350 feet away. Volatile organic compounds (VOCs) in the plume include trichloroethene (TCE); 1,1- and 1,2-dichloroethenes (DCEs); and chloroethene (CE). Total VOC concentrations are over 100,000 µg/L upgradient of the peat and approach 1 µg/L in the peat. Geochemical results are indicative of reductive dechlorination in the peat. As the mass of VOCs decrease, the relative concentrations of degradation products to parent compounds increase. Ethene, the last step in the dechlorination process, was observed at concentrations of up to 870 µg/L. Geochemical indicators showing anaerobic conditions, required for reductive dechlorination, included low reduction-oxidation potential (redox), dissolved oxygen (DO), nitrate, and sulfate concentrations and high dissolved iron and methane concentrations. Microbiological investigations performed using phospholipid fatty acid (PLFA) analysis and denatured gradient gel electrophoresis (DGGE) revealed a natural selection of the *in situ* bacterial population. Biomass estimates, determined by the total PLFA concentration, showed a significant increase in the center of the plume and directly downgradient in a sample from the peat (as compared to background biomass content). PLFA profiles also show a shift to a more active Gram-negative bacterial community in the center of the plume and in the peat when compared to background. Results from the DGGE profiles provided more specific identification of the bacterial communities, and revealed the presence of several organisms capable of degrading chloroethenes (*Dechloromonas, Actinomycetes* and *Geobacter*).

INTRODUCTION

Objective. Biodegradation of chlorinated solvents has shown promise to address remediation of impacted groundwater. This study documents naturally occurring reductive dechlorination of one such plume in peat at a former industrial facility in the upper Midwest. The unique aspect of this site is that dissolved concentrations of VOCs up to 100,000 µg/L decrease to concentrations of less than 10 µg/L over a distance of less than 350 feet. Most of the dramatic reduction occurs where soil transitions from mineral-rich ablation till and outwash to organic-rich peat.

MATERIALS AND METHODS:

Field Installations. Soil borings were drilled using a combination of hollow-stemmed auger direct-push techniques. Groundwater samples were collected from permanent PVC monitoring wells and temporary stainless-steel drive-points. Following development, samples were collected with a peristaltic pump using low-flow purging and sampling protocols. In-field measurements included pH, specific conductance, redox, DO, and temperature.

Chemical Analyses. Groundwater samples were analyzed for chloroethenes using standard U.S. Environmental Protection Agency (USEPA) Method 8260 (gas chromatography/mass spectroscopy, GC/MS). Dissolved gases (ethene, ethane and methane) were analyzed by RSK SOP-175, a gas chromatography headspace equilibration technique. Iron was analyzed using Method SW846 6010B. Other inorganics (chloride, total organic carbon [TOC]) were analyzed by various USEPA 300 and 400 series methods.

Microbiological Analyses. Microbial communities were assessed using PLFA and DGGE methods. These nonculture-based methods are not subject to the inherent limitation of more classical techniques. The PLFA analysis is a biochemical analysis performed by GC/MS that provides insight into the viable biomass, community structure, and metabolic activity of a sample (White et al., 1997). DGGE is a DNA-based technique that produces banding patterns representative of the numerically dominant organisms (Muyzer et al., 1993). Sequence analysis of prominent bands is used to infer species identity.

RESULTS AND DISCUSSION

Site Description and History. The site is an undeveloped 11-acre parcel located in a small urban area in the upper Midwest (Figure 1). At the upgradient end of the parcel, treated wastewater from an adjacent manufacturing facility was discharged to a ¼-acre wastewater lagoon for a period of approximately 15 to 20 years (1950's through the 1970's). The facility used TCE as a degreasing solvent along with petroleum-based cutting and lubricating oils. While there are no records on waste management practices, it appears that liquids (including chloroethenes and oils) left the lagoon by subsurface infiltration and (to a lesser extent) evaporation.

Geologic and Hydrogeologic Setting. The site is underlain by unconsolidated Pleistocene-age deposits. Regional information and site-specific investigations reveal a complex setting with up to four units occurring in the upper 25 feet below the site. A brief description of the units, from deepest to shallowest, follows:

- ***Basal till*** underlies the entire site and is a very stiff clay or silty clay till with some sand and gravel. It occurs at depths of 25 to 50 feet. This till functions as an aquitard and supports a dense nonaqueous-phase liquid (DNAPL) pool beneath part of the manufacturing facility.
- ***Ablation till*** is found above the basal till. It is typically softer, coarser-grained, and more permeable than the basal till.

- ***Outwash*** blankets the basal and ablation tills. The hydraulic conductivity of the outwash ranged from $1x10^{-3}$ to $6x10^{-3}$ cm/s and averaged $3x10^{-3}$ cm/s.
- ***Peat,*** an organic-rich unit, underlies the wetland, and surrounding area. With distance away from the wetland, it transitions into the outwash.

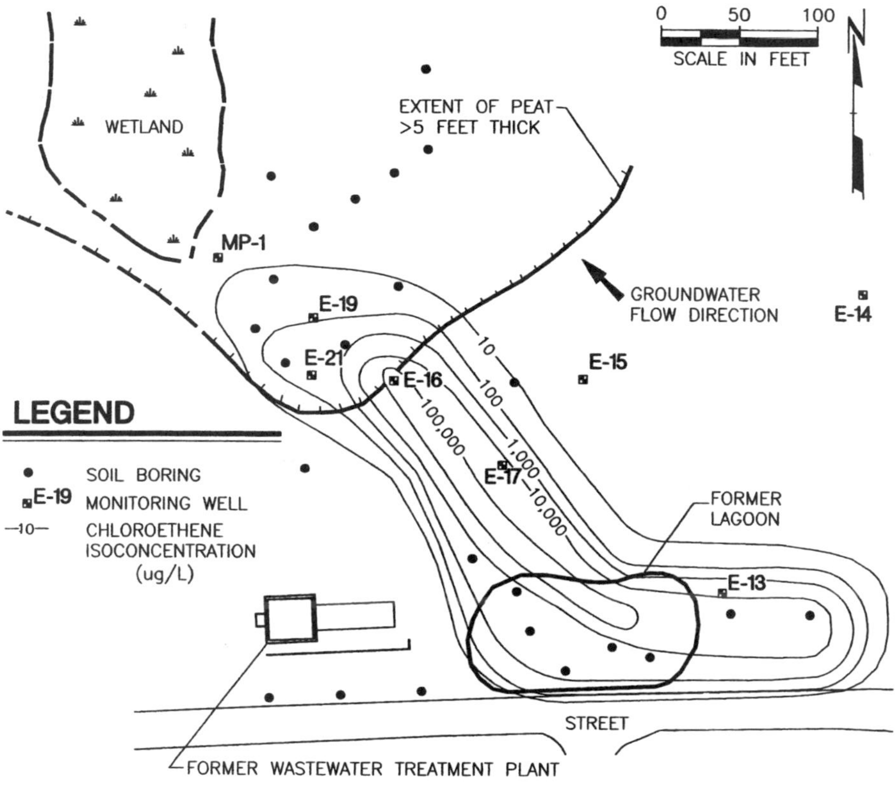

FIGURE 1. Site layout and Chloroethene Distribution

The outwash and peat were deposited in depressions within the surface of the tills. In some cases, the depressions were channels carved into the basal till.

The water table occurs in the outwash, peat, and till at a depth of 0 to 5 feet. Groundwater in the outwash and peat flows to the northwest, from the lagoon toward the wetland. The horizontal gradient of approximately 0.03 produces an estimated average linear groundwater velocity of 100 to 600 ft/yr.

Shallow groundwater discharges into a 2-acre wetland located 350 feet northwest from the lagoon. Shallow monitoring points installed in the wetland show a consistent upward gradient of approximately 0.08 from the underlying sediment into the wetland. Surface water drains from the wetland through a ditch on the western side of the wetland.

Occurrence of Chloroethenes. Based on the time frame within which the lagoon was in operation, the chloroethenes should have migrated to the wetland. Figure 1 shows that the chloroethenes form a narrow plume, which begins at the vicinity of the lagoon and extends to the area where the peat thickens to more than 5 feet and predominates the shallow flow system. Concentrations over 100,000 μg/L were found in E-16, approximately 230 feet from the lagoon and 120 feet from the wetland (Table 1). About 80 percent of the VOCs at this location are TCE, and less than 5 percent are CE. This downgradient limit of the high VOC concentrations is near the boundary where the outwash and ablation till transition into peat.

Groundwater samples from wells and borings about 25 to 60 feet downgradient of the peat/outwash boundary contained concentrations of DCEs and CE between 10 and 90 μg/L. TCE was detected at less than 5 μg/L. Moving further downgradient, a small groundwater seep adjacent to the wetland (MP-1) produced only 4.6 μg/L of CE and 0.63 μg/L of DCEs. Several phases of investigation failed to identify the discharge of the VOC plume into the wetland. Surface water samples in the wetland showed no detectable concentrations of chloroethenes (< 5 μg/L).

Ethene, the product of complete dechlorination of chloroethenes, is often difficult to detect because it is readily biodegraded. However, ethene was reported at a concentration of 870 μg/L at E-16 (upgradient of the peat), with lower concentrations in the peat at E-19 (61 μg/L) and E-21 (1.9 μg/L). Ethene was not detected in the background sample.

Natural Biodegradation Assessment. Lorah and Olsen (1999) presented evidence of reductive dechlorination of 1,1,2,2-tetrachloroethane (1,1,2,2-TCA) and TCE in organic-rich (peat) deposits adjacent to a tidal freshwater wetland. Initial concentrations of 1,1,2,2-TCA on the order of 2,000 μg/L and of TCE on the order of 300 μg/L were reduced by factors of about 1,000 and 100, respectively, (nondetectable levels) as the groundwater flowed through the highly anaerobic and methanogenic portions of the wetland.

The concentrations of chloroethenes on either side of the peat deposits encountered show 100,000-fold reductions. Concentration reductions alone suggest either degradation or simply sorption in the organic-rich peat deposits. However the relative increase of degradation products to the parent TCE suggests that the chloroethenes were being naturally biodegraded through reductive dechlorination (USEPA, 1998) and not just sorbed by the organic matter.

At this site, groundwater samples collected from wells located in the background (E-14), the center of the plume (E-16), and downgradient in the peat (E-19 and E-21) were analyzed to assess whether the geochemical conditions in the peat support degradation by reductive dechlorination. Samples were field-analyzed for pH, dissolved oxygen (DO), specific conductance, and temperature; and were laboratory-analyzed for dissolved ferrous iron, ethene, ethane, and methane. The analytical results are summarized in Table 1.

Several lines of geochemical evidence show that groundwater in the peat is strongly anaerobic, a condition required for reductive dechlorination. DO at the edge of the peat (E-16) is only 0.6 mg/L and decreases to 0.2 mg/L in the peat.

TABLE 1. Field and Laboratory Analytical Summary

Well Location	Background	Plume Center	Plume Within Peat	
Well Designation	E-14	E-16	E-19	E-21
Field Parameters				
pH (standard units)	7.0	6.8	6.7	7.0
Eh (mvolts)	57	-69	-139	-121
Dissolved oxygen (mg/L)	0.4	0.6	0.2	0.2
Conductivtiy (mmhos/cm)	150	1030	720	680
Temperature (celcius)	14.5	14.9	13.3	13.6
VOCs				
Trichloroethene, µg/L	<1.0	78000 D[1]	<1.0	<1.0
1,2-Dichloroethenes, µg/L	<2.0	23000	9.8	<2.0
Vinyl chloride, µg/L	<2.0	3400	22	<2.0
Dissolved Gases				
Ethene, µg/L	<0.50	870D	61	1.9
Ethane, µg/L	<0.50	33	1500D	490D
Methane, ug/L	87D	1600D	13000D	11000D
Total Organic Carbon	2.8	4.5	30	17
Inorganic Chemicals				
Iron, mg/L	<0.050	2.7	19	21
Potassuim, m/L	0.66	1.8	3.4	3.0
Chloride, mg/L	<20	110	35	44
Nitrogen, ammonia, mg/L	0.19	0.47	6.4	2.9
Nitrogen, tot. kjeldhal, mg/L	<0.50	0.89	<0.20	<0.20
Nitrogen, nitrate, mg/L	<0.20	<0.20	<0.05	<0.05
Nitrogen, NO2 + NO3, mg/L	<0.05	<0.05	6.9	3.9
Sulfate, mg/L	<0.20	40	<0.20	<0.20
PLFA Microbiology				
pmols PLFA/ml filtered	0.3	7	6	0.4
Cells/ml filtered	6.34E+03	1.35E+05	1.11E+05	8.69E+03
picomoles prokaryote PLFA	0.3	5	5	0.4
picomoles eukaryote PLFA	0.1	1	0.5	0.04
ratio prokaryote/eukaryote	5	4	11	9
Growth/Turnover Rate[2]				
cy17:0/16:1w7c	1.82	0.43	0.16	0.32
cy19:0/18:1w7c	0.34	0.17	0.40	0.00
Total	2.16	0.60	0.56	0.32
Environmental Stress[3]				
16:1w7t/16:1w7c	0.00	0.16	0.03	0.00

Notes 1. D = sample dilution outside calibration range

2. ratios < 0.1 log phase, 0.1 to 5.0 Stationary Phase, > 5.0 Decline Phase

3 ratios > 0.1 adapting to environmentally induced stress

Redox shows a strong negative shift from the background to the center of the plume to downgradient in the peat. Nitrate was not detected in any of the wells.

Where nitrogen species are present at downgradient locations, they are in reduced forms as ammonium and nitrite. Soluble iron increases from background (<0.05 mg/L), to the edge of the peat (2.7 mg/L), to downgradient within the peat (19 to 21 mg/L). Sulfate concentrations are reduced from 40 mg/L upgradient of the peat to nondetectable levels in the peat. Methane concentrations in the peat (13,000 and 11,000 μg/L) approach 50 percent of the methane solubility limit (24,000 μg/L). These results, showing iron, nitrogen and sulfate reduction and methane generation, indicate methanogenic conditions, an environment in which chloroethenes can be reductively dechlorinated through cometabolic pathways (Chapelle, 1993).

Microbiological Assessment. Microbial populations were assessed by a combined PLFA and DGGE approach. Samples were again collected from wells located in the background (E-14), the center of the plume (E-16), and downgradient in the peat (E-19 and E-21). Biomass concentrations, determined by total PLFA content, were considerably higher in the center of the plume (E-16) and directly downgradient (E-19) as compared to that found upgradient at E-14 (Table 1). Biomass at downgradient well E-21, which is not located along the center-line of the plume, was similar to the background well.

The PLFA profiles showed that the upgradient sample contained a very diverse community structure that was more evenly distributed than the other samples. Profiles from samples E-16, E-19, and E-21 all contained higher proportions of Gram negative bacteria; at least 12 percent higher than the background well. Evidence of metal-reducing bacteria and *Actinomycetes* were detected in all four samples.

Metabolic activity of the Gram negative community was assessed from ratios of different biomarkers. Ratios of these fatty acids (cy17:0/16:1w7c and cy19:0/18:1w7c, Table 1) showed that samples within the contaminated zone had faster turnover rates. The Gram negative community at the center of the plume (E-16) also showed evidence of adapting to the environment (i.e., the presence of VOCs) by decreasing the permeability of its cellular membrane. This is reflected in the summed ratio of trans/cis PLFA (16:1w7t/16:1w7c, Table 1).

The DGGE analysis allowed greater specificity in identifying specific bacterial populations. The DGGE bacterial profiles were complementary to the PLFA results revealing unique populations within each of the four sampling locations. The profiles within the plume revealed several prominent members, which appeared to be tolerant of the environmental conditions. Overall, several types of bacteria identified from the groundwater samples are capable of reductive dechlorination. These organisms include the following:

- *Actinomycetes* was found in the plume upgradient of the peat. This bacteria can decompose complex or resistant organic compounds.

- Species closely related to *Geobacter* were found in the peat. *Geobacter* falls within the Desulfomonas subgroup, and several members of this subgroup are capable of reductive dechlorination.
- Dechloromonas was found in samples from the peat. This bacteria has been associated with biodegradation of pentachlorophenol.

CONCLUSION

The investigation documented a plume of chloroethenes migrating in groundwater and undergoing natural biodegradation as the groundwater plume moves from a mineral-rich outwash/ablation till aquifer, into an organic-rich peat. Decreasing mass of chloroethenes and as a shift from parent to degradation products (including ethene) indicated that complete dechlorination was occurring. Geochemical conditions within the peat show favorable conditions for reductive dechlorination in an anaerobic environment.

Microbiological analysis show bacterial communities that are in an adaptive state with a number of species that can degrade organic compounds such as the chloroethenes.

REFERENCES

Chapelle, F. H., *Groundwater-Microbiology and Geochemistry,* 424 pp., John Wiley, New York, 1993.

Lorah, M.M. and L.D. Olsen. Natural attenuation of chlorinated organic compounds in a freshwater tidal wetland; Field evidence of anaerobic biodegradation. *Water Resour. Res.,* 35(12):3811-3827.

Muyzer, G., E.C. De Waal, and A.G. Uitterlinden, "Profiling of Complex Microbial Populations by Denaturing Gradient Gel Electrophoresis Analysis of Polymerase Chain Reaction-Amplified Genes Coding for 16S rRNA". *Applied and Environmental Microbiology,* 1993. 59(3):695-700.

USEPA. 1998. Technical protocol for evaluating natural attenuation of chlorinated solvents in groundwater. Office of Res. And Development, Washington, D.C., EPA/600/R-98/128.

White, D.C., H.C. Pinkart, and D.B. Ringelberg, "Biomass measurements: Biochemical approaches," in *Manual of Environmental Microbiology,* C.J. Hurst, et al., Editors. 1997, *ASM Press*: Washington. p. 91-101.

INTEGRATED ASSESSMENT OF MONITORED NATURAL ATTENUATION AT A PCE CONTAMINATED SITE

Mark A. Widdowson and ***Heather V. Rectanus*** (Virginia Tech, Blacksburg, VA)
Mathew J. Higgins (Bucknell University, Lewisburg, PA)
John T. Novak and Duane F. Berry (Virginia Tech, Blacksburg, VA)

ABSTRACT: The goal of this study was to interpret microcosm experimental results as part of assessment of monitored natural attenuation (MNA) at a tetrachloroethene (PCE) contaminated site. Models were developed using the SEAM3D (Sequential Electron Acceptor Model, 3 Dimensional) reductive dechlorination package to simulate microcosm experiments. The results of the numerical modeling study suggested that while the aquifer provided the reduced environment for reductive dechlorination, different reductive pathways were exhibited in each microcosm. In one microcosm, PCE was reduced sequentially to vinyl chloride (VC) while in the other microcosm partial reduction of PCE to *cis*-1,2-dichloroethene (*cis*-1,2-DCE) occurred with limited observed trichloroethene (TCE) production. Slow *cis*-1,2-DCE degradation rates from the first microcosm and the limited ability to degrade *cis*-1,2-DCE in the latter microcosm may stem from sparse microbial populations capable of reducing *cis*-1,2-DCE, different affinities of the microbes to degrade higher chlorinated compounds over lesser chlorinated compounds, or both.

INTRODUCTION

The widespread use of tetrachloroethene (PCE) in dry cleaning and industrial processes has led to PCE as an ubiquitous groundwater contaminant. While recalcitrant in aerobic environments, PCE has been shown to degrade by indigenous microorganisms under anaerobic conditions via sequential reductive dechlorination (Bradley, 2000; Fetzer, 1998). This ability of microorganisms to intrinsically biodegrade PCE promulgated the discussion of monitored natural attenuation (MNA) as a remediation strategy at PCE-contaminated sites. Biodegradation usually is the major contributing mechanism in the reduction of the contamination in MNA (Wiedemeier et al., 1998). However, many times groundwater data indicate that PCE is partially dechlorinated and its daughter products, *cis*-1,2-dichloroethene (*cis*-1,2-DCE) and/or vinyl chloride (VC), accumulate in the environment (Bradley, 2000). This buildup of the more reduced chlorinated solvents could be a result of decreased energy yield in the reductive dechlorination reaction or insufficient amounts of organic carbon to drive the reductive dechlorination reaction. Therefore, before MNA can be recommended for use at a site, the long-term ability of the site to support complete reductive dechlorination of PCE must be assessed.

Objective. The objective of the study is the interpretation of microcosm experiment data through the use of numerical modeling. The numerical microcosm models were designed to simulate microcosm experimental data with the goal being an improved model for predicting the microbiological processes and their affect on contamination in the field.

Site Description. The study site is located at a military facility in Virginia Beach, VA, and was home to a former exchange laundry/dry cleaning facility. From 1973 to 1978, the dry cleaning facility discharged dry cleaning waste containing approximately 200 gallons of PCE and 1100 gallons of soaps, sizing, and dyes into the storm sewer. In preparation for the construction of a new base commissary, the main portions of the storm sewer were removed in 1987, and the remaining portions of the storm sewer were removed in 1992. The plume is controlled hydraulically by a leaky sanitary sewer line.

MICROCOSMS EXPERIMENTS

Aquifer sediment samples were collected in 1999 from MLS12 and MLS22 at a depth of 1.8 m below ground surface (6 ft bgs) to provide aquifer sediment from an established active microbial location (MLS12) and from a shallow location adjacent to the leaky sewer line (MLS22). Along with the sediment samples, groundwater parameters such as chloroethenes and redox indicators were sampled at the MLS12 and MLS22 locations. Chloroethene concentrations at MLS12 and MLS22 are shown in Figure 1. At MLS12, PCE and each daughter product are observed in the groundwater. In the groundwater at MLS22, PCE, TCE and ethene were observed. Dissolved hydrogen and chloride concentrations are shown in Figure 2. Hydrogen concentrations indicated the terminal electron accepting process (TEAP) was sulfate reduction at MLS12 and methanogenesis at MLS22. Dissolved redox species, dissolved oxygen, sulfate, methane and iron, supported these findings (Rectanus, 2000).

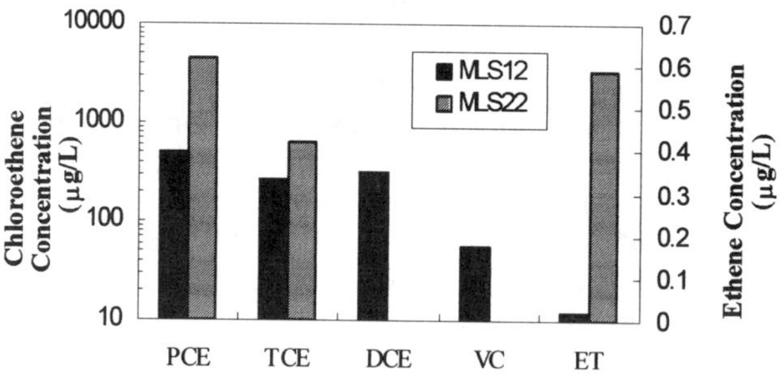

FIGURE 1. Chloroethene and ethene groundwater concentrations at MLS12 and MLS22.

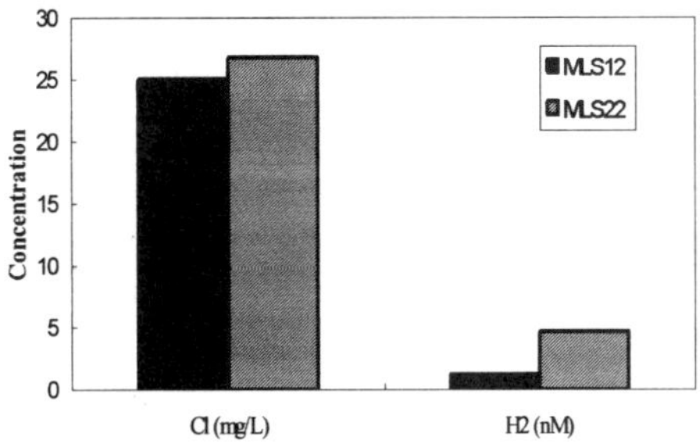

FIGURE 2. Groundwater parameters at MLS12 and MLS22.

The microcosms were prepared under anaerobic conditions using aseptically collected aquifer sediment samples in the laboratory. Each microcosm contained 150 g aquifer sediment, 200 mL autoclaved MLS12 groundwater and 30 μM of either PCE or trichloroethene (TCE). The microcosms were constructed in Erlenmeyer flasks that were retrofitted to house a Mininert™ valve. After construction, the microcosms were inverted and covered at 20°C under nitrogen until sampled (See Rectanus, 2000 for analytical methods). The controls were prepared in a similar manner to the live microcosms using autoclaved aquifer sediment and autoclaved groundwater.

The microcosms established that biological transformation of the chlorinated solvents occurs in aquifer sediment sampled from the site at both MLS12 and MLS22. Degradation of PCE, TCE and *cis*-1,2-DCE were observed in the MLS12 experiments, as well as, production of TCE, *cis*-1,2-DCE, and VC. In the MLS22 microcosm, partial degradation of the chloroethenes (PCE and TCE) to *cis*-1,2-DCE was observed.

MICROCOSM MODEL

Conceptual Model. The microcosms were modeled using the contaminant fate and transport model SEAM3D (Sequential Electron Acceptor Model, 3 Dimensional) developed by Waddill and Widdowson (1998) to estimate the biodegradation rates of PCE and its daughter products. SEAM3D utilizes the cell centered, finite difference approach to simulate contaminant fate and transport. A single layer finite difference grid consisting of four rows and four columns was constructed to simulate the experimental microcosm configuration. Based on the experimental microcosm configuration, a uniform distribution of the initial chloroethene concentration was required throughout the microcosm model. To simulate these experimental conditions, each cell in the model grid received an identical initial chloroethene concentration. Concurrent static groundwater conditions and uniform chloroethene concentrations eliminated the effects of advection and dispersion on contaminant fate. Based on previous sorption experiments that demonstrated the effects of sorption were negligible compared to the amended concentration of the chloroethene, the microcosm model did not include sorption. Therefore, biodegradation via reductive dechlorination remained as the only mechanism responsible for the reduction of the chloroethene concentrations over time.

Mathematical Model. The reductive dechlorination package (beta v 2.0) of SEAM3D was used to model the degradation of the chloroethenes occurring in the microcosm experiments. The reductive dechlorination package uses Monod kinetics as it sequentially reductively dechlorinates the chloroethene based on the following general equation,

$$\frac{dC_C}{dt} = -R_C^{bio} + \xi_{PC} R_{PC}^{bio} - R_{CDO}^{bio}, \tag{1}$$

where C_c is the concentration of the contaminant (such that the subscript c = 1-5, for PCE, TCE, *cis*-1,2-DCE, VC or ethene, respectively), R_C^{bio} is the biodegradation term for the contaminant, ξ_{PC} is the production coefficient of the contaminant generated by the parent compound (PC=c-1), R_{PC}^{bio} is the biodegradation term for the parent compound, and R_{CDO}^{bio} is the biodegradation term from direct oxidation. The biodegradation term, R_C^{bio}, can be represented by

$$R_C^{bio} = \frac{M_C}{\Theta} v_C = \frac{M_C}{\Theta} v_C^{\max} \left[\frac{C_C}{C_C + K_C} \right] I_C, \tag{2}$$

where M_C is the biomass of the contaminant, Θ is the effective porosity, v is the reaction rate of the contaminant, $v_C^{\max}$ is the maximum specific utilization rate of the contaminant, K_C is the half saturation coefficient of the contaminant, and I_C is the inhibition function of the contaminant. The biomass in SEAM3D is divided into two microbial population sets (Harkness et al., 1999; Magnuson et al., 1998). The first set incorporates the PCE and TCE degrading microorganisms, and the second set incorporates the DCE and VC degrading microorganisms. The two microbial populations can be presented in the forms,

$$\frac{1}{M_1} \frac{dM_1}{dt} = Y_{PCE} v_{PCE} + Y_{TCE} v_{TCE} - k_{d1}, \tag{3}$$

$$\frac{1}{M_2} \frac{dM_2}{dt} = Y_{DCE} v_{DCE} + Y_{VC} v_{VC} - k_{d2}, \tag{4}$$

where M is the biomass, Y is the yield coefficient, k_d is the decay coefficient. The inhibition function is represented by

$$I_C = \prod_{C=1}^{C-1} \left[\frac{K_{PC/C}}{K_{PC/C} + C_C} \right], \tag{5}$$

where $K_{PC/C}$ is the coefficient of the contaminant inhibition by the parent compound. Direct oxidation as discussed by Bradley and Chapelle (1998) of the contaminants, *cis*-1,2-DCE and VC, can be expressed as

$$R_{CDO}^{bio} = \sum^{TEAP} \frac{M}{\Theta} v_{CDO}^{\max} \left[\frac{C_{CDO}}{C_{CDO} + K_{CDO}} \right] \left[\frac{E}{E + K_E} \right], \tag{6}$$

where M is the microbial population of the direct oxidizers for a specific TEAP, $v_{CDO}^{\max}$ is the maximum specific utilization rate of the contaminant for a specific TEAP, K_{CDO} is the half saturation coefficient for a specific TEAP, E is the electron acceptor concentration associated with the TEAP, and K_E is the half saturation coefficient of the electron acceptor.

Model Parameters. A numerical microcosm model was developed for each microcosm. The degradation parameters were calibrated individually for each chlorinated compound and sequentially (starting with PCE) such that the chloroethene under investigation was simulated as the sole amended compound. Then, the inhibitory functions were fit as the model sequentially degraded the chloroethene beginning with PCE. The production coefficient, ξ_C, was set to 1:1 stoichiometery for sequential reductive dechlorination. If a chloroethene was not modeled, ξ_C was set equal to zero. To model steady state microbial kinetics, the initial microbial biomass terms were set equal to the porosity. The subsequent parameters of microbial growth, such as the yield coefficient and decay coefficient, were set equal to zero. The concentration versus time profile of the MLS22 microcosm was shifted to remove the 44-day lag phase that occurred at the beginning of the microcosm experiment. The initial concentration of PCE for the model was set equal to the original amended concentration. Concentration versus time profiles were used in calibrating the model to the experimental data and then in determining the reductive dechlorination rate parameters for biodegradation rate determination. In this study, the direct oxidation option was omitted in the microcosm model given the design of microcosm experiments did not allow for quantification of CO_2 production by the dechlorinating microorganisms. Given the aforementioned input parameters, the models were run over a period of 200 days at a 0.5-day time step with the explicit finite difference solution technique.

RESULTS AND DISCUSSION

PCE and TCE degradation. The maximum substrate utilization rate for PCE, $v_{PCE}^{\max}$, and the PCE half saturation coefficients, K_{PCE}, were equivalent between the two locations (Table 1). The MLS12 microcosm followed the conventional reductive dechlorination process where PCE is converted to TCE stoichiometerically (Figure 3). The TCE maximum substrate utilization rate, $v_{TCE}^{\max}$, for MLS12 was similar to $v_{PCE}^{\max}$. The K_{TCE} was smaller than the K_{PCE} by an order of magnitude, which indicates that low levels of TCE could be degraded faster than low levels of PCE. The TCE degradation parameters differed between the microcosms as a result of low levels of TCE observed in MLS22 (Figure 4). Due to the low levels of observed TCE, the MLS22 TCE-amended microcosm was used to estimate the TCE degradation rates for the MLS22 microcosm. These degradation rates were then applied to the MLS22 PCE-amended microcosm. However, manipulation of the inhibitory function of PCE on TCE degradation, $K_{PCE/TCE}$, could not simulate the observed TCE concentrations in the PCE-amended microcosm (Rectanus, 2000).

This inability of the model to simulate the degradation of TCE indicated that the TCE degradation in the MLS22 microcosm does not follow the conceptual pathway of sequential reductive dechlorination. Rather, PCE was degraded to *cis*-1,2-DCE before the reduced chloroethene was released to the aqueous environment. Holliger and Schumacher (1994) proposed that PCE degrading organisms could produce *cis*-1,2-DCE through intracellular TCE production. Upon intracellular *cis*-1,2-DCE production, the cell releases the *cis*-1,2-DCE to the aqueous environment. In order to simulate the observed TCE concentrations in MLS22, v_{TCE}^{max} was set at an artificially high value. Therefore, the reported values for TCE degradation in Table 1 do not reflect an aqueous phase degradation rate but could indicate an intracellular rate of TCE degradation.

***cis*-1,2-DCE degradation.** For MLS12, the maximum substrate utilization rate for *cis*-1,2-DCE, v_{DCE}^{max}, was lower in comparison to v_{PCE}^{max} and v_{TCE}^{max} by an order of magnitude. Since the production of VC was not observed in MLS22, the decrease in *cis*-1,2-DCE over time could not be substantiated as a result of reductive dechlorination. Nevertheless, estimated *cis*-1,2-DCE degradation rates by the numerical model were similar to the MLS12 *cis*-1,2-DCE degradation rates and were limited to comparative purposes. Haston and McCarty (1999) reported lower degradation rates for *cis*-1,2-DCE and VC in comparison to PCE and TCE. One explanation for the slower degradation rates could be due to insignificant numbers of *cis*-1,2-DCE degrading microbes present in the microcosms. Another explanation could stem from a preferential selection of the microbes to degrade higher chlorinated ethenes.

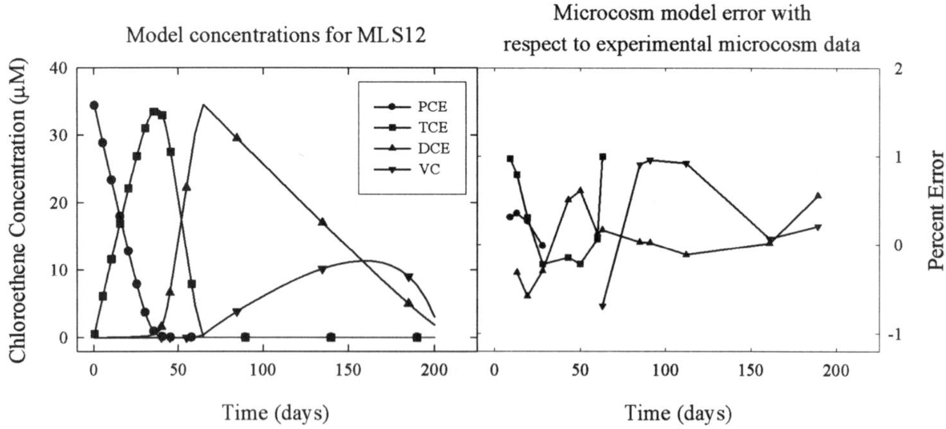

FIGURE 3. Microcosm model results for MLS12.

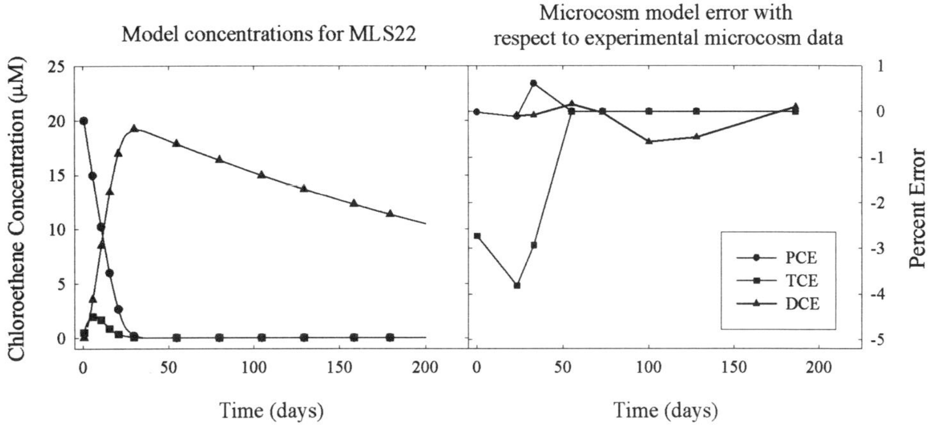

FIGURE 4. Microcosm model results for MLS22.

TABLE 1. Biodegradation parameters for the chloroethenes.

Microcosm	v^{max}_{PCE} (day^{-1})	v^{max}_{TCE} (day^{-1})	v^{max}_{DCE} (day^{-1})	K_{PCE} (mg/L)	K_{TCE} (mg/L)	K_{DCE} (mg/L)	$K_{PCE/TCE}$ (mg/L)	$K_{PCE/DCE}$ (mg/L)	$K_{TCE/DCE}$ (mg/L)
MLS12	0.40	0.42	0.05	0.41	0.01	0.07	0.01	0.01	0.01
MLS22	0.40	1.00*	0.06	0.57	0.20*	1.30	1.90	0.10	0.60

* - Artificially high rate to simulate the direct conversion of PCE to DCE.

CONCLUSIONS

SEAM3D was used to develop microcosm models to simulate the reductive dechlorination process occurring in experimental microcosms. The degradation parameters were based on Monod kinetics and illustrated either different affinities of the microbes to degrade higher chlorinated compounds over lower chlorinated compounds or insignificant numbers of *cis*-1,2-DCE degrading microbes present in the microcosms. Most importantly, the higher degradation rates of PCE and TCE corroborate the observations of PCE, TCE, and *cis*-1,2-DCE and intermittent observations of VC and ethene seen in the groundwater analysis. The results of the microcosm modeling study suggest that while the subsurface environment provides the reduced condition essential for reductive dechlorination, the site is not an efficient system for reductive dechlorination. The lack of the efficiency may stem from sparse microbial populations capable of reducing *cis*-1,2-DCE, high levels of PCE not conducive to reductive dechlorination beyond *cis*-1,2-DCE, or both.

ACKNOWLEDGEMENTS

Partial support for this research came from the National Science Foundation Graduate Fellowship program and the Charles E. Via, Jr. Foundation.

REFERENCES

Bradley, P.M. 2000. "Microbial Degradation of Chloroethene in Groundwater Systems". *Hydrogeology Journal. 8*(1):104-111.

Bradley, P.M., and F.H. Chapelle. 1998. "Microbial Mineralization of VC and DCE under Different Terminal Electron Accepting Conditions". *Anaerobe. 4*:81-87.

Fetzer, S. 1998. "Bacterial Dehalogenation". *Appl. Microbiol. Biotechnol. 50*:633-657.

Harkness, M.R., A.A. Bracco, M.J. Brennan, K.A. Deweerd, and J.L. Spivack. 1999. "Use of Bioaugmentation to Stimulate Complete Reductive Dechlorination of Trichloroethene in Dover Soil Columns". *Environ. Sci. Technol. 33*(7):1100-1109.

Haston, Z.C. and P.L. McCarty. 1999. "Chlorinated Ethene Half-Velocity Coefficients for Reductive Dechlorination". *Environ. Sci. Technol. 33*(2):223-226.

Hollinger, C. and W. Schumacher. 1994. "Reductive Dehalogenation as a Respiratory Process". *Antonie van Leeuwenhoek. 66*:239-246.

Magnuson, J.K., R.V. Stern, J.M. Gosset, S.H. Zinder, and D.R. Burris. 1998. "Reductive Dechlorination of Tetrachloroethene to Ethene by a Two-Component Enzyme Pathway". *Appl. Environ. Microbiol. 64*(4):1270-1275.

Rectanus, H.V. 2000. "Assessment of Intrinsic Bioremediation at a PCE-Contaminated Site". M.S. Thesis, Virginia Polytechnic Institute and State University, Blacksburg, VA.

Waddill, D.W. and M.A. Widdowson. 1998. "Three-Dimensional Model for Subsurface Transport and Biodegradation". *Journal of Environmental Engineering. 124*(4):336-344.

Wiedemeier, T.H., M.A. Swanson, D.E. Moutoux, E.K. Gordon, J.T. Wilson, B.H. Wilson, P.E. Haas, R.N. Miller, J.E. Hanson, and F.H. Chapelle (Eds.). 1998. *Technical Protocol for Evaluating Natural Attenuation of Chlorinated Solvents in Groundwater.* U.S. Environmental Protection Agency Technical Report, EPA/600/R-98/128, National Risk Management Research Laboratory, Cincinnati, OH.

SUITABILITY OF BIOREMEDIATION AT A FORMER SOLVENT RECYCLING SITE

Pamela J. Morrill, Jolyn C. Leslie, Mary M. Trute, and Patrick J. Evans
(Camp Dresser & McKee Inc., Bellevue, Washington)

ABSTRACT: The shallow aquifer at a former solvent recycling site is contaminated with mineral spirits, a variety of volatile aromatic and chlorinated solvents, and semivolatile organic compounds. Field monitoring was conducted to evaluate existing natural attenuation processes. Monitoring results and evaluation of historical data determined that the site is methanogenic, reductive dechlorination is actively occurring, and the plume is at steady state. However, an unacceptable time frame would be required to meet cleanup levels required for clean closure of this former RCRA treatment, storage, and disposal facility. Monitoring results further indicate that anaerobic biodegradation processes are unlikely to be enhanced by amendments such as electron donors. Laboratory treatability studies were conducted on other *in situ* alternatives, including surfactant flushing, chemical oxidation, and thermal treatment. Surfactant flushing appeared successful in contaminant removal, however treatment of the extracted surfactant-laden water would be very costly. Site contaminants, especially in soil, were not effectively degraded by chemical oxidation. Thermal treatment at high enough temperatures was successful in reducing contaminant concentrations to required cleanup levels. Although thermal treatment is very costly it appears to be the most promising of the technologies reviewed that would result in a clean closure for the site within a reasonable time frame.

BACKGROUND

Various remediation technologies were evaluated as a part of a feasibility study for cleanup of a former solvent recycling facility, as required by an Agreed Order issued by the regulatory agency.

For about 13 years the facility operated as a treatment, storage, and disposal (TSD) facility under Subtitle C of the Resource Conservation and Recovery Act (RCRA). The majority of spent solvents were mineral spirits generated from parts washing, but a variety of chlorinated solvents are also present as secondary contaminants. About 380,000 liters of spent solvent were processed per month. The facility currently operates as a chemical and petroleum storage, blending, and distribution facility.

SUBSURFACE CONDITIONS

The site lies on reclaimed tidal flats. Subsurface materials are characterized by dredge fill overlying tideflat sediments. The first 0.3 to 1.2 meters (m) is structural fill. This fill is underlain by dredge fill referred to as the "Upper Sand," which consists of clean to silty fine to medium sand (0.6 to 2.4 m thick). Beneath the Upper Sand is the "Upper Silt," which is the original tideflat material and consists of soft, clayey silt/silty clay with abundant organic material. Upper Silt occurs at about 1.8 to 3 m below ground surface (bgs) and ranges from 3.9 to 4.5 m thick. The silt is underlain by "Second Sand," which is a clean to silty fine sand. The thickness of this sand is unknown, but typically begins at about 5.8 to 6.4 m bgs.

Perched water occurs in the Upper Sand, beginning at about 0.9 to 1.5 m bgs. The Upper Silt acts as an aquitard between the perched water in the Upper Sand and the aquifer in the Second Sand.

CONTAMINATION TYPE AND EXTENT

The remedial investigation (RI) determined that contamination is essentially limited to the structural fill, Upper Sand, and perched water. The groundwater plume extends in a north-south direction across the site. The aquifer in the Second Sand is not affected. Site soil and groundwater contain a wide variety of contaminants of concern (COCs) as listed below. Light nonaqueous-phase liquid (LNAPL) is found sporadically at various locations.

Petroleum Hydrocarbons – mineral spirits (C_9-C_{11} aliphatic hydrocarbon mix).
Volatile Organic Compounds (VOCs) – tetrachloroethene (PCE), trichloroethene (TCE), 1,1,1-trichloroethane (1,1,1-TCA), 1,2-dichlorobenzene (1,2-DCB), and 1,4-dichlorobenzene (1,4-DCB), benzene, ethylbenzene, toluene, xylene (BETX), acetone, and 2-butanone.
Polycyclic Aromatic Hydrocarbons (PAHs)-benzo(a)anthracene, benzo(a)pyrene, chrysene.
Semivolatile Organic Compounds (SVOCs)–pentachlorophenol (PCP), 2-methylphenol, 4-methylphenol, and bis(2-ethylhexyl)phthalate (BEHP).

EVALUATION OF CURRENT BIOLOGICAL ACTIVITY

An evaluation of current biological activity was conducted by sampling wells along a north-south transect of the site. The purpose was to evaluate natural attenuation and enhanced biological degradation as remediation technologies for this site.

Methods. Groundwater samples were collected on two occasions: December 1999 and June 2000. In December 1999, ten wells were sampled. During the June sampling, three of the previously sampled wells were eliminated from the sampling program and two additional wells were added to provide a better profile across the center of the plume. The wells were purged and sampled by low-flow methods using a peristaltic pump. EPA protocols were used for sample collection and handling. A Hydrolab MiniSonde was used for field-measured parameters, except for ferrous iron, which was measured using a Hach test kit. Dissolved hydrogen was measured using the bubble strip method in combination with a reduction gas analyzer. EPA and standard protocols were used for laboratory analyses of organic and inorganic compounds.

Findings. The following table summarizes results of field-measured parameters and inorganic chemical analyses with an interpretation regarding potential degradation processes that would be suitable under the conditions indicated by the measured values.

TABLE 1. Summary of field-measured parameters and inorganic analyses.

Parameter	Plume Edge	Center of Plume	Interpretation
Oxygen (mg/L)	0.15 to 0.27	0.07 to 0.25	Less than 0.5 mg/L is ideal for reductive dechlorination.
Iron II (mg/L)	20 to 22.5	15 to 25	Concentrations are high, indicating a substantial amount of iron reduction has occurred.
Methane (mg/L)	7.0 to 10.32	9.93 to 10.84	Concentrations indicate highly methanogenic conditions.
Oxidation Reduction Potential (mV)	-19 to –109	-60 to –100	Not within the optimal range for reductive dechlorination, but low enough for it to occur. However, ORP data is not always reliable.
Nitrate (mg/L)	<0.005 to 0.011	<0.005 to 0.044	Very low concentrations, suggesting nitrate reduction is of little significance with respect to biological transformation.
Sulfate Sulfide (mg/L)	5.5 to 13 <0.05	9 to 12 <0.05 to 0.06	Not suggestive that sulfate reduction is actively occurring, but sulfate concentrations are within an optimal range to allow this process.
Hydrogen (nM)	1.2 to 18	1.3 to 7.6	Concentrations are supportive of reductive dechlorination
Chemical Oxygen Demand (mg/L)	16 to 82	180 to 400	Concentration within the plume is much higher than at the edges due to high concentrations of contaminants.
Total Organic Carbon (mg/L)	4.2 to 21	36 to 110	Optimal concentration for carbon as an electron donor and energy source is greater than 20 mg/L. Concentrations at the edge of the plume are less, because of less contamination.

Organic chemical analyses further show that reductive dechlorination is occurring throughout the site, as evidenced by substantial quantities of the "daughter" products of TCE and 1,1,1-TCA. TCE's first daughter product, *cis*-1,2-dichloroethene (DCE), is present at concentrations that are significantly higher than TCE concentrations. Even vinyl chloride (VC) concentrations are high with respect to TCE. None of these daughter products are known to have been present in significant quantity as secondary contaminants of the mineral spirits. Ethene was also detected in all wells that contain significant concentrations of chlorinated ethenes, suggesting complete dechlorination is occurring. Figure 1 compares concentrations of TCE and its daughter products across the site.

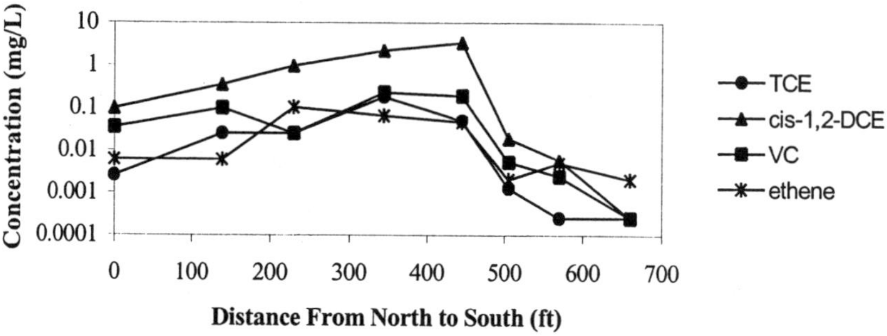

FIGURE 1. Chlorinated ethene concentrations across the site.

For 1,1,1-TCA, both the daughter products of biotic (1,1-dichloroethane [1,1-DCA]) and abiotic (1,1-dichloroethene) degradation are present in significant quantities. 1,1-DCA is higher in all instances than 1,1,1-TCA. Further, significant concentrations of ethane were present in nearly all wells, suggesting that complete dechlorination is occurring. Figure 2 compares 1,1-TCA and its daughter products of biotic degradation across the site.

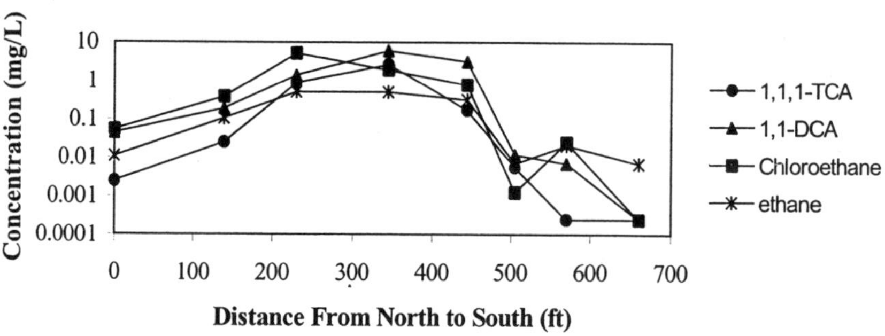

FIGURE 2. Chlorinated ethane concentrations across the site.

Petroleum hydrocarbon concentrations are high within the chlorinated solvent plume and decline to nondetectable concentrations about 55 m south of the site. The chlorinated solvent plume also ends at this location. In the north half of the site, the hydrocarbon plume appears to extend beyond the site boundaries. Additional source areas in the north half of the site complicate evaluation of natural attenuation.

Even with an active system of natural attenuation, the data and site modeling suggest cleanup to state and federal mandated levels for clean closure would require several hundred years. This is not consistent with current requirements to attain clean closure within a short timeframe. The data further suggest that attempting to enhance natural degradation processes would be impractical. For example, the site is heavily contaminated with mineral spirits; thus, adding an electron donor such as HRC® or molasses would provide no additional benefit. In addition, microcosm studies were conducted using various electron donors. The data did not show enhancement of reductive dechlorination (data not shown). Ferrous iron concentrations are high enough that attempting to increase aerobic conditions would be difficult due to the high demand for oxygen and would also likely result in iron fouling. Given these conditions, additional treatment methods were evaluated, including surfactant flushing, chemical oxidation, and thermal treatment. The following sections summarize the results of those studies.

SURFACTANT STUDIES

Surfactants are a class of chemical agents that show promise as an *in situ* technique for removing LNAPL from soil and aquifers. A literature search was conducted to select specific surfactants that have been used with some success under similar circumstances. Next, four phases of laboratory treatability studies were conducted. An initial screening was conducted on numerous surfactants using Sudan dye to evaluate NAPL removal efficiencies and determine the most effective concentrations. The second phase consisted of flask studies on select surfactants. The third phase consisted of a column study on the most promising surfactants. The fourth phase consisted of evaluating the effectiveness of various antifoaming agents on the surfactant/water solution.

Initial screening by dye testing and flask studies indicated a wide variation in contaminant removal by different surfactants. Three surfactants that performed the best during the initial screening tests were subjected to the column study; Min-Foam 2X, Brij 97, and a mix of Alcodet SK & 260. The column study also showed that some surfactants adversely react in the subsurface, as was the case for the Alcodet SK & 260 mix, which gelled and broke the column. Table 2 summarizes results of the column study for Min-Foam 2X and Brij 97. Both the Min-Foam 2X and the Brij 97 performed well. The overall performance of the Min-Foam 2X appears better than the Brij 97. Brij 97 also costs about twice as much as the Min-Foam 2X. Min-Foam 2X achieved nearly 100 percent removal of the VOCs and 86 percent removal of the TPH, as shown below. Removal efficiency of the SVOCs was not well defined due to elevated detection limits and inconsistency between the baseline and water control sample.

TABLE 2. Summary of surfactant column study results.

	Baseline	Water Control	Brij 97	Min-Foam 2X
Total VOCs (mg/kg)	4.5	5.4	0.37	0.062
TPH (mg/kg)	4,600	4,300	940	590

Analysis of the surfactant effluent from the columns and review of the chromatograms indicates that TPH removal is likely higher than indicated by the column studies. This is because some of the surfactant was incidentally quantified with the motor oil range hydrocarbons.

The most promising aspect of this treatability study was the apparent success of surfactant flushing in removal of the heavy petroleum hydrocarbons. Since the heavy end petroleum hydrocarbons and SVOCs should respond similarly, removal of SVOCs should also be substantial. In addition, much of the SVOCs are likely to be associated with the petroleum LNAPL. Removal of the LNAPL and the majority of contaminants adsorbed to the soil by surfactant flushing, as this study indicates is possible, would allow for later implementation of natural attenuation to remove residual contaminant concentrations.

Unfortunately, treatment of the extracted contaminated groundwater would be complicated by the surfactant. An interim groundwater treatment system that has been installed on the site includes an air stripper and activated carbon beds. Unless the air stripper is modified for operation at a very high pressure, excessive foaming will occur. Carbon is only intended for treatment of contaminants in the parts per billion/low parts per million range. At a 1-percent solution, the surfactant would overwhelm the activated carbon.

CHEMICAL OXIDATION

In situ chemical oxidation involves the injection of strong oxidants into the subsurface, in some cases with other chemicals that function with catalysts, to remediate soil and groundwater. This study consisted of three phases of in-house laboratory treatability studies, as well as laboratory treatability studies by two vendors. The first phase was an initial screening performed on hydrogen peroxide, Fenton's Reagent (hydrogen peroxide with an iron catalyst), and potassium permanganate at various concentrations. The second phase consisted of flask studies on select treatments performed in-house and by two vendors. The third phase consisted of a column study on the most promising chemical oxidizer.

The results of the flask studies indicated that significant VOC and SVOC degradation in site soil and groundwater could be achieved by chemical oxidation using Fenton's reagent, but TPH removal was practically nonexistent. The column study further showed that the effectiveness of chemical oxidation on soil treatment is significantly reduced when applied *in situ*. The effectiveness of SVOC removal during the column study was not well defined due to elevated concentrations of BEHP in the control and treated samples and lack of other SVOCs in the control and treated samples. In addition, acetone appeared to be created due to partial oxidation of alkanes. Based on screening measurements during the column study, there is no indication that better contaminant reduction would be achieved if treatment were to continue for a longer duration.

THERMAL

In situ thermal technologies involve the application of heat to the subsurface, which in turn causes an increase in contaminant volatility, thereby allowing removal by vapor extraction techniques. At high enough temperatures, contaminants may also be destroyed. Boiling points of the various COCs for this site vary widely. For 1,1,1-TCA, it is 74.1°C, for PCP it is 309.8°C, and for chrysene it is 448°C. However, physical and chemical changes induced during the heating process, such as generation of steam, may lower the temperatures required to achieve removal of some of the higher boiling point contaminants.

Several methods of *in situ* thermal treatment methods have recently become commercially available, including steam injection, six-phase electrical heating, radio frequency (RF) and thermal conduction, although they are generally still in the emerging technology stage. Each of the various methods of *in situ* heating has specific limitations that may make it unsuitable for use on any given site. For example, six-phase heating achieves its maximum temperature within a short period of time, but is limited to a maximum temperature of 100°C, too low for many COCs at this site. Steam injection works well in highly permeable soils, but poorly in low-permeability soil layers and lenses, leaving some areas untreated. Both RF heating and thermal conduction require that the area to be heated is dewatered; otherwise, a substantial amount of energy will be lost while attempting to heat the subsurface. RF heating theoretically can reach temperatures much higher than 100°C. However, information to date does not support the conclusion that RF heating can reach temperatures much higher than 100°C within a reasonable time frame.

Two laboratory treatability studies were carried out to evaluate heating technologies and the required temperature for treatment of site soils. Both studies were conducted as soil pan studies using 250 grams of homogenized soil in each pan. In the first study, four soil subsets were heated to 100°C, 150°C, 200°C, and 250°C in an aerated oven for one week. A second sample heated to 100°C had moisture added continually to simulate the six-phase heating process. One sample was subjected only to ambient conditions as a control. In the second laboratory study, the six-phase simulated heating process was repeated, but the test was carried out for a month. The results of these studies indicated that temperatures less than 150°C would not be suitable to attain treatment at this site. Temperatures between 100°C and 200°C resulted in the creation of ketones due to partial oxidation of alkanes. A temperature of 250°C achieved over 90 percent reduction in all contaminants as shown below. Table 3 compares contaminant removal with treatment temperatures for the one-week treatability study.

TABLE 3. Summary of thermal treatment test results.

	Baseline	Control	100°C*	100°C	150°C	200°C	250°C
Acetone (mg/kg)	<0.059	0.029	0.061	0.91	2.2	1.6	0.16
2-Butanone (mg/kg)	<0.059	0.015	0.016	0.35	0.7	0.33	0.02
Total VOCs (mg/kg)	2.9	0.045	0.085	1.3	2.9	1.9	0.19
Total SVOCs (mg/kg)	4.0	2.4	0.98	1.4	1.3	0.35	Not detected
TPH (mg/kg)	2,800	3,400	2,200	1,500	550	39	Not detected

* Water added during heating process.

Results of the laboratory treatability study indicate that thermal treatment would be an effective form of treatment. Based on the temperature required for effective treatment of site contaminants and nature of the subsurface materials, thermal conduction was considered the only method that would be effective for this site.

CONCLUSIONS

Natural attenuation processes are actively occurring at the site, as evidenced by field parameters, inorganic analytical results, and the presence of substantial concentrations of contaminant degradation products. The data suggest that the system is already optimized and that bioremedation would not be enhanced by addition of amendments or oxygen. In addition, review of monitoring data over the past 10 years indicates that the contaminant plume is stabilized. However, given the high contaminant concentrations at this site, attainment of remediation cleanup levels will not occur for hundreds of years, which is not acceptable to the regulatory agency.

The variety of chemical contaminants at this site and geology make it particularly challenging for implementation of other *in situ* remediation methods. As indicated in Table 4, thermal treatment is the most promising of the alternative technologies reviewed.

TABLE 4. Comparison of treatment technologies.

Treatment Method	Ability to Attain Cleanup Levels	Relative Cost	Comments
Natural Attenuation	Soil – No GW – No	$	▪ Reductive dechlorination and other biodegradation processes are active at this site. ▪ Contaminant plume has stabilized.
Surfactant	Soil – Probable GW – Probable	$$$	▪ Requires specialized treatment & disposal of extracted surfactant-laden water. ▪ May leave low permeability layers untreated.
Chemical Oxidation	Soil – No GW – Possible	$$	▪ Health and safety issues of using strong oxidizers.
Thermal	Soil – Yes GW - Yes	$$$$	▪ Possible destabilization issues when soils become desiccated. ▪ Methane must be removed prior to implementing treatment. ▪ Requires dewatering.

NAPHTHALENE AND PHENANTHRENE MINERALIZATION COUPLED TO FE(III) REDUCTION AND MECHANISMS OF ACCESSING INSOLUBLE FE(III)

J. Ramsay, K. Robertson, S. Meylan, S. Brown, Y.-S. Luu and P. Lee
(Queen's University, Kingston, Ontario, Canada K7L 3N6)
B. Ramsay (Polyferm Canada Inc., Harrowsmith, Ontario, Canada K0H 1V0)

ABSTRACT: A mixed anaerobic culture from a coal-tar contaminated sediment was found to mineralize naphthalene and phenanthrene when Fe(III) was provided as the major terminal electron acceptor. Naphthalene disappearance of about 100% in 20 days was measured by fluorescence spectrometry in non-radiolabeled experiments. There was a maximum of 53% mineralization detected in radiolabeled naphthalene microcosms in 147 days. Some of the naphthalene that was not mineralized, disappeared due to sorption but the majority of unmineralized naphthalene appeared to have accumulated as metabolic intermediates. The microorganisms at this site appeared to use oxygen, nitrate, Fe(III) and sulphate to degrade naphthalene. Experiments in which ferrihydrite was separated from the microorganisms inside dialysis membranes indicated that direct contact was necessary for extensive Fe(III) reduction, but a compound able to pass through the dialysis membrane also played a role in Fe(III) reduction. The presence of siderophore producers was demonstrated on agar plates containing an Fe(III)-chrome azurol S complex.

INTRODUCTION

Some polyaromatic hydrocarbons (PAHs) are toxic, mutagenic, and carcinogenic to humans (Cerniglia, 1992) and are known to persist in the environment. They are listed as priority pollutants by the US Enivronmental Protection Agency (Keith and Telliard, 1979). The aerobic degradation of aromatic hydrocarbons has been well studied (Cerniglia, 1993; Crawford, 1991). It has been shown that the degradation of monoaromatic and polyaromatic hydrocarbons can occur under nitrate-reducing (Al-Bashir et al., 1990; Burland and Edwards, 1999; MacRae and Hall, 1998, Mihelcic and Luthy, 1988; Rockne et al., 2000), sulphate-reducing (Coates et al. 1997; Edwards et al., 1992; Lovley et al. 1995; Meckenstock et al., 2000; Zhang and Young, 1997), methanogenic (Genthner et al., 1997; Parker, 1995) or Fe^{3+}-reducing (Anderson and Lovley, 1999; Lovley and Phillips, 1988) conditions. Iron is the second most abundant element present in the earth. However, under the conditions that microbial activity would occur, Fe(III) is poorly water soluble (e.g. 10^{-19} M of soluble Fe(III) in a saturated $Fe(OH)_3$ solution at neutral pH) (Seeliger et al., 1998).

The first pure culture capable of anaerobically oxidizing aromatic hydrocarbons using Fe(III) as a terminal electron acceptor (TEA) was isolated by Lovley and Phillips (1988). This bacterium, GS15, (later called *Geobacter metallireducens*) completely mineralized toluene, phenol and *p*-cresol heterotrophically using Fe^{3+} as the electron acceptor (Lovley and Lonergan, 1990). Although naphthalene degradation has been shown to occur under Fe(III)-reducing conditions in a petroleum-contaminated aquifer (Anderson and Lovley, 1999), a direct link between naphthalene degradation and Fe(III) reduction has not been proven.

This study examines whether a mixed microbial culture from a sediment at a former coal gasification site is capable of using a variety of TEAs and, in particular, Fe(III) in the degradation of PAHs such as naphthalene and phenanthrene.

MATERIAL AND METHODS

Source of Inoculum. The inoculum used in these experiments were either a coal-tar-contaminated sediment obtained from a former coal gasification plant (Kingston, Ontario) or a one-step enrichment culture of the sediment which had been added to a defined mineral salts medium (MSM) (Lovley and Phillips, 1986) containing naphthalene and ferrihydrite (Lovley and Phillips, 1986).

Microcosm Studies with Various TEAs. Duplicate biotic and abiotic microcosms were prepared in 125 mL amber bottles covered with a butyl rubber septum secured by a screw cap with a hole in the middle of the cap. Three times the stoichiometric amount of each terminal electron acceptor required to mineralize 10 mg/L of naphthalene was added in MSM.

Naphthalene Mineralization in Excess Fe^{3+}. Duplicate biotic and abiotic microcosms were prepared in 125-mL glass, serum bottles sealed with a butyl rubber stopper and aluminium crimp. A stoichiometric excess of Fe(III) was added to 100 mL of MSM containing 20 or 50 mg/L of either naphthalene, phenanthrene, anthracene, fluoranthrene, pyrene, and benzo(a)pyrene spiked with about 100,000 counts per minute (cpm) of the corresponding ^{14}C-isotope dissolved in toluene or methanol. CO_2 was trapped in 1N KOH and radioactivity measured with a Beckman LS 6500 Scintillation Counter.

Anaerobic Conditions. Anaerobic conditions were created by purging with oxygen-free nitrogen gas.

Naphthalene Analysis by Fluorescence Spectrophotometer. Prior to sampling, microcosms were shaken vigorously for 10 seconds and the sediment allowed to settle for 2 hours. A volume of 0.5 mL from the upper aqueous phase was added to a vial containing 2.5 mL of ethanol and 2.0 mL of deionized water. All samples were stored in the freezer until time of analysis. Prior to analysis, all samples were placed in a sonicator for 10 min. Naphthalene concentration was measured using a fluorescence spectrophotometer (Photon Technologies International) with the source of light coming from a 75 watt xenon lamp. A synchronous scan was run for each sample at the appropriate excitation (265 nm) and emission (312 nm) wavelengths. Results from the scans were recorded online using Felix 1.11 software from Photon

Technologies International. Using a calibration curve, peak areas for each sample were used to calculate naphthalene concentration. Between samples, the cuvette was rinsed with ethanol and dried with nitrogen gas.

Analyses. Methane in the headspace was analysed using a Varian Star 3400 gas chromatograph equipped with a flame ionization detector. The injector and oven temperatures were at 50°C while the detector temperature was at 150°C. Helium was used as the carrier gas at a flow rate of 50 mL/min. Methane was analyzed at the maximum sensitivity possible and had a retention time of 0.15 min. Nitrate and sulphate were analysed by ion chromatography using a Dionex DX-300 with a conductivity detector. Samples were filtered through a 0.45 μm filter, then injected into Ionpac AS9-SC guard and analytical columns with 1.7 mM sodium bicarbonate and 1.8 mM sodium carbonate solution as mobile phase. Total iron was analysed with an Atomic Absorption Spectrometer, Varian Spectra AA 20 plus, using an air and acetylene flame with absorption measured at 248.3 nm.

RESULTS AND DISCUSSION

Naphthalene Disappearance. In separate, non-radiolabeled microcosms, an excess of either oxygen, nitrate, Fe(III) or sulphate theoretically necessary to mineralize 10 mg/L of naphthalene was added. Naphthalene concentration was measured by fluorescence spectrometry and its disappearance due exclusively to biotic processes, as reported in Table 1, was obtained by subtracting the abiotic losses from the biotic microcosms. When no TEA was added, there was little (12%) disappearance of naphthalene in 16 days (Table 1). Naphthalene disappeared fastest when nitrate was provided as the TEA compared to O_2, Fe(III) and SO_4^{2-}. The latter three had similar rates of disappearance.

No TEA. Although no TEA was added, there were 38 mg/L of SO_4^{2-} and at least 26 mg/L of Fe(III) transferred with the inoculum. The 12% disappearance of naphthalene due to biotic processes in 16 days (Table 1) was probably the result of using only Fe(III) as the TEA as there was no reduction in the SO_4^{2-} concentration (Table 2). The lack of TEA resulted in a slow rate of naphthalene disappearance.

O_2. The headspace of these microcosms was filled with pure oxygen resulting in about five times the minimal amount of O_2 required for the mineralization of 10 mg/L naphthalene. Under these conditions, the rate of naphthalene disappearance was similar to when Fe(III) or sulphate was provided as TEA (Table 1).

NO_3^-. Naphthalene disappearance was fastest when nitrate was provided as a TEA (Table 1). By day 60, all the naphthalene had disappeared (data not shown). With three times the stoichiometric amount of NO_3^-, 211 mg/L of nitrate was consumed and about the same amount of nitrite was produced after 60 days of incubation (Table 2). The theoretical amounts of nitrate required to mineralize 10 mg/L naphthalene with N_2 or nitrite as the end product of nitrate reduction are 46 and 116 mg/L respectively (last column of Table 2). The amount of nitrate converted to nitrite was in excess of the theoretical amount necessary to mineralize 10 mg/L of naphthalene. This may have been due to the presence of other organics transferred with the inoculum.

TABLE 1: Naphthalene disappearance as measured by fluorescence spectrometry in the presence of various terminal electron acceptors (TEAs). Initial naphthalene concentration was 10 mg/L with at least three times the stoichiometric amount of TEA for mineralization. (* % disappearance in biotic - % disappearance in abiotic microcosms)

TEA added	% disappearance at day 16 due to biotic processes only*	Time for 90% naphthalene disappearance (days)
none	12	41
oxygen	31	23
nitrate	71	15.5
Fe(III)	36	25
sulfate	28	25

Fe(III). Since Fe(III) ions are essentially insoluble at neutral pHs, it is difficult to measure directly the change in Fe(III) concentrations. When Fe(III) is reduced, the more soluble Fe(II) ions are formed. In this study, the change in the total soluble Fe concentration was determined by Atomic Absorption and assumed to represent Fe(II) concentration. Thus the change in total soluble Fe should represent the amount of Fe(II) produced from Fe(III) reduction or from the solubilization of Fe(III). More than the theoretical amount of Fe(III) required to mineralize 10 mg/L naphthalene (last column of Table 2) was measured as a change in total soluble Fe at day 60 when there was no detectable naphthalene present. Trace amounts of SO_4^{2-} was transferred with the inoculum and the change in SO_4^{2-} concentration was much less than the amount required to mineralize the naphthalene added. Based on the Fe(III) and SO_4^{2-} data, Fe(III) was likely used as a TEA. Furthermore, there was no significant change in the soluble iron concentration in the abiotic control.

SO_4^{2-}. After 60 days of incubation, there was a disappearance of 40 mg/L of sulphate (Table 2). This was quite close to the theoretical amount needed to mineralize 10 mg/L of naphthalene (i.e 45 mg/L SO_4^{2-}).

CO_2. Methane was not detected in the headspace of any of the microcosms tested indicating that either CO_2 was not utilized as a TEA or that the amount of CH_4 produced was too low to be measured.

^{14}C-Microcosom Studies in Excess of Fe(III). In separate experiments with excess Fe(III) and no other TEA added and at an initial concentration of 20 or 50 mg/L of a single PAH, the coal-adapted consortium was able to mineralize up to 50% ^{14}C-

TABLE 2: TEA concentrations at day 0 and 60 at an initial concentration of 10 ppm naphthalene. The actual amount of TEA consumed and the theoretical amount of each TEA needed to mineralize 10 ppm naphthalene are also shown.

	Day 0			Day 60				TEA consumed			Theoretical TEA to mineralize 10 mg/L naphthalene
	NO_3^- (mg/L)	soluble Fe (mg/L)	SO_4^{2-} (mg/L)	NO_3^- (mg/L)	NO_2^- (mg/L)	soluble Fe (mg/L)	SO_4^{2-} (mg/L)	NO_3^- (mg/L)	Fe(III) (mg/L)	SO_4^{2-} (mg/L)	(mg/L)
TEA added											
none	tr	45 ± 9	38 ± 8	tr	-	71 ± 14	42 ± 8	0	26	-	
nitrate	294 ± 29	32 ± 6	24 ± 5	83 ± 8	249	56 ± 11	30 ± 6	211	24	-	46 (NO_3^- to N_2) 116(NO_3^- to NO_2^-)
Fe(III)	tr	120 ± 24	41 ± 8	tr		390 ± 78	31 ± 6	0	270	-	209 (Fe(III))
sulfate	tr	-	178 ± 18	tr		nd	138 ± 14	0	-	40	45 (SO_4^{2-})

naphthalene, 6% ^{14}C-phenanthrene, 0.8% ^{14}C-anthracene and trace amounts of ^{14}C-pyrene and ^{14}C-benzo(a)pyrene in 147 days (Table 3). Nitrate and sulphate analyses at day 0 indicated the presence of trace or nondetectable amounts. The $^{14}CO_2$ mineralization results together with the TEA data indicate that this consortium is capable of mineralizing various PAHs including naphthalene using oxygen, nitrate, Fe(III) and sulphate as TEA.

TABLE 3: Mineralization of ^{14}C-PAHs in the presence of excess Fe(III) by coal-tar adapted consortium

PAH	Initial conc. (mg/L)	% mineralized at day 147	
		Abiotic	Biotic
naphthalene	20	5.1	53
	50	5.1	4.9
phenanthrene	20	0.2	6.9
anthracene	20	0	0.8
fluoranthene	20	0	0.1
pyrene	20	0	0.2
benzo(a)pyrene	20	0	0.3

These results indicate that if TEAs such as O_2, NO_3^-, Fe(III) and SO_4^{2-} were present at this site that intrinsic biodegradation of PAHs, particularly naphthalene and phenanthrene, would occur. It would appear that methanogenic biodegradation of naphthalene would be least likely.

CURRENT INVESTIGATIONS

In these studies, the Fe(III) is provided as ferrihydrite which is quite insoluble and hence, perphaps, not bioavailable. We are currently investigating the mechanism(s) by which the microorganisms at this site access insoluble Fe(III). Preliminary experiments in which the ferrihydrite was separated from the microorganisms inside dialysis membranes indicate that although direct contact was necessary for extensive iron reduction, and a compound able to pass through the dialysis membrane also played a role in Fe(III) reduction. The presence of siderophore producers was demonstrated on agar plates containing an Fe(III)-chrome azurol S complex (Schwyn and Neilands, 1987). Although complete naphthalene disappearance by fluorescence spectrometry was observed, a maximum of 53% naphthalene mineralization was obtained. Identification of the intermediates is currently being attempted.

CONCLUSIONS

1) The microbial consortium from coal-tar contaminated sediment used oxygen, nitrate, Fe(III) and sulphate as terminal electron acceptors to degrade naphthalene. CO_2 was probably not used as a TEA as methane was not detected in any microcosm

tested.
2) In the presence of excess Fe(III), the consortium mineralized up to 50% naphthalene, 6% phenanthrene, 0.8% anthracene and trace amounts of pyrene and benzo(a)pyrene in 147 days.

ACKNOWLEDGEMENTS
The authors thank Environmental Science and Technology Alliance Canada and Queen's University for their financial support.

REFERENCES

Al-Bashir, B., T. Cseh, R. Leduc, and R. Samson. 1990. "Effect of soil/contaminant interactions on the biodegradation of naphthalene in flooded soil under denitrifying conditions". *Appl. Microbiol. Biotechnol. 34*:414-419.

Anderson, R.T., and D.R. Lovley. 1999. "Naphthalene and benzene degradation under Fe(III)-reducing conditions in petroleum-contaminated aquifers". *Bioremediation J. 3*:121-135.

Burland, S.M., and E.A. Edwards. 1999. "Anaerobic benzene biodegradation linked to nitrate reduction". *Appl. Environ. Microbiol. 65*:529-533.

Cerniglia, C.E. 1992. "Biodegradation of polycyclic aromatic hydrocarbons". *Biodegradation. 3*:351-368.

Cerniglia, C.E. 1993. "Biodegradation of polycyclic aromatic hydrocarbons". *Curr. Opinion Biotechnol. 4*:331-338.

Coates, J.D., J. Woodward, J. Allen, P. Philp, and D.R. Lovley. 1997. "Anaerobic degradation of polycyclic aromatic hydrocarbons and alkanes in petroleum-contaminated marine harbor sediments". *Appl. Environ. Microbiol. 63*:3589-3593.

Crawford, R.L. 1991. "Bioremediation of groundwater pollution". *Curr. Opin. Biotechnol. 2*:436-439.

Edwards, E.A., L.E. Wills, M. Reinhard, and D. Grbic-Galic. 1992. "Anaerobic degradation of toluene and xylene by aquifer microorganisms under sulfate-reducing conditions". *Appl. Environ. Microbiol. 58*:794-800.

Genthner, B.R.S., G.T. Townsend, S.E. Lantz, and J.G. Mueller. 1997. "Persistence of polycyclic aromatic hydrocarbon components of cresote under anaerobic enrichment conditions". *Arch. Environ. Contam. Toxicol. 32*:99-105.

Keitn, L.H., and W.A. Telliard. 1979. "Priority pollutants I - A perspective view". *Environ. Sci Technol. 13*:416-423.

Lovley, D.R., J.D. Coates, J.C. Woodward, and E.J. P. Phillips. 1995. "Benzene oxidation coupled to sulfate reduction". *Appl. Environ. Microbiol. 61*: 953-958.

Lovley, D.R., and D.J. Lonergan. 1990. "Anaerobic oxidation of toluene, phenol, and p-cresol by the dissimilatory iron-reducing organism, GS-15". *Appl. Environ. Microbiol. 56*:1858-1864.

Lovley, D.R., and E.J.P. Phillips. 1986. "Organic matter mineralization with reduction of ferric iron in anaerobic sediments". *Appl. Environ. Microbiol. 51*:683-689.

Lovley, D.R., and E.J.P. Phillips. 1988. "Novel mode of microbial energy metabolism: organic carbon oxidation coupled to dissimilatory reduction of iron or manganese". *Appl. Environ. Microbiol. 54*:1472-1480.

MacRae, J.D., and K.J. Hall. 1998. "Biodegradation of polycyclic aromatic hydrocarbons (PAH) in marine sediments under denitrifying conditions". *Wat. Sci. Tech. 38*:177-185.

Mihelcic, J.R., and R.G. Luthy. 1988. "Degradation of polycyclic aromatic hydrocarbons under various redox conditions in soil-water systems". *Appl. Environ. Microbiol. 54*:1182-1187.

Meckenstock, R.U., E. Annweiler, W. Michaelis, H.H. Richnow, and B. Schink. 2000. "Anaerobic naphthalene degradation by a sulfate-reducing enrichment culture". *Appl. Environ. Microbiol. 66*:2743-2747.

Parker, W.J. 1995. "Fate of polynuclear aromatic hydrocarbons during anaerobic digestion of municipal wastewater sludges". *Water Environ. Res. 67*:1052.

Rockne, K.J., J.C. Chee-Sanford, R.A. Sanford, B.P. Hedlund, J.T. Staley, and S.E. Strand. 2000. "Anaerobic naphthalene degradation by microbial pure cultures under nitrate-reducing conditions". *Appl. Environ. Microbiol. 66*:1595-1601.

Schwyn, B., and J.B. Neilands. 1987. "Universal chemical assay for the detection and determination of siderophores". *Anal. Biochem. 160*:47-56.

Seeliger, S., R. Cord-Ruwisch, and B. Schink. 1998. "A periplasmic and extracellular c-type cytochrome of Geobacter sulfurreducens acts as a ferric iron reductase and as an electron carrier to other acceptors or to a partner bacteria". *J. Bacteriol. 180*:3686-3691.

Zhang, X., and L.Y. Young. 1997. "Carboxylation as an initial reaction in the anaerobic metabolism of naphthalene and phenanthrene by sulfidogenic consortia". *Appl. Environ. Microbiol. 63*:4759-4764.

INVESTIGATION OF PAHS INTRINSIC DEGRADATION IN SOILS OF FORMER COKE FACILITY SITES

Christine Lors, Jean-Remi Mossmann, Pascal Barbé, Agnès Laboudigue
(Centre National de Recherche sur les Sites et Sols Pollués, Douai, France)

ABSTRACT: Both microcosm and respirometric experiments have been used toevaluate PAHs intrinsic degradation in a soil from a former coke site. The study has been carried out in solid phase over a period of 8 months, at 20°C, in aerobic conditions, on a soil sample which contained about 1 g/Kg dry soil of the 16 PAHs from the US EPA list and a metallic pollution in Zn, Pb, Cu and Cd, reaching about 1 g/Kg dry soil. Microbial degradation activity was assessed by monitoring the CO_2-production and O_2-consumption as well as by following the decrease in PAHs content in the soil samples. Experiments on sterilised samples were also carried out, in order to evaluate the part of abiotic processes in the degradation. Part of the most volatil and soluble PAHs (2 and 3 ring PAHs) are rapidly eliminated (within 7 days) by abiotic processes. Afterwards, biologic phenomena become dominant in the PAHs degradation. They contribute alone to a 25% decrease in total PAH content. This concerned mainly 3- and 4- cycle PAHs, which were degraded by respectively 30 and 25%. For the 3- cycle PAHs, the most reduction was for anthracene, phenanthene and fluorene, whereas for the 4- cycle PAHs, fluoranthene significantly decreased. 5- and 6- cycle PAHs hardly or not biodegraded even after 8 months. O_2-consumption and CO_2-production revealed a high microbial activity which is associated with the decrease in PAH concentrations. Long term experiments in laboratory are thus suitable to estimate the potential to natural attenuation.

INTRODUCTION

Polycyclic aromatic hydrocarbons (PAHs) are the most important organic pollutants in natural media, with petroleum hydrocarbons and chlorinated solvents. Environmental agencies take interest in these compounds for their recognised toxic, mutagenic and carcinogenic effects on humans. These pollutants are present since many decades in soils of former sites from coal pyrolysis activities. In order to manage these sites according to a risk assessment approach, it is important to have a better knowledge on PAH behaviour in soils. Among processes involved in their transformation, biodegradation is mainly responsible for the decrease of PAH concentrations in soils.

Objectives. The general goal of this study is to demonstrate PAH's intrinsic degradation in a soil of a former coke facility site. The first step of this study was to determine if environmental conditions were favourable to the PAHs biodegradation through a complete characterisation of a soil sample from the site. The second step used laboratory experiments to estimate PAH's natural biodegradation by evaluating kinetics of biodegradation of these compounds and by following the evolution of physico-chemical and microbiological parameters likely to influence this microbial process.

MATERIALS AND METHODS

The investigations were carried out on a soil sampled from a former coke site. The degradation study was conducted in a first step by microcosm experiments (250 mL glass beakers, which contained 50 g of soil sieved to 2 mm). The soil moisture content was to 80% of field capacity. The microcosms were incubated at 20°C, in aerobic conditions for 8 months. Recurrent samples were prepared for analysis after 0, 7, 28, 63, 98, 147 and 224 days. Degradation was evaluated by following the decrease in PAHs concentration over time. Concentrations of the 16 PAH compounds (US-EPA) were quantified using High Performance Liquid Chromatography coupled with a UV photodiode array detector (HPLC - UV). Physico-chemical (pH, organic carbon, total nitrogenous and phosphoric compounds) and microbiological (total bacterial and PAH degrading microflorae) parameters were also investigated during the experiment.

For bacterial enumeration, microflorae were extracted from 5 g of soil after shaking in 50 mL sterile Ringer solution (Ramsay, 1984). Enumeration of microorganisms was performed on the supernatant. The most probable number (MPN) of specific PAH-degrading microorganisms was determined on 96-well microplates according to Stieber et al. (1994) with the following PAH used as sole carbon and energy source: naphthalene, phenanthrene, fluorene, and fluoranthene. For total microflora, the growth substrate was nutrient broth.

Experiments with sterilized soil were also carried out in order to evaluate the part of strict abiotic processes.

Beside these microcosm experiments, respirometric studies were carried out to evaluate the O_2-consumption and CO_2-production over time for the same soil sample. Experiments were carried out in a respirometer (BI-16, Bioscience, USA), consisting of eight 250 mL flasks placed in a water bath maintained at 20°C. Each flask, containing 50 g of soil, was connected to an electrolytic oxygen producing cell, which maintained a constant oxygen pressure in the flask by defined electric impulses whereas CO_2 produced was trapped in a 5 mL solution of KOH (1 M). O_2-consumption was measured continuously by the respirometer equipment. PAHs concentration, PAHs degraders, chemical and toxicity parameters were also determined in the reactors at different periods of time, corresponding to the different phases of the degradation as previously identified in microcosm experiments.

RESULTS AND DISCUSSION

Soil characterisation at the initial state

The physico-chemical characteristics of the soil are reported in Table 1. This soil contains about 1 g/Kg dry material of the 16 PAHs from the US EPA. Anthracene and fluoranthene are the main components among the 16 PAHs (Figure 1). It is also polluted in metallic compounds, particularly Zn, Pb, Cu and Cd, reaching about 1g/Kg dry soil. This soil shows a high toxicity (2083 T.U.), which is explained by the effect of both organic and inorganic pollution. The physicochemical characteristics (pH close to 7 and moisture close to 20%) and nutritive quality (Total C, N and P) of this soil are favourable to facilitate biodegradation.

TABLE 1. Physico-chemical characteristics of the soil sample.

Texture	Moisture (%)	pH	Toxicity (T.U.)	T.O.C. (%)	Total N	Total P	16 PAHs	Zn	Pb	Cd	Cu
					(mg/Kg of dry soil)						
Silty sand	19	8	2083	13	2595	2970	1124	537	128	15	224

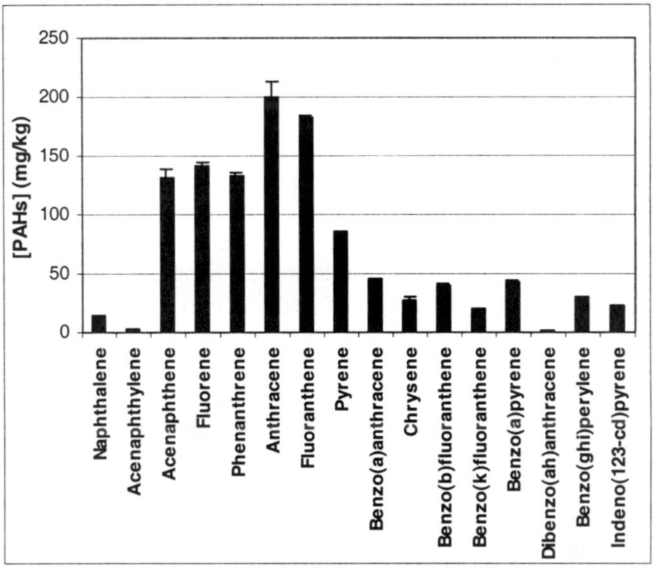

FIGURE 1. Initial concentrations of 16 PAHs in the soil sample.

In the same way, the microbiological investigations revealed the occurrence of good biological activity; the total microflorae is in the range usually found in unpolluted soil (10^8 bacteria and 10^5 fungi per g of dry soil). The high proportion of 3- and 4- cycle PAHs degraders (10^6 à 10^7 bacteria per g of dry soil) highlights adaptation of this microflora to the aged pollution (Kästner et al., 1994).

DEGRADATION EXPERIMENTS IN MICROCOSMS

Kinetics of PAH's degradation. 2-cycle PAHs (mainly naphthalene) which occur in very small amount in the beginning of the experiments (~ 1%), follow similar evolution pattern either in biotic or in abiotic conditions (Figures 2a et 2b). Disappearance of naphthalene is thus mainly due to abiotic processes such as volatilization.

3-cycle PAHs disappear following abiotic processes at the beginning of the experiment (between day 0 and day 7); then physico-chemical evolution becomes negligible with respect to biodegradation (Figure 2a et 2b). The latter is then the main responsible in 3-cycle PAH degradation. About 25% of 3-cycle PAHs disappear during the 21st following days after the abiotic stage (Figure 2a). A new stage of degradation occurs after a 70 days lag time. After 8 months incubation, about 53% of the 3 cycle PAHs are biologically degraded. Anthracene alone, which is the dominant contaminant in the 3-cycle PAHs, contributes to 29% of the degradation observed, followed by fluorene (Figure 3). Phenanthrene is degraded by about 62% (biotic and abiotic processes) and contributes to about 20% to 3-cycle PAH disappearance. The pattern of fluorene disappearance is similar to that of anthracene (60% off the initial fluorene).

4-cycle PAHs show very similar trend to 3-cycle PAHs, but in a lower range (only 24% off after 8 months incubation time) (Figure 2a). The decrease observed after 28 days is mainly due to fluoranthene disappearance which is the most prevalent component in this class (Figure 1). Kinetics pattern of fluoranthene decrease is very comparable to that observed for 3-cycle PAHs: two degrading stages, interspaced by a stabilization stage between days 28 and 98 (Figure 4). Pyrene and benzo(a)anthracene also take part to this result, but to a lesser extent. Chrysene shows a smaller decrease during the 8 months incubation.

5- cycle PAHs only show a slight reduction of their concentration (about 16% loss after 8 months) (Figure 2a), mainly due to benzo(a)pyrene degradation. No evolution of 6- cycle PAHs concentrations occurred during this period.

Those results agree well with those of Bidaud (1998), that evidenced a significant biodegradation of 3- and 4- cycle PAHs whereas 5- and 6- cycle PAHs were less or not degraded.

Bacterial concentration. The total bacterial microflora steady concentration indicates that biological activity remained optimal and able to favor biodegradation during the experiment (Figure 5).

The first step of 3- and 4- cycle PAHs degradation is linked to a significant increase in these PAHs degraders in relation to the amount and availability (solubility) of these PAHs in soil. During the second step of degradation, the level of the PAH's degrading microflora, which reached an optimum at the end of the first phase, remained constant. These results suggest the occurrence of different degradation mechanisms. The first step should be linked to an access to dissolved substrate in the aqueous phase (Bouchez et al., 1997) whereas the second step should correspond to an interfacial accession (Ortega-Calvo and Alexander, 1994).

Respirometric study. The microbial activity was monitored in the course of the experiments by measuring O_2-consumption and additionally CO_2-production under aerobic conditions. The soil sample shows high O_2-consumption, CO_2-production and PAHs degradation. The kinetics of O_2-consumption show an initial phase of exponential increase without any lag period (Figure 7). The decrease in PAHs concentration is concomittent with CO_2 production and O_2-

consumption (Figures 6 et 7). CO_2-production, O_2-consumption and their kinetics are good indicators of microbial activity and PAH biodegradation. The decrease in PAH concentrations well agrees with both respirometric and microcosm experiments, showing similar kinetics (Figure 7). To 224 days, degradation of total and 3- to 6- cycles PAHs led to a similar reduction of their concentration, to about respectively of 46, 70, 25, 15 and 0%.

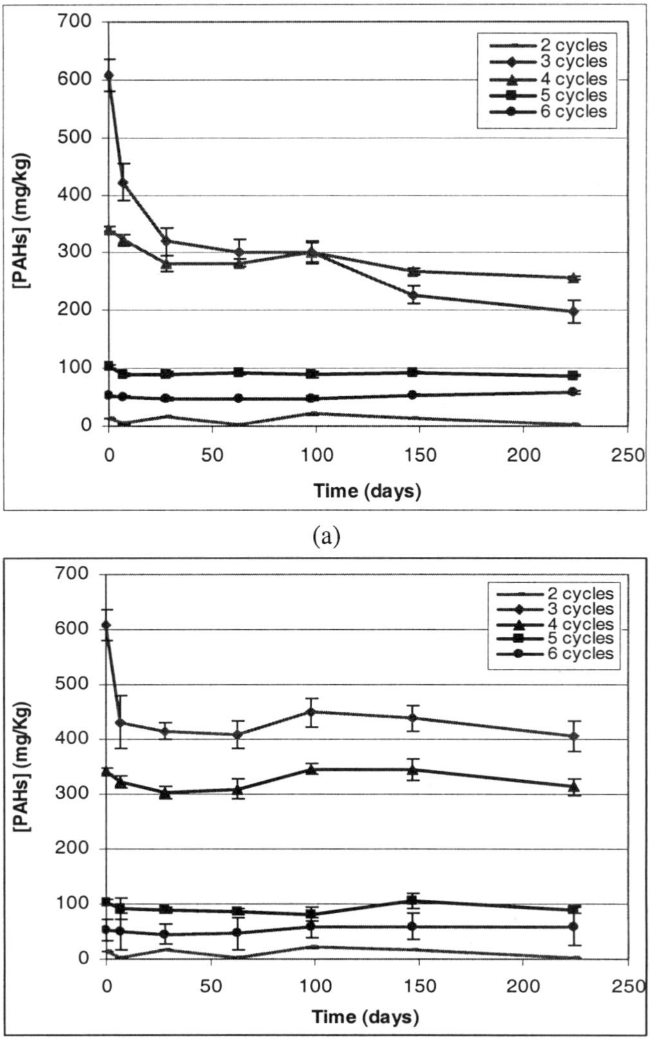

FIGURE 2. PAHs concentrations in the unsterilized (a) and sterilized (b) soil sample during the degradation experiment.

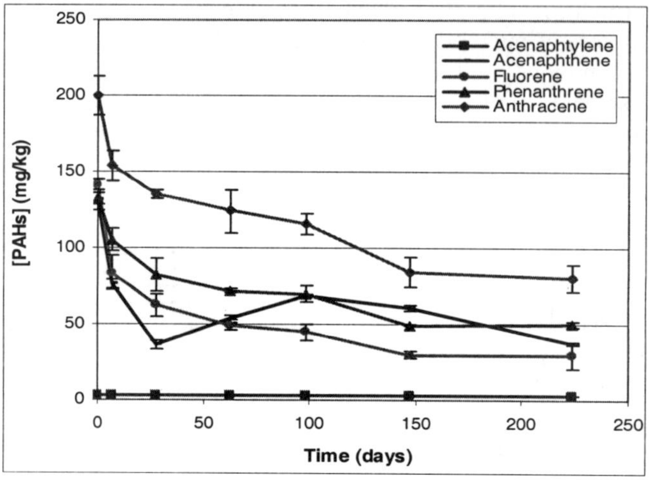

FIGURE 3. 3- cycle PAHs concentrations in the unsterilized soil during the degradation experiment.

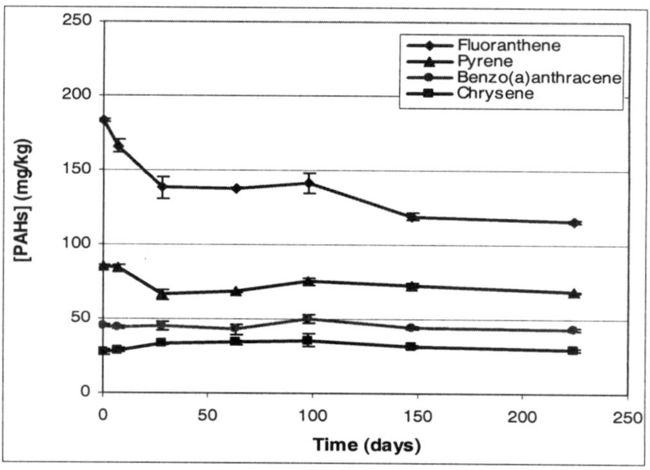

FIGURE 4. 4- cycle PAHs concentrations in the unsterilized soil during the degradation experiment.

CONCLUSION

The experiments carried out in this study show a significant reduction of the 3- and 4- cycle PAH concentration over 8 months. The 5- cycle PAHs present a lesser and much slower degradation whereas the 6-cycle PAHs undergo no evolution. They also evidence that observed degradation is mainly due to biological processes. Part of the reduction of the 16 PAHs concentration over 8 months is associated with bacterial degrading activity. O_2-consumption and CO_2-production are good indicators of microbial activity and PAHs biodegradation.

Degradation experiments thus demonstrate the occurrence of PAHs intrinsic biodegradation, which concerned more particularly 3- and 4- cycle PAHs.

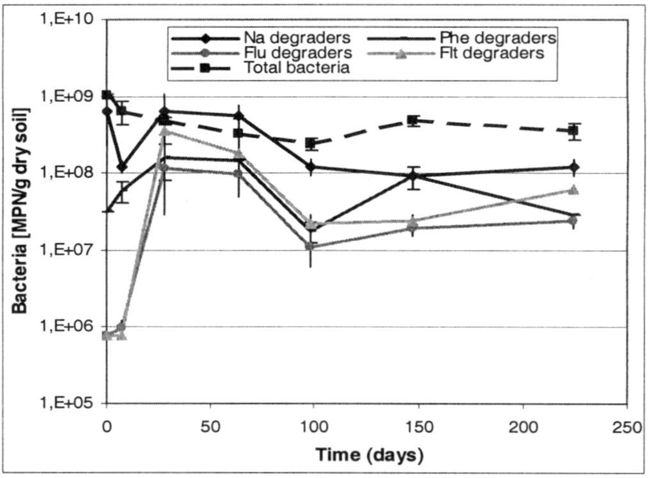

FIGURE 5. Concentration of bacteria during the degradation experiment.

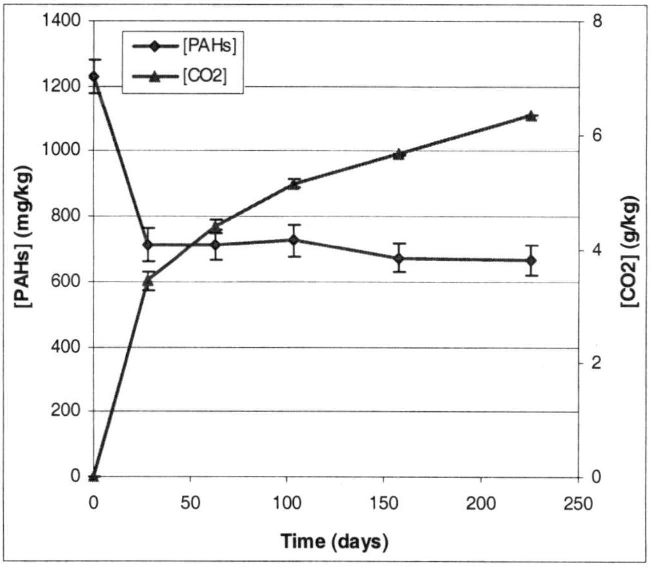

FIGURE 6. 16 PAHs (EPA) concentration and CO_2-production during the degradation experiment

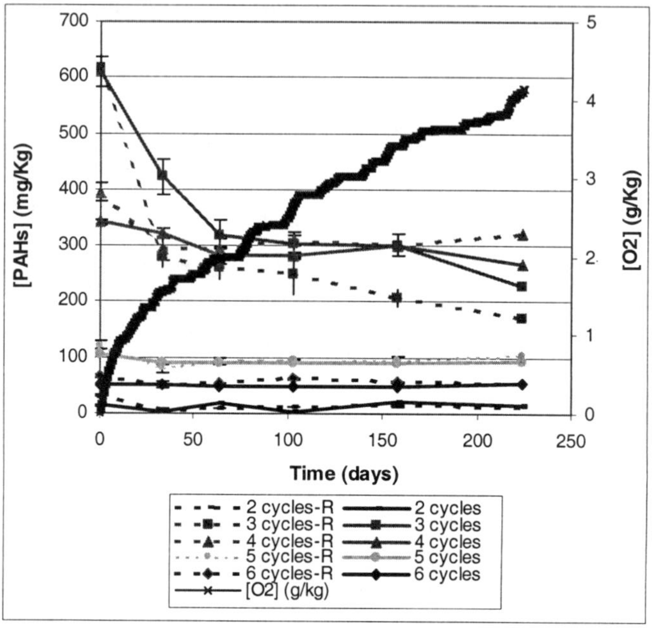

FIGURE 7. Comparison of PAHs concentration trends (by cycle) in microcosm and in respirometric (R) experiments – O_2 consumption during the respirometric experiment

REFERENCES

Alsaleh E. S., Smith S., Mason J., 1999. Investigations into the factors limiting bioremediation of PAH contaminated sites. *Bioremediation Technologies for Polycyclic Aromatic Hydrocarbon Compounds*, 5(8): 111-122, Leeson A., Alleman B. C. (eds), Batelle Press, San Diego, USA.

Bidaud C., 1998. Biodegradation de Hydrocarbures Aromatiques Polycycliques. Approche Microbiologique et Application au Traitement d'un Sol Pollué. Thèse de doctorat, Institut National Polytechnique de Grenoble.

Bouchez M., Blanchet D., Besnainou B., Leveau J.Y., Vandecasteele J.Y., 1997. Kinetic studies of biodégradation of insoluble compounds by continuous determination of oxygen consumption. *J. Appl. Microbiol.* 82: 310-316.

Kästner M., Breuer-Jammali M., Mahro B., 1994. Enumeration and characterisation of the soil microflora from hydrocarbon-contaminated soil sites able to mineralise polycyclic aromatic hydrocarbons (PAH). *Appl. Enviro. Microbiol.* 41 : 297-273.

Lundstedt S., Haglund P., Van Bavel B., Öberg L., 1999. Analysis of PAHs and metabolites during treatment of contaminated soil. *Bioremediation Technologies for Polycyclic Aromatic Hydrocarbon Compounds*. 5(8): 135-146, Leeson A., Alleman B. C. (eds), Batelle Press, San Diego, USA.

Ortega-Calvo J.J., Alexander M., 1994. Roles of bacterial attachment and spontaneous partitioning in the biodegradation of naphthalene initially present in nonaqueous-phase liquids. *Appl. Environ. Microbiol.* 60: 2643-2646.

Ramsay A.J., 1984. Extraction of bacteria from soil : efficiency of shaking or ultrasonification as indicated by direct counts and autoradiography. *Soil Biol. Biochem.*, 16: 475-481.

Stieber M., Haeseler F., Werner P., Frimmel F.H., 1994. A rapid screening method for microorganisms degrading polycyclic aromatic hydrocarbons in microplates. *Appl. Microbiol. Biotechnol.*, 40: 753-755.

NATURAL ATTENUATION IN A PAH/BTEX PLUME: THE IMPACT OF MICROBIOLOGICAL PROCESSES

S. Schulze (Water Technology Center, Karlsruhe, Germany),
B. Barczewski, N. Klaas (VEGAS, University of Stuttgart, Germany),
R. Ebner (GMF, Karlsruhe, Germany),
M. Birkle (Fraunhofer Institute, Karlsruhe, Germany),
A. Tiehm (Water Technology Center, Karlsruhe, Germany)

ABSTRACT: The investigations focus on the microbial Natural Attenuation processes in the plume of a gasworks residues disposal site. Twenty multi-level-wells covering the whole plume were installed to examine the three-dimensional distribution of microorganisms, pollutants, electron acceptors, and nutrients. First results of the field monitoring programme indicate different microbial respiration processes in the plume, i.e. nitrate, Fe(III) and sulfate reduction. However, the reduction of polycyclic aromatic hydrocarbons (PAH) and BTEX (benzene, toluene, ethylbenzene and xylenes) in the plume could not be related to specific metabolic processes on the basis of the field data alone. Additional microcosm studies are recommended and currently performed.

INTRODUCTION

In Germany, Natural Attenuation (NA) is discussed at present as an alternative to classical remediation techniques. Naturally occurring processes retaining and reducing pollutants in the aquifer are increasingly examined and integrated in remediation concepts. Whereas the physicochemical processes such as adsorption are mainly leading to a redistribution of the contaminants, microbial degradation processes resulting in pollutant mineralization are of major interest.

In the field, the availability of suitable electron acceptors, nutrients, or auxiliary substrates have to be considered important factors affecting pollutant biodegradation. Most low molecular weight PAH and the BTEX are easily degradable in the presence of oxygen, but their biodegradability significantly varies under anaerobic conditions. With nitrate as an electron acceptor, biodegradation has been observed for ethylbenzene, toluene, and xylene, but not for benzene (Hutchins 1991, Schink 2000). Under denitrifying conditions, biodegradation of PAH was reported in a few cases (Leduc et al. 1992, McNally et al. 1998, Milhelcic and Luthy 1988, Rockne et al. 2000). Field studies indicate that degradation of BTEX and PAH is possible in the presence of Fe(III) (Lynkilde and Christensen 1992). Utilization of Mn(IV) as electron acceptor was observed during biodegradation of naphthalene (Langenhoff et a. 1996). Degradation of benzene (Lovley 1995) and naphthalene (Meckenstock et al. 2000) was shown in the presence of sulfate. Under methanogenic conditions, biotransformation of benzene and toluene was observed (Grbic-Galic and Vogel 1987, Grbic-Galic 1990). Not only the availability of suitable electron acceptors, but also the lack of nutrients can limit microbial NA processes (Althoff et al.

2001). Furthermore, the presence of easily degradable compounds was reported to affect the degradation of low concentrated pollutants (Providenti et al. 1993).

Objective. The aim of our project is to study the microbial NA processes in a PAH and BTEX plume. Extended field investigations are carried out to identify the predominating zones of biodegradation. The spatial distribution of pollutants, microorganisms, electron acceptors, and nutrients is determined. Additionally, biodegradation is studied in microcosms and sediment columns in the laboratory. This paper presents first results of the field monitoring programme.

Site description. The investigations focus on the plume of a small landfill named „Stürmlinger Sandgrube", located in the northern part of Karlsruhe, Germany. The aquifer consists of porous sand and gravel (average K_f value: 2×10^{-3} m sec^{-1}). The water table is in 8 – 10 m depth. From 1925 to 1956, the exploited pit was filled with municipal waste, later on with gasworks residues, waste products of benzene production, and other unknown chemicals. Currently, the plume length is much shorter than the length calculated from contamination mass and groundwater flow data, indicating Natural Attenuation processes are occuring.

MATERIALS AND METHODS

Multilevel wells. Downgradient the contaminated area a net of twenty multilevel wells was installed. These multilevel wells consist of 5 pipes ending in about 9, 13, 17, 21, and 25 m depth. The wells are arranged in planes with increasing distance (5, 20, 40, 110m) to the disposal site (see fig. 1). Two additional wells are installed in the center line of the plume at 170 m (E5/21) and 260 m (E6/22) (not shown in fig. 1).

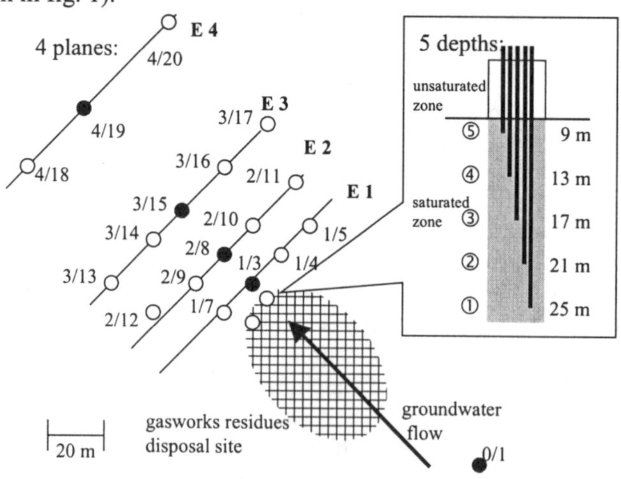

FIGURE 1. Monitoring of the plume: multilevel wells are arranged in four planes (E1 to E4).

Sampling. Sediment samples for PAH analysis were taken in glass bottles with PE cups. The PE surface was covered with aluminum foil to minimize losses by sorption. Samples for bacterial enumeration were collected in sterilized 1L-tins and immediately flushed with argon gas to remove oxygen.

Groundwater samples were taken after intensive flushing. Sterilized bottles were used for MPN (Most Probable Number)-samples. All samples were stored at 4°C for no longer than 24 hours.

Microbial Enumeration. Bacterial numbers were determined by the MPN-microplate technique (Stieber et al. 1994). Total heterotrophs were determined under aerobic, denitrifying, iron reducing, and sulfate reducing conditions. PAH-, BTEX- and mineral oil degrading bacteria were determined under aerobic conditions. Sediment samples were extracted with 0.9% NaCl.

RESULTS AND DISCUSSION

Pollutant concentrations. The spatial distribution of PAH in the sediment is shown in figure 2. Pollutant concentration decreases with increasing distance from the disposal site. About 40 m downgradient the contamination source (plane E3), PAH were not detectable in the sediment material (detection limit 20 µg/kg).

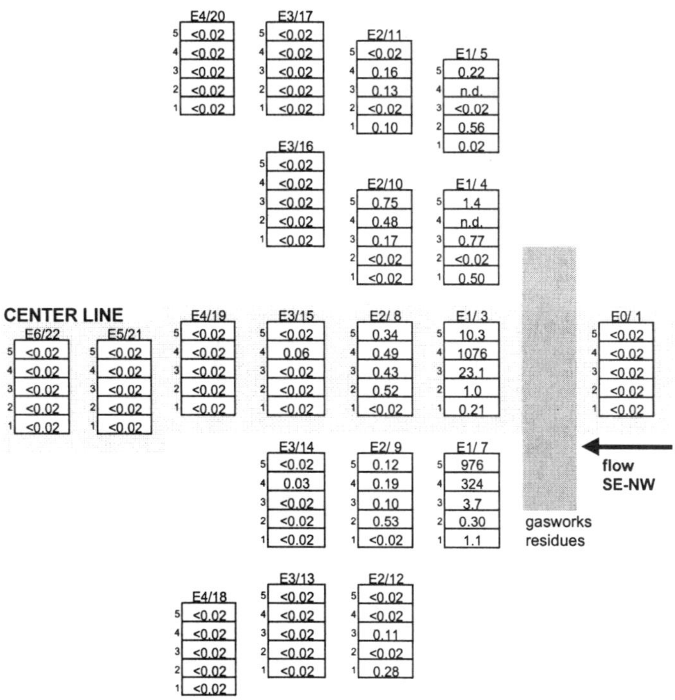

FIGURE 2. Spatial distribution of 16 EPA-PAH in sediment material [mg/kg dry weight]. The location of the sampling points is shown in figure 1.

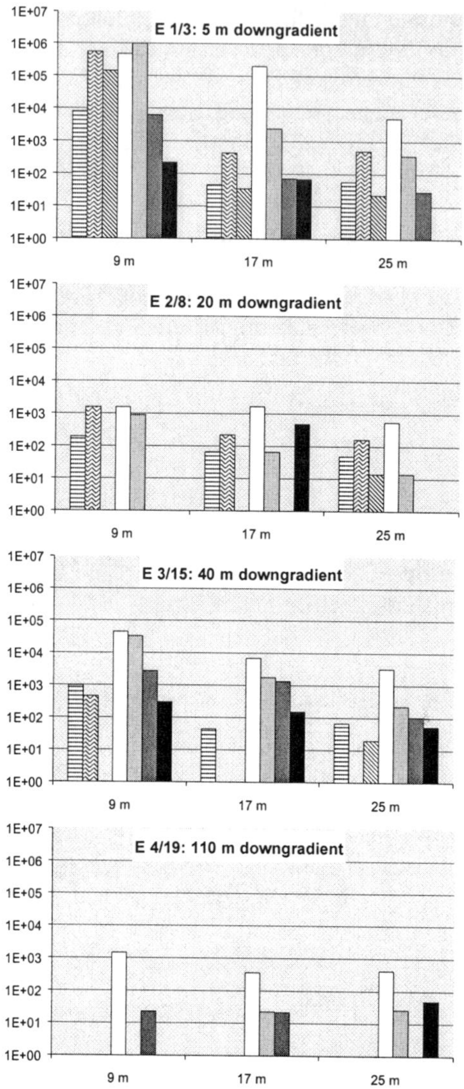

FIGURE 3. Distribution of microorganisms [MPN/g dry weight] in the center line (9, 17 and 25 m depth)

pollutant degrading microorganisms:	total heterotrophs:
BTEX	aerobic
2-and 3-ring PAH	denitrifying
3- and 4-ring PAH	iron-reducing
	sulfate-reducing

In the groundwater, PAH concentrations up to 12 mg/L (16 EPA-PAH) were found in plane E1, 5 m downgradient of the contamination source. In particular, high concentrations of naphthalene were found, especially in the upper layers. Aqueous phase concentrations of naphthalene decreased more rapidly than the 3- and 4- ring PAH. For example, naphthalene was below the detection limit (0.01 µg/L) in plane E5, whereas phenanthrene was still detectable.

BTEX concentrations up to 60 mg/L were found in plane E1, with benzene as the predominating compound. In plane E5, the monoaromatics were below the detection limit (1 µg/L). According to the field data, a preferential degradation of a particular BTEX was not yet observed.

Microbiological Parameters. Bacterial numbers were determined in sediment and water samples. Fig. 3 presents the spatial distribution of bacteria in the aquifer material. Data are shown for 4 wells along the centerline of the plume. Three samples were taken at each well in different depth. High numbers of total heterotrophs and pollutant degrading bacteria (up to 10^6/g sediment dry weight) were found. Bacteria utilizing oxygen, nitrate, iron(III), or sulfate were detected in most of the samples. No predominating microbial respiration process could be identified on the basis of MPN results. However, elevated ratios of PAH and BTEX degrading to total heterotrophic bacteria in the high contaminated areas indicate active biodegradation.

Electron acceptors and nutrients. The redox potential upgradient of the landfill (E0) was between +30 and +240 mV. In plane E1, values between –200 and -300 mV were detected. Downgradient the redox potential increased again up to values between –170 mV and +40 mV at plane E6. Oxygen was below the detection limit of 0.5 mg/L upgradient of the landfill (E0) as well as in the plume. Only in the upper capillary zone, oxygen was detectable in concentrations up to 4.2 mg/L.

Nitrate [mg/L]

Level	E6/22	E5/21	E4/19	E3/15	E2/ 8	E1/ 3	gasworks residues	E0/ 1
5	7.3	1.1	38.2	21.2	11.9	<1		23
4	5.9	<1	<1	<1	<1	n.d.		9.7
3	4.7	<1	<1	<1	<1	<1		9.5
2	2.7	<1	<1	<1	<1	<1		8.3
1	<1	<1	<1	<1	<1	<1		<1

flow SE-NW

Nitrite [mg/L]

Level	E6/22	E5/21	E4/19	E3/15	E2/ 8	E1/ 3	E0/ 1
5	0.01	0.02	0.04	0.04	0.12	<0.01	<0.01
4	0.02	0.01	<0.01	<0.01	<0.01	n.d.	0.02
3	0.02	<0.01	0.02	<0.01	<0.01	<0.01	<0.01
2	0.01	<0.01	<0.01	<0.01	<0.01	<0.01	<0.01
1	<0.01	<0.01	<0.01	<0.01	<0.01	<0.01	<0.01

flow SE-NW

Sulfate [mg/L]

Level	E6/22	E5/21	E4/19	E3/15	E2/ 8	E1/ 3	E0/ 1
5	104	134	37	83	106	35	128
4	111	124	24	50	49	n.d.	135
3	115	129	134	135	134	139	135
2	121	102	128	119	126	128	133
1	126	120	89	90	123	125	115

flow SE-NW

Sulfide [mg/L]

Level	E6/22	E5/21	E4/19	E3/15	E2/ 8	E1/ 3	E0/ 1
5	<0.1	<0.1	<0.1	0.1	<0.1	0.5	<0.1
4	<0.1	<0.1	<0.1	1.6	1.6	n.d.	<0.1
3	<0.1	<0.1	<0.1	<0.1	0.6	<0.1	<0.1
2	<0.1	<0.1	<0.1	0.1	<0.1	<0.1	<0.1
1	<0.1	<0.1	<0.1	0.3	<0.1	0.3	<0.1

flow SE-NW

Phosphate [mg/L]

Level	E6/22	E5/21	E4/19	E3/15	E2/ 8	E1/ 3	E0/ 1
5	0.06	0.02	0.03	0.08	0.28	n.d.	0.01
4	0.05	0.04	0.11	0.75	0.86	n.d.	0.02
3	0.04	0.04	0.04	0.1	0.09	0.03	0.02
2	0.04	0.02	0.03	0.06	0.17	0.04	0.02
1	0.04	0.04	0.04	0.06	0.02	0.09	0.29

flow SE-NW

FIGURE 4. Spatial distribution of nitrate, nitrite, sulfate, sulfide and phosphate in the center line of the plume (n.d.= not determined).

Nearly all along the plume, nitrate (fig. 4) was detected only in the upper layer near the water table, indicating a vertical input, e.g. from agriculture or nitrification. Whereas the inflowing groundwater contained nitrate down to about 18 m depth, no nitrate was detected in the corresponding deeper horizons of the plume up to plane E5. These findings and the detection of nitrite near the water table (fig. 4) indicate that denitrification occurs in the plume.

About 130 mg/L sulfate were detected upgradient. Up to E4, areas of significantly reduced sulfate concentration were observed in the upper layers (fig. 4). In the deeper layers, sulfate decrease was less pronounced. Sulfide as the product of sulfate reduction was detected predominately in the first three planes.

Although a consistent pattern could not be observed, elevated Fe(II) concentrations were detected in an extended plume area up to E5 (data not shown). Therefore Fe(III) reduction has to be considered an important process for this site.

Methane was found in concentrations up to 140 μg/L close to the contamination source (data not shown). Highest concentrations were observed in the layers between 13 m and 21 m depth. Near the water table, significantly lower methane concentrations were detected.

Phosphate concentrations are low in the inflow groundwater and all along the plume (fig. 4). Potential growth limitations are obvious. Preliminary lab experiments with dosage of naphthalene give evidence for nutrient limitations (data not shown).

CONCLUSION

The field data indicate different microbial respiration processes in the plume. Denitrification, Fe(III) and sulfate reducing activity, as well as methanogenesis were observed. Pollutant decline in the plume was clearly demonstrated, but cannot yet be assigned to a specific metabolic process on account of the interactions of different abiotic and biotic reactions. Currently, additional lab experiments are performed with authentic field samples. Microcosm studies are required to understand the microbial attenuation processes and to predict contaminant degradation in the plume.

ACKNOWLEDGEMENT

The authors gratefully acknowledge financial support by the German ministry of education and research (BMBF 02 WT 9957/9).

REFERENCES

Althoff K., Mundt M., Eisentraeger A., Dott W., and Hollender J. 2001. "Microcosm-experiments to assess the potential for natural attenuation of contaminated groundwater." *Wat. Res. 35*(3): 720-728.

Grbic-Galic, D., and Vogel T.M. 1987. "Transformation of toluene and benzene by mixed methanogenic cultures." *Appl. Environ. Microbiol. 53*: 254-260.

Grbic-Galic D. 1990. "Methanogenic transformation of aromatic hydrocarbons and phenols in groundwater aquifers." *Geomicrobiol. J. 8*: 167-200.

Hutchins S. R. 1991. "Biodegradation of monoaromatic hydrocarbons by aquifer microorganisms using oxygen, nitrate, or nitrous oxide as terminal electron acceptor." *Appl. Environ. Microbiol. 57* (8): 2403-2407.

Langenhoff A.A.M., Zehnder A.J.B., and Schraa G. 1996. "Behaviour of toluene, benzene and naphthalene under anaerobic conditions in sediment columns." *Biodegradation 7*: 267-274.

Leduc R., Samson B., Al-Bashir B., Al-Hawari J., and Cseh T. 1992. "Biotic and abiotic disappearence of four PAH compounds from flooded soil under various redox conditions." *Wat. Sci. Tech. 26*: 51-60.

Lovley D.R.. 1995. "Bioremediation of organic and metal contaminants with dissimilatory metal reduction." *J. Ind. Microbiol. 14*: 85-93.

Lynkilde J., and Christensen T.H. 1992. "Fate of organic contaminants in the redox zones of a landfill leachate pollution plume (Vjeen, Denmark)." *J. Contaminant Hydrology 10*: 291-307.

Meckenstock R. U., Annweiler E., Michaelis W., Richnow H. H., and Schink B. 2000. "Anaerobic naphthalene degradation by a sulfate-reducing enrichment culture." *Appl. Environ. Microbiol. 66* (7): 2743-2747.

McNally D. L., Mihelcic J. R., and Lueking D. R. 1998. "Biodegradation of three- and four-ring polycyclic aromatic hydrocarbons under aerobic and denitrifying conditions." *Environ. Sci. Technol. 32*: 2633-2639.

Mihelcic J.R., and Luthy R.G. 1988. "Degradation of polycyclic aromatic hydrocarbon compounds under various redox conditions in soil-water systems." *Appl. Environ. Microbiol. 54*: 1182-1187.

Providenti M. A., Lee H., and Trevors J.T. 1993. "Selected factors limiting the microbial degradation of recalcitrant compounds." *J. Ind. Microbiol. 12*: 379-395.

Rockne K. J., Chee-Sanford J. C., Sanford R. A., Hedlund B. R., Staley J. T., and Strand S. E. 2000. "Anaerobic naphthalene degradation by microbial pure cultures under nitrate-reducing conditions." *Appl. Environ. Microbiol. 66* (4): 1595-1601.

Schink B. 2000. "Principles of anaerobic degradation of organic compounds." In Rehm H.-J. and Reed G. (Eds.); Biotechnology, 2nd edition, Vol 11b, pp. 169-192. Wiley.

Stieber M., Haeseler F., Werner P., and Frimmel F. 1994. "A rapid screening method for micro-organisms degrading polycyclic aromatic hydrocarbons in microplates." *Appl. Microbiol. Biotechnol. 40*: 753-755.

EXPERIENCE WITH NATURAL ATTENUATION OF CHLORINATED SOLVENTS AT LANDFILLS

Robert D. Norris (Brown and Caldwell Denver, Colorado)
Robert D. Mutch (Brown and Caldwell Mahwah, New Jersey)

ABSTRACT: For many decades landfills were constructed without provisions for mitigating environmental impact. Typically wastes including solvents were either poured into landfill wastes or waste containing drums were placed in unlined excavations or gravel pits, etc. As a result of environmental awareness and strong regulations, landfills designed in the last twenty years are much less likely to release organic solvents to the environment. Still, there is a large legacy of landfills that have released contaminants to groundwater. Most of these landfills were constructed more than twenty years ago and thus plumes emanating from these landfills have frequently either reached a steady state or are shrinking in part due to the groundwater conditions associated with landfills. The distribution of chlorinated compounds within, beneath, and downgradient of landfills is quite variable because of the haphazard ways in which materials were added to landfills, the range of co-contaminants present in the landfills, and the types of geological settings where the landfills were constructed. Experience shows that conditions favorable to attenuation of chlorinated solvents typically occurs within and downgradient of landfills, e.g., the presence of electron donors within the landfill mass and associated groundwater as well as the redox shadow created by the presence of the landfill. Important considerations are the impacts of drum deterioration, transport mechanisms including landfill gas and groundwater advection, and geochemical conditions associated with landfills.

INTRODUCTION

Many of the landfills that have organic contamination plumes associated with them were constructed prior to regulations concerning the use of impermeable bottom layers, caps, leachate collection systems, etc. In fact many guidance documents had suggested that as long as there were six or more feet of soil beneath the waste, natural attenuation mechanisms would protect the environment. Solvents and fuels were frequently disposed of at the landfills. In some cases these liquids were directly added to the landfill mass or placed in drums which leaked or were damaged. Today there are large numbers of former industrial and co-disposal landfills that have significant groundwater contamination plumes emanating from them as a result of previous practices.

One might expect that drums would eventually corrode and release their contents, which would then migrate to the soils and water table as a result of gravity. An important question is how long might the drums remain intact? This has important implications with regards to whether there is benefit to excavating drums that have been in the landfill for years or decades. It is also important for considering monitored natural attenuation or other remedies for the impacted groundwater (Mutch and Norris, 2001).

FATE OF DRUMS IN LANDFILLS

Drums deposited in a landfill lose their integrity and thus release their contents, in several ways, some occurring immediately after placement in the landfill and others over a somewhat longer time through corrosion. Immediate spillage resulted from drums being dropped off trucks, missing or lose bungs, or intentionally puncturing to avoid pressure buildup or explosion, often at the recommendation of regulatory agencies. Compaction along with municipal solid or other waste resulted in crushing of drums. As a result leachate could reflect the contents of the wastes within days and groundwater could be impacted in a very short time.

Corrosion of drums results in the release of liquid wastes, generally, over a much shorter time than many people think. Corrosion of metal placed in landfills or the subsurface occurs in two forms – uniform oxidation of the metal and pitting. Pitting is the greatest concern with respect to maintaining the integrity of drums because the rate of penetration is much faster than the general rate of oxidation. Pitting first allows the contents to directly escape. As pitting advances, then percolating water can leach or displace the original liquid further transferring the contents to the landfill mass and eventually to groundwater.

Corrosion is a function of many factors including minor damage to drums that is likely to occur during use, transport to the landfill, or during placement of the drum. Other factors include, the drum material and coating, moisture levels, pH, temperature presence of salts, and aerobic versus anaerobic conditions. Thus the rate of corrosion varies considerably (Frankel, 1982). Studies have shown that in most cases drums not immediately crushed or otherwise damaged can be expected to have failed in less than ten years with a fifty- percent failure rate after five to six years. Exceptions are where drum placement provided protection from moisture and physical damage. Our experience at one landfill is that of over ten thousand drums excavated after more than twenty years in a landfill, none were intact.

The use of uncontrolled or mixed waste landfills was ended more than twenty years ago. Thus the importance of drum corrosion rates to monitored natural attenuation is twofold: 1) excavation of drums is unlikely to provide any real source control, and 2) the rate at which organic chemicals might be entering the groundwater is likely to have reached a steady state or is declining.

FATE AND TRANSPORT OF RELEASED SOLVENTS

Solvents released from landfill drums first migrate in the form of a non-aqueous phase liquid (NAPL). Chlorinated solvents behave as dense, non-aqueous phase liquids (DNAPL), while non-chlorinated solvents such as benzene or toluene behave as light, non-aqueous phase liquids (LNAPL). Above the groundwater table, the behavior of DNAPL and LNAPL chemicals would be fairly similar (Cohen and Mercer, 1993; Poulsen and Kueper, 1992). The behavior of NAPL within a co-disposal landfill should be similar in many respects to behavior within a porous geologic medium with several distinctions attributable to the unique properties of solid waste such as wettability issues. The landfill materials will behave differently as fresh, fairly dry or more aged or wetter material affect sorption characteristics differently. Landfill materials, depending on the type of materials and degree of compaction may create channels for NAPL flow or may provide traps. Residual NAPL saturation may be as high as 17 percent.

It can be anticipated that the migration as NAPL will result in a zone of residual saturation of solvent within the waste and, potentially, within the underlying geologic strata, depending on the depth of penetration, the geology, and the presence or absence of liner systems. The NAPL within the zone of residual saturation will thereafter be subject to solubilization into percolating moisture from rainfall or as byproduct water from biodegradation within the landfill mass or into groundwater if the landfill is below the zone of saturation. The NAPL will also be subject to volatilization into landfill gas. Solvents volatilized into landfill gas would also be partitioned, to some extent, into infiltrating moisture and, ultimately, into leachate from the landfill. Some volatilization from the dissolved phase solvents in groundwater or leachate would also occur. The dissolved phase constituents would be subject to migration in groundwater and natural attenuation. Gas phase constituents would migrate with the landfill gas and in the absence of an effective landfill gas collection system, migrate outside the landfill through the unsaturated zone.

LNAPL reaching the groundwater surface will tend to pancake out creating a separate floating liquid phase. DNAPL reaching the groundwater surface will first tend to spread laterally, and then as the mass of DNAPL accumulates will continue downward over a tortuous path until reaching an impermeable layer where it will again spread laterally. Once reaching the groundwater, organic constituents will dissolve in the groundwater at rates depending upon solubility of the specific compounds, the groundwater seepage velocity, and the extent to which the distribution of the NAPL facilitates contact with the flowing groundwater. The dissolved constituents will then migrate along the flow of the groundwater. The extent of migration will be mitigated by attenuation mechanisms.

NATURAL ATTENUATION

Natural attenuation occurs downgradient of a landfill as a result of several mechanisms. There are two primary types of mechanisms: (1) physical mechanisms including advection, dispersion, diffusion, and adsorption that either dilute or retard movement of dissolved phase constituents but do not reduce the mass of constituents, and (2) degradation mechanisms that result in lower dissolved phase concentrations and reduction in migration as a result of reducing the total mass of organic constituents

DEGRADATION MECHANISMS

Biodegradation is the process where microorganisms convert constituents to other compounds. In both soils and groundwater, ubiquitous microorganisms use petroleum hydrocarbons, ketones, alcohols, etc., as a food and energy source, ultimately converting what are considered contaminants to carbon dioxide and water. For example:

Benzene + oxygen → carbon dioxide + water

Biodegradation occurs because microorganisms, mostly bacteria, benefit from using many organic compounds as food and energy sources. During biodegradation, the organic chemical compound is either oxidized or reduced. When one species is oxidized, another compound must necessarily be reduced. This occurs when electrons are transferred from one species to another. By way of definition, we refer to the compound that is being oxidized as the electron donor and the compound that is being reduced as the electron acceptor. In most microbial processes, oxidation/reduction reactions release energy for use by the microorganisms and provide carbon for new cell material.

Biodegradation of petroleum hydrocarbons (alcohols, ketones, etc.) occurs most rapidly in the presence of oxygen, which serves as the electron acceptor. Downgradient of landfills groundwater is nearly always oxygen limited, however, microorganisms are able to use other electron acceptors typically used in a roughly sequential manner depending upon the oxidation/reduction (redox) potential range over which various microorganisms utilize them, typically in the order oxygen, nitrate, manganese (IV), iron (III), and then sulfate. As electron acceptors are used, the oxidation/reduction potential trends downward, favoring reductive dechlorination of chlorinated solvents.

Commonly used solvents, such as PCE, TCE, 1,1,1-TCA, 1,2-DCA, carbon tetrachloride, chloroform, and methylene chloride, have been identified in landfill leachate and in groundwater associated with landfills. These and other chlorinated compounds can be degraded through chemical and biological processes. Chemical degradation is not generally important except for a few compounds such as 1,1,1-TCA. Although aerobic biodegradation and a process called cometabolism occur at appreciable rates for some chlorinated compounds, e.g., methylene chloride, under certain conditions, anaerobic degradation that results in reduction of the chlorinated solvent compounds is by far the most important microbial process for natural attenuation. These processes, sometimes called reductive dechlorination, involve the sequential replacement of chlorine atoms with hydrogen atoms. For example, PCE can be converted sequentially to ethene and ethane, which can then be oxidized to carbon dioxide and water:

PCE → TCE → DCE → vinyl chloride → ethene and ethane

DCE produced in these reactions consists primarily (~90%) of *cis*-1,2-DCE with some trans-1,2-DCE and some trace amounts of 1,1-DCE. The degradation of cis-1,2-DCE also occurs by other mechanisms as documented in several recent articles published by USGS (Bradley and Chapelle, 2000). These researchers have shown that cis-1,2-DCE can be converted to methane and carbon dioxide. One might expect these reactions to occur around the edges of the plume where dissolved oxygen levels are typically highest.

Reductive dechlorination processes require a sequence of reactions. Microorganisms utilize electron acceptors such as oxygen to degrade natural organic materials in soils as well as introduced chemicals such as petroleum hydrocarbons, food products, etc., which can be degraded through a series of steps to produce carbon dioxide and water and lower the oxidation/reduction potential. Some of the intermediate products undergo fermentation reactions that release low concentrations of hydrogen as a byproduct. The hydrogen can be used in reductive dechlorination. Other conditions important to these biological processes include a near neutral pH, temperature, presence of nutrients, especially nitrogen and phosphorus sources, and an absence of toxins.

LANDFILL EFFECTS ON NATURAL ATTENUATION

Landfills commonly create uniquely favorable environments for natural attenuation of chlorinated solvents. Landfills are typically anaerobic environments as evidenced by the production of large quantities of methane in landfill gases. As a result, reductive dechlorination occurs within the landfill mass resulting in many landfill gases and leachates containing reductive dechlorination byproducts in greater concentrations than the parent compounds. The presence of a landfill cap or other object that interferes with infiltration of aerated water (precipitation) reduces the influx of oxygen. This results in lower dissolved oxygen content and lower redox potential. Generally, the most important effect of landfills on reductive dechlorination is the presence of electron donors. Common electron donors found in landfill leachate and associated groundwater include petroleum hydrocarbons, alcohols, ketones, etc. At co-disposal sites, food related compounds, which are excellent electron donors, are also frequently present but not always analyzed for. Critically, the degradation intermediates undergo fermentation to produce hydrogen that is necessary for the reductive dechlorination process.

The environment of the groundwater beneath and downgradient of co-disposal landfills is generally quite conducive to reductive dechlorination. Occurrence of reductive dechlorination is typically supported by low DO, low redox potential, high alkalinity (evidence of carbon dioxide formation), high dissolved organic matter, presence of volatile fatty acids, and elevated dissolved hydrogen. Increased dissolved iron and manganese concentrations are also common. Chloride levels may also be elevated but are typically obscured by other sources of chloride. The presence of degradation intermediates such as 1,2-DCE, vinyl chloride, and ethene provide the most direct evidence of reductive dechlorination. Fate and transport modeling using readily available models such as BIOSCREEN, BIOPLUME III, BIOCHLOR, etc. can provide support that biodegradation is limiting the extent of migration and estimates of degradation rates. Descriptions of these models as well as the models themselves can be downloaded from the USEPA Ada, Oklahoma Web page (USEPA, Ada, Oklahoma). Where historical groundwater quality data are available for a number of years (i.e., five years of longer), steady state or shrinking plumes provide strong direct evidence of the degree of benefit of natural attenuation.

IMPLICATIONS TO REMEDIAL ACTION

The feasibility of drum excavation from a co-disposal landfill would seem to be best judged by the percentage of original waste mass that can be recovered. If the majority of the solvent has already been released from the drums into the waste matrix, that fraction of the original solvent mass is largely unrecoverable, at least without wholesale excavation of the landfill. This calls into question the merits of attempting to recover the remaining solvent still within the drums. Perhaps more importantly, if the majority of the solvent has already been released from the drums, then leachate or groundwater quality should already reflect the maximum impact of the drummed solvents. Concern for additional impact from additional release from drums would therefore be unfounded. Moreover, the question of the extent to which natural attenuation can mitigate releases of drummed solvents from the landfill should be answerable based upon observations of current conditions.

Furthermore, excavation is very expensive, disruptive to the landfill, and can produce short-term public health impacts and nuisances during implementation. These include volatile emissions, odor, the possibility of landfill fires, and spills during transportation. Both the theoretical and experiential evidence suggest that, if 20 or 25 years have elapsed since disposal of drummed solvents within a landfill, any expectations that drum excavation can recover a significant fraction of the original solvent mass should be low.

Capping, leachate collection, increased landfill gas collection, etc. are much more likely to be beneficial and cause fewer problems. Our experience is that these remedies have been sufficient to substantially mitigate plumes emanating from landfills. Active remedies immediately downgradient of the landfill may also provide substantial benefit where monitored natural attenuation is not sufficient.

Case Study No. 1: Shope's Landfill Superfund Site, Girard, Pennsylvania. The Shope's Landfill site near Erie, Pennsylvania, operated from the 1950s until 1979. A variety of wastes, including plant trash, scrap rubber, and drummed waste solvents were disposed of in the unlined landfill. The liquid wastes included chlorinated solvents, alcohols, and monoaromatic hydrocarbons. In the early 1980s, an initial remediation was implemented. Surficial drums were removed. In addition, a geomembrane cap and an upgradient, soil-bentonite, slurry-trench cut-off wall were installed. Following completion of a Feasibility Study (FS), a soil vapor extraction and a source area groundwater recovery and treatment system were implemented in the early 1990s. The groundwater recovery system was designed to control the higher concentrations near the landfill.

A review of the historical data indicated that the plume had reached its maximum extent before the mid-1980s. The plume contained anaerobic biodegradation products of chlorinated solvents including ethene, representing more than 99 percent of the chlorinated ethenes. Their presence was considered evidence of biodegradation as a contributing mechanism of natural attenuation and that the degradation process was going to completion. The geochemical data collected at the site supported the occurrence of reductive dechlorination as an attenuation mechanism. Even though a substantial number of drums remained within the landfill, the combination of partial drum removal, a cap, and a cut-off wall appears to have contributed to a shrinking plume. By 1990, contaminants in wells located downgradient of the landfill property boundary were largely gone and the agencies allowed nearly all of the off-site wells to be decommissioned.

Table 1 shows the great preponderance of daughter products over TCE and, possibly, PCE, the presumed parent compounds. The table also shows the significant decrease in VOCs along the apparent flow path as well as the dramatic decease in vinyl chloride levels at the property boundary over the last few years. Additionally the geochemical data suggests that nitrate, manganese, and iron are consumed as part of the process. The high sulfate in the plume may have resulted from sulfate addition containing compounds in the landfill. Consumption of sulfate may have also occurred. Chloride data, not shown, implies that chloride is produced within the plume but can not be clearly distinguished from the presence of chloride containing compounds within the materials disposed of in the landfill.

Case Study No. 2: Grenada, Mississippi Landfill. An off-site landfill at a manufacturing facility near Granada, Mississippi is an example of attenuation of a plume where the selected remedial activity has yet to occur pending resolution of property issues. This wooded disposal area received buffing compound wastes that contained chromium and TCE as well as some drummed wastes. After discovery of the landfill during a remedial investigation of the plant and associated lands, a few surficially located drums were removed. Investigation of soil and groundwater identified the extent of the impact. The extent of groundwater quality impacts was determined to be relatively limited. The presence of TCE daughter products indicated the role of anaerobic biodegradation in attenuation of the plume. Continued groundwater sampling showed that the plume dimensions were not changing appreciably over time and that the groundwater geochemistry was indicative of conditions favoring continued anaerobic degradation of the chlorinated ethenes present. Fate and transport modeling indicated that the plume would never reach the downgradient receptor, Grenada Lake. The Mississippi Department of Environmental Quality (MDEQ) accepted the remedy identified in the FS, which was consolidation and capping of the landfill materials and monitored natural attenuation for the groundwater plume.

Table 2 provides a summary of the MNA data. The concentrations of both TCE and *cis*-1,2-DCE are shown to be stable or possibly decreasing for *cis*-1,2-DCE. On a molar basis *cis*-1,2-DCE exceeds the parent compound by 20-30 percent. Vinyl chloride is generally present but at low ug/L levels. The table also shows that dissolved oxygen decreases within the plume and while manganese (II) increases within the plume. The other electron acceptors appear to be less involved.

Case Study No. 3: Upstate New York Landfill. Largely municipal wastes were placed in this unlined landfill. Investigations by the new owners determined the presence of PCE and other chlorinated ethenes and ethanes. Migration in the bedrock aquifer remained almost entirely within the landfill owner's property boundaries. The extent of the plume was much less than anticipated in the absence of biodegradation as an attenuation mechanism. In portions of the plume, degradation products represented more than 90 percent of the released compounds. Historical data indicated a relatively steady state condition. Geochemical data indicated conditions supportive of reductive dechlorination. The portion of the plume that showed the highest proportion of daughter products underlies a former food waste lagoon. In order to achieve closure in a timely fashion, NYSDEC has proposed enhanced reductive dechlorination as a groundwater remedy in addition to capping and leachate collection within the landfill.

Table 3 shows the decreases in VOC concentrations with distance to the east and to the south of the apparent source area. The ratios of daughter products to parent compounds show extensive degradation to the east. Less extensive degradation is observed to the south, however, somewhat higher dissolved oxygen levels and oxidation/reduction potentials would be consistent with cis-1,2-DCE being degraded through oxidative processes. Ranking of individual wells according to the USEPA protocol for MNA also suggest more reducing conditions to the east than the south

CONCLUSIONS

Natural attenuation plays a pivotal role in controlling the fate and transport of solvents including drummed solvents from a co-disposal landfill. The conditions within a co-disposal landfill are uniquely suited to foster the biodegradation of chlorinated solvents. In most cases, monitoring of leachate or perimeter groundwater quality should reflect the impact of drummed solvents on overall leachate or groundwater within a few years of disposal. Excavation and recovery of drummed solvents from a co-disposal landfill will likely yield a low solvent mass recovery percentage unless undertaken within the first few years after drum deposition.

REFERENCES

Bradley and Chapelle, 2000. *Environmental Science and Technology*, Vol. 34, pp. 221-222

Cohen and Mercer, 1993. *DNAPL Site Evaluation*, C.K. Smokley, Boca Raton, Florida.

Fried, J.J., P.Muntzer and L. Zilliox, 1979. "Groundwater pollution by transfer of oil."

Frankel, I, 1982. "Estimating the life span of a buried steel drum," *Industrial Wastes* 19, January-February.

Mutch, R.D. and R. D. Norris, 2001. "The Fate of Drummed Organic Solvents in a Landfill Environment: Implications to Remedial Action", presented at the Waste Tech 2001 Conference, San Diego, California.

Poulsen, M.M. and B. H. Kueper, 1992. "A field experiment to study the behavior of tetrachloroethylene in unsaturated porous media", *Environmental Science and Technology*, Vol. 26, No. 5, p 889-895.

USEPA Ada, Oklahoma (current) Web Page: http://www.epa.gov/ada/csmos.html

TABLE 1. Summary of Data from Shopes-Landfill Site

Area of site	TCE ug/L	Cis-1,2-DCE Ug/L	Vinyl Chloride Ug/L	Ethene ug/L	Nitrate Mg/L	Manganese Mg/L	Iron (II) Mg/L	Sulfate Mg/L
Source	ND	1,500	42,600	17,000	1.9	0.4	1.8	38
Side Plume	ND	ND	409	110	0.6	1.2	3.7	201
Background	ND	ND	ND	ND	.03	2.2	0.8	40
Property Boundary 4/98	ND	ND	920					
11/98	ND		183					
5/99	ND		27					
12/99	ND		19					
5/2000	ND		49	4,600				
12/2000	ND		6					

ND: Not Detected

TABLE 2. MNA Data from the Grenada, Mississippi Landfill

Area	TCE 1993 (ug/L)	TCE 1996 (ug/L)	*cis*-DCE 1993 (ug/L)	*cis*-DCE 1996 (ug/L)	Dissolved Oxygen (mg/L)	Manganese (II) (mg/L)
Background	ND	ND	ND	ND	8.0	3.8
Source	767	700	947	670	2.0	32
Source	1030	920	820	660	2.2	44
Plume	NS	280	NS	330	3.0	77
Downgradient	20	26	27	23	7.0	16
Side Gradient	5	0	ND	ND	6.9	3.7
Perimeter	1	1.5	ND	ND	7.5	2.1

ND: Not Detected

TABLE 3. Upstate New York Landfill MNA Data

Distance from Source (feet)	VOCs East of Source (ug/L)	VOCs South of Source (ug/L)	East: Ratio Daughter to Parent Compound	South: Ratio Daughter to Parent Compound	East: USEPA Protocol Ranking Points	South: USEPA Protocol Ranking Points
Source	2,500-3,500	2,500-3,500	5	5	21	21
750-900	1,000-2,000	400-1,200	13	0.8	17-20	10-11
1500-1600	00-600	ND-40	37-67	1.1	NC	NC
2400-2800	ND	ND	NA	NA	NC	NC

ND: Not Detected

NA: Not Applicable

NC: Not Calculated

PROBABILISTIC APPLICATION OF BIOCHLOR FOR POINT OF COMPLIANCE

Christine L. Harwood (Michael Baker Corporation, Coraopolis, PA, USA)

ABSTRACT: The popular model, BIOCHLOR, for determining movement of chlorinated solvents in a simple aquifer with reductive dechlorination is coupled with the point estimate method to produce a normal probability distribution of the expected distance to a risk-based concentration of vinyl chloride. The probability distributions of the concentrations of trichloroethene, dichlorethene (total) and vinyl chloride at the property boundary are also found. The input variables that were assumed to be random variables were hydraulic conductivity, hydraulic gradient, longitudinal dispersivity, ratio of transverse to longitudinal dispersivity, source concentration, and half-lives in the environment. A typical "worst case scenario" was calculated using BIOCHLOR and the probability that the distance to the risk-based concentration will be exceeded was found.

INTRODUCTION

The widespread use of the BIOCHLOR model (Aziz et al., 1999)for evaluating groundwater contaminant transport with degradation of chlorinated compounds in a simplified aquifer setting has led to deterministic solutions of "distance to point of (regulatory) compliance" of chlorinated contaminants of concern. It is widely recognized that several of the input parameters to the BIOCHLOR model have variation. That is, they have a probability distribution, and are not easily characterized by one value. Instead of using all the available information for each of these parameters, often several randomly chosen scenarios are modeled and the "worst case" scenario with the most conservative input parameters is presented as a deterministic result.

Probabilistic solutions to problems in the environment have the ability to incorporate uncertainty into models and are advancing in their use, as more people understand their value and application. Some alternatives to deterministic analyses include the Monte Carlo method, first-order second moment analyses, and the point estimate method.

The Monte Carlo method is by far the most popular application of probabilistic modeling, and examples of its application are frequent in the literature (Bonomo et al., 2000; Chrostowski et al., 2000; Copty and Findikakis, 2000). Because this method relies on many realizations of a deterministic solution, its application is usually limited to a computer model coupled to a Monte Carlo simulation package. While this is not technically difficult to accomplish, it is computationally intensive.

First order and second order probabilistic applications are increasing in use and have been published by the U.S. EPA for use in groundwater contaminant transport and remediation (Hamed and Benient, 1999). Smith and Charbeneau

(1990) used first order uncertainty analysis to determine the concentration of a contaminant at an exposure point (water supply well).

The point estimate method was first developed by Emilio Rosenblueth in 1975. It has been applied to problems of health risk (Mishra, 2000), geotechnical design (Thevanayagan et al., 1996), and others. The point estimate method was used to determine the probability distribution of the distance to the point of regulatory compliance for vinyl chloride. In effect, how far away from the source is the concentration of vinyl chloride within the risk-based concentration or RBC? Is this location off the property boundary? What is the concentration at the property boundary or other compliance point? What is the probability that the regulatory RBC is exceeded at that point? These are questions that come to mind when a site is assessed for risk to human health and the environment. Naturally, in the context of a subsurface environment where a very small amount of actual data is available, uncertainty is inherent. Although this is widely recognized, in the vast majority of cases, no uncertainty analyses are done, and the uncertainty is not quantified.

METHODS

Because of the sequential degradation of chlorinated solvents (e.g. trichloroethene → *cis* 1,2-dichloroethene → vinyl chloride → ethene/ethane) in the groundwater under specialized aquifer conditions, the computer model BIOCHLOR has been widely used to model the movement of the chlorinated solvents in the groundwater. This model, in the form of an Excel spreadsheet, provides an analytical solution to the groundwater contaminant transport equation with a constant, planar contaminant source. It also accounts for the increase in daughter products of degradation, and their subsequent degradation under idealized conditions. The reader is referred to the model documentation (Aziz et al., 1999) for further information on this model. All model runs were done for a one hundred-year scenario.

The generalized point estimate method for a function "y" with "n" independent (not correlated) random variables can be written as follows:

$$y_{\pm 1, \pm 2 \ldots, \pm n} = f[\overline{x_1} \pm \sigma(x_1), \overline{x_2} \pm \sigma(x_2), \ldots \overline{x_n} \pm \sigma(x_n)]$$

where $\overline{x_i}$ is the mean of variable x_i, and $\sigma(x_i)$ is the standard deviation of the variable x_i. This equation has 2^n permutations of the +'s and -'s. The mean and standard deviation of the 2^n outcomes of the function "y" are simply found from this set of outcomes.

The random variables, x_i, chosen for the BIOCHLOR (i.e. "y") application were those that are not typically known with any degree of certainty at a site. They are hydraulic conductivity, hydraulic gradient, longitudinal dispersivity, ratio of transverse to longitudinal dispersivity, source concentration, and half-lives in the environment. When applying the point estimate method, the mean and standard deviation of these random variables must be known or estimated. The site used for these calculations was an actual site, and all available data is actual data. However, because its location is irrelevant, it will remain anonymous.

For the case of hydraulic conductivity, there were results from seven slug tests available. The mean and standard deviation of these values was found. Even though hydraulic conductivity is usually assumed to have a log-normal probability distribution, the values obtained from the slug tests did not yield a log-normal distribution significantly different from the normal distribution. Therefore, a normal distribution was assumed.

Several quarters of groundwater elevation data were available to determine an estimate of hydraulic gradient. From the resulting contour maps, eleven values of hydraulic gradient were found and the mean and standard deviation of these values were used. The longitudinal dispersivity was first estimated using the equation of Xu and Eckstein (Aziz et al., 1999). Then an educated guess as to the coefficient of variation of this parameter was chosen to be 50 percent. That is, the standard deviation divided by the mean is 0.5. The mean ratio of transverse to longitudinal dispersivity was chosen to be 0.1 with a coefficient of variation of 50 percent.

In a typical BIOCHLOR application, the source concentrations of trichloroethene (TCE), *cis* 1,2 dichloroethene (DCE), and vinyl chloride (VC) are usually assumed to be the maximum value of the contaminants found in the existing monitoring wells. These are known values. Therefore, the source concentrations cannot be any less than those. However, they may be much higher, depending on whether the monitoring well is anywhere near the source. To simplify this problem, the actual data for source concentrations is assumed to be the "-" value of concentrations. If a coefficient of variation of 25 percent is assumed for the normal distribution of the source concentration, the mean and the "+" value can be found. Even though TCE, DCE, and VC are all three input as source concentrations, they were assumed to be one variable.

The half-lives in the environment are input in the biodegradation section of BIOCHLOR. The values found in Mackay et al. (1999) for TCE, DCE, and VC in groundwater were highly variable. The mean was taken to be the value approximately in the middle of the given values, and the coefficient of variation was assumed at 80 percent.

Table 1 summarizes the input variables for the BIOCHLOR runs. The parameters used in BIOCHLOR that were assumed to be "constants" for this study were porosity (0.25), ratio of vertical to longitudinal dispersivity (0.01), soil bulk density (1.6 kg/L), fraction organic carbon (0.0018), and K_{oc} (TCE=25 L/kg, DCE=50 L/kg, and VC=3 L/kg). Note that the K_{oc} values used are considered conservative. Higher values, which would model more sorption capacity of the soil, are readily found in the literature (Mackay et al., 1999).

The "worst case scenario" that was assumed for this site used the maximum hydraulic conductivity and hydraulic gradient at the site. In addition, the dispersivity was calculated using the method of Xu and Eckstein, the other dispersivity ratios were assumed at 0.1 and 0.01, and the sorption and biodegradation constants were kept at the most conservative levels. The source concentrations were those observed at the site. These parameters are also found in Table 1.

TABLE 1. Summary of random variables input values

Random Variable	Mean	Standard Deviation	Coefficient of Variation	"+"(1)	"-" (2)	Worst Case Scenario
Hydraulic Conductivity (ft/day)^	1.913	1.73	~90%	3.642	0.183	5.04
Hydraulic Gradient^	0.046	0.016	34%	0.062	0.031	0.073
Longitudinal Dispersivity (feet)	25	12.5	50%	37.5	12.5	21.6
Ratio of Transverse to Longitudinal to Longitudinal Dispersivity	0.1	0.05	50%	0.15	0.05	0.1
Initial Concentration(s) (mg/L)						
TCE	0.533	0.133	25%	0.666	0.4^	0.4
DCE	0.16	0.04	25%	0.2	0.12^	0.12
VC	0.124	0.032	25%	0.16	0.095^	0.095
Half-Lives in Groundwater (years)						
TCE	2.5	2	80%	4.5	0.5	4.5
DCE	4.5	3.6	80%	8	0.9	8
VC	4.5	3.6	80%	8.1*	1.0*	8.1*

^Based on actual data

* Can not have two rate constants the same in BIOCHLOR

(1) Mean "+" Standard Deviation

(2) Mean "-" Standard Deviation

RESULTS AND DISCUSSION

The results of the sixty-four (2^6) BIOCHLOR simulations can be found in Table 2. The concentration of trichloroethene, dichloroethene, and vinyl chloride at the property boundary 300 feet (91.44 m) downgradient of the source area was found for each simulation. In addition, the distance to the risk-based concentration (RBC) for vinyl chloride (in this case, 0.002 mg/L) was found. These four results were tabulated. After completing the sixty-four simulations, the mean and standard deviations of the results were found using an Excel spreadsheet. Table 2 contains the probabilistic outcomes, the mean and standard deviation, for the concentrations at the property boundary and the distance to the RBC for vinyl chloride.

The worst case scenario outcomes are also shown on this table. If the worst case scenario outcomes are placed on the cumulative normal distributions of the probabilistic outcomes, the probability that they will be exceeded can be found. This was done and the resulting exceedance probabilities are also shown in Table 2. As can be seen, the probability that the "worst case scenario" concentration for vinyl chloride at the property boundary will be exceeded is 44.3%. The probability that the distance to the point of compliance for the RBC of vinyl chloride will exceed the worst case scenario distance is 51.3%. These are high probabilities, considering that the "worst case scenario" is supposed to model an unlikely event. As can be inferred from these probabilities, the "worst case" is more likely to be the average case.

CONCLUSIONS

The point estimate method was easily applied to the use of BIOCHLOR. It is easily applied to any model in which random variables are used as input parameters, including BIOSCREEN, MT3D, RT3D and many others. Because the subsurface is rarely, if ever, known with any degree of certainty, probabilistic models should be used whenever modeling is done to evaluate groundwater contaminant concentrations or distance to compliance of regulatory contaminant levels. Its ease of use and its simplicity lends itself to understanding by clients, regulators, and the general public in this application of the point estimate method to a model of in-situ biodegradation of chlorinated compounds.

TABLE 2. Results of BIOCHLOR runs using + and - values of random variables.

Biochlor Run	Concentration at Point of Compliance (mg/L) * TCE	DCE (total)	VC	Distance to RBC (ft) for VC**
++++++[1]	0.016	0.007	0.004	525
+++++-	0.006	0.009	0.005	675
++++--	0.004	0.005	0.003	375
++++-+	0.009	0.004	0.003	375
+++-++	0.027	0.011	0.007	960
+++-+-	0.010	0.015	0.009	1050
+++--+	0.016	0.007	0.004	600
+++---	0.006	0.009	0.005	675
++-+-+	0.027	0.012	0.007	960
++--++	0.076	0.032	0.021	3250
++-++-	0.016	0.026	0.015	1200
++-+--	0.009	0.015	0.009	1050
++--+-	0.026	0.043	0.025	1500
++-+++	0.046	0.019	0.013	1800
++---+	0.045	0.019	0.012	1800
++----	0.016	0.026	0.015	1400
+---++	0.066	0.037	0.021	3150
+---+-	0.009	0.031	0.025	1000
+----+	0.040	0.022	0.012	1920
+-----	0.005	0.018	0.015	900
+-+++-	0.002	0.006	0.005	525
+-++++	0.014	0.007	0.004	600
+--+++	0.040	0.022	0.013	2000
+-+---	0.002	0.007	0.005	525
+--+-+	0.024	0.013	0.008	1050
+--++-	0.005	0.019	0.015	900
+--+--	0.003	0.011	0.009	675
+-++-+	0.008	0.004	0.003	375
+-++--	0.001	0.004	0.003	375
+-+--+	0.014	0.008	0.004	600
+-+-++	0.023	0.013	0.007	1000
+-+-+-	0.004	0.001	0.008	675
-+++++	0.002	0.006	0.004	450
-++++-	0.000	0.000	0.000	150
-+++--	0.000	0.000	0.000	140
-+++-+	0.001	0.004	0.003	375
-++-++	0.003	0.010	0.008	600
-++-+-	0.000	0.000	0.000	200

* 300 feet (91.4 m) downgradient of source
** 0.002 mg/L
(1) Indicates the value used (+ or -) of each random variable listed in order in Table 1.

TABLE 2. Results of BIOCHLOR runs using + and - values of random variables (con't).

Biochlor Run	Concentration at Point of Compliance (mg/L) * TCE	DCE (total)	VC	Distance to RBC (ft) for VC**
- + + - - +	0.002	0.006	0.005	450
- + + - - -	0.000	0.000	0.000	150
- + - + - +	0.003	0.010	0.008	600
- + - - + +	0.007	0.028	0.022	840
- + - + + -	0.000	0.000	0.000	175
- + - + - -	0.000	0.000	0.000	150
- + - - + -	0.000	0.000	0.000	205
- + - + + +	0.004	0.017	0.014	750
- + - - - +	0.004	0.017	0.013	750
- + - - - -	0.000	0.000	0.000	175
- - - - + +	0.001	0.009	0.012	525
- - - - + -	0.000	0.000	0.000	140
- - - - - +	0.000	0.005	0.007	450
- - - - - -	0.000	0.000	0.000	135
- - + + + -	0.000	0.000	0.000	138
- - + + + +	0.000	0.003	0.003	375
- - - + + +	0.000	0.005	0.007	450
- - + - - -	0.000	0.000	0.000	138
- - - + - +	0.000	0.003	0.004	375
- - - + + -	0.000	0.000	0.000	135
- - - + - -	0.000	0.000	0.000	125
- - + + - +	0.000	0.002	0.002	300
- - + + - -	0.000	0.000	0.000	112.5
- - + - - +	0.000	0.003	0.003	375
- - + - + +	0.001	0.004	0.005	450
- - + - + -	0.000	0.000	0.000	142.5
Average	0.010	0.010	0.007	702.984
Standard Dev.	0.016	0.010	0.007	649.790
Worst Case Scenario	0.032	0.012	0.008	675
Cumulative Normal Distributio	0.915	0.618	0.557	0.483
Probability of Exceedance (%)	8.5	38.2	44.3	51.7

* 300 feet (91.4 m) downgradient of source

** 0.002 mg/L

(1) Indicates the value used (+ or -) of each random variable listed in order in Table 1.

REFERENCES

Aziz, C.E., C.J. Newell, J.R. Gonzales, P. Haas, T.P. Clement, and Y. Sun. 1999. *BIOCHLOR, Natural Attenuation Decision Support System, Beta Version 1.0, User's Manual.*

Bonomo, L., Caserini, S., Pozzi, C, and Uguccioni, D.A. 2000. "Target Cleanup Levels at the Site of a Former Manufactured Gas Plant in Northern Italy: Deterministic versus Probabilistic Results." *Environmental Science and Technology*. 34(18): 3843-3848.

Chrostowski, P.C., Preziosi, D.V., Pearsall, L.J., Foster, S.A. 2000. "A Probabilistic Model of Time-To-Clean-Up by Natural Attenuation." *Natural Attenuation Considerations and Case Studies: Remediation of Chlorinated and recalcitrant Compounds.* Wickramanayake, G.B., Gavaskar, A.R. and Kelley, M.E. (eds.) Battelle Press.

Copty, N.K., and Findikakis, A.N. 2000. "Quantitative Estimates of the Uncertainty in the Evaluation of Ground Water Remediation Schemes." *Ground Water*. 38(1): 29-37.

Hamed, M.M., Bedient, P.B., 1999. *Reliability-Based Uncertainty Analysis of Groundwater Contaminant Transport and Remediation.* U.S. Environmental Protection Agency Technical Report, EPA/600/R-99/028, National Risk Management Research Laboratory, Cincinnati, OH.

Harr, M.E. 1987. *Reliability-Based Design in Civil Engineering*. McGraw-Hill Book Company, New York, NY.

Mackay, D., W. Shiu, K. Ma. 1999. *Physical-Chemical Properties and Environmental Fate Handbook* (CD-ROM). Chapman & Hall/CRCnetBASE, Boca Raton Florida.

Mishra, S. 2000. "Uncertainty Propogation Using the Point Estimate Method." *Calibration and Reliability in Groundwater Modelling, ModelCARE 99.* IAHS Publication No. 265, pp. 292-296.

Rosenblueth, E. 1975. "Point Estimates for Probability Moments." *Proceedings of the National Academy of Science*. 72(10). USA. 3812-3814.

Smith, V.J. and Charbeneau, R.J., 1990. "Probabilistic Soil Contamination Exposure Assessment Procedures." *Journal of Environmental Engineering*. 116(6): 1143-1163.

Thevanayagam, S., Kavazanjian, E., Jacob, A., Nesarajah, S. 1996. "Deterministic and Probabilistic Analyses of Preload Design at a Hydraulic Fill Site." *Uncertainty in the Geologic Environment: From Theory to Practice*. Shackelford, C.D., Nelson, P.P., and Roth, M.J.S. (eds.). pp. 1417-1431.

A GIS TECHNIQUE FOR ESTIMATING NATURAL ATTENUATION RATES AND MASS BALANCES

Neal D. Durant, P. Srinivasan, Charles R. Faust, Daniel K. Burnell, and Katrina L. Klein (GeoTrans, Inc., Sterling, Virginia)
David S. Burden (U.S. EPA, Ada, Oklahoma)

ABSTRACT: Regulatory approval of monitored natural attenuation (MNA) as a component for site remediation often requires a demonstration that contaminant mass has decreased significantly over time. Successful approval of MNA also typically requires an estimate of past and future natural attenuation rates. Calculation of electron acceptor and/or donor mass budgets is also useful in establishing the potential effectiveness of MNA. In this paper we present the application of a GIS program (TINMASS) for quantifying and visualizing contaminant and electron acceptor/donor mass-in-place. TINMASS uses a triangulated irregular network (TIN) to interpolate dissolved contaminant mass between monitoring points. The technique is best suited to sites where the plume is receding or has reached a steady-state, and the existing network of monitoring wells includes points located at or outside the perimeter of the plume, both horizontally and vertically. Natural attenuation rates can be estimated from the slope of the line formed by a log-linear plot of contaminant mass versus time. We present an example application illustrating use of the method for mass-balance analysis of sequential decay of trichloroethene. A second example is presented illustrating use of the method for evaluating fuel hydrocarbon degradation by quantifying masses of biodegradation end-products.

INTRODUCTION

In April 1999, the U.S. Environmental Protection Agency (EPA) issued a policy directive regarding the use of monitored natural attenuation (MNA) at Superfund, RCRA corrective action, and underground storage tank sites (OSWER Directive 9200.4-17P). The directive described lines of evidence that are typically investigated for determining the adequacy of MNA as a site remedy. One line of evidence is historical groundwater and/or soil data that demonstrate a meaningful trend of decreasing contaminant mass and/or concentration over time (U.S. EPA, 1999). Another line of evidence is hydrogeologic and geochemical data that demonstrate indirectly the types and rates of natural attenuation processes that are active at a site. In most cases, biodegradation is recognized as the most important natural attenuation process (for organic contaminants). As such, MNA evaluations are typically designed to estimate natural biodegradation rates. Rate estimates are used in analytical and/or numerical models to predict the time required to attain cleanup standards under a MNA remediation scenario.

Establishing a decreasing trend in contaminant mass is sometimes difficult because it can require multiple sampling events over several years, as well as a monitoring network that is sufficient to define the limits of the plume vertically and horizontally. Even at sites where the monitoring network is adequate, synthesis and interpretation of historical groundwater monitoring data can be complicated by temporal and spatial variation. For example, declining trends may be evident in some wells, but absent in others. Even if multiple wells at a site show decreasing concentration trends, the local persistence of contamination can sometimes jeopardize the approval of MNA as a site remedy.

A variety of methods are available for estimating site-specific intrinsic biodegradation rates, most of which have been described in existing EPA guidance manuals (U.S. EPA, 1998; Wiedemeier et al., 1998). The conservative tracer normalization (dilution-correction) and Buscheck and Alcantar (1995) methods are two of the more commonly employed techniques for estimating biodegradation rates. Both of these techniques require the installation of a series of monitoring points distributed along a linear transect that extends from the contaminant source to downgradient, along the length of the plume. An alternative approach to estimating attenuation/biodegradation rates is to quantify the contaminant mass-in-place trend over time (Dupont et al., 1997).

In this paper we describe the development and application of TINMASS, a simple GIS tool for visualizing natural attenuation trends and estimating natural attenuation rates. TINMASS provides a means for estimating past and present contaminant mass-in-place, as well as an effective graphical framework for visualizing mass reduction trends. Unlike the normalization and Buscheck and Alcantar (1995) methods, TINMASS does not require the placement of wells along a linear transect extending from source to downgradient. Not only can the method be used to estimate mass loss trends for the contaminant, it can also be used to estimate increases in the mass of inorganic end-products (e.g., Fe^{II}, H_2S, and CH_4), thereby facilitating a mass balance analysis. In addition, the method allows for specification of spatial variation in porosity and aquifer thickness.

DESCRIPTION OF TINMASS

TINMASS is a MapInfo® for Windows application that can be used to calculate the mass of a dissolved constituent in groundwater based on water quality data contained in a database (e.g., Microsoft Excel® or Access®). Input data for TINMASS include the following parameters at each sample location: (1) aquifer or interval thickness; (2) concentration of the constituent or chemical; and (3) porosity. Coordinates of each sample are taken directly from the MapInfo® object representing the sample location. An example input parameter window for TINMASS is shown in Figure 1.

Computational Technique. A triangulated irregular network (TIN) algorithm is used to interpolate constituent mass between monitoring points. The technique is a simple numerical integration approach commonly used to estimate volumes in civil engineering applications. In the first step, an optimum network of triangles connecting all sample locations is generated using the Dulauney Triangulation procedure. Dulauney Triangulation connects points to form a network of triangles that have as nearly equal angles at their vertices as possible. The procedure was based on an algorithm described by Watson (1982).

In the second step, an estimate of the mass of dissolved contaminant within each triangle is determined based on the assumption that aqueous concentration (C_w), porosity (n), and aquifer/interval thickness (b) vary linearly between the sample locations. The dissolved mass within the aquifer bounded horizontally by each triangle is determined by:

$$M_w = \int_{area} nC_w b(dxdy)$$

where n, C_w, and b vary linearly across the triangle. Details on the interpolation and integration may be found in Istok (1989). The total mass of dissolved constituent in the aquifer unit equals the sum of the mass from each triangle.

Running TINMASS will create a TIN network as a MapInfo® table and calculate mass in units of pounds or kilograms. Data from the model run (input data fields or values entered, number of nodes, TIN triangles generated, and total mass) are stored in another MapInfo® summary table indexed by a Run ID. The Run ID may be used as an identifier of the contaminant and the sampling period. Thus, results from multiple periods and different chemicals may be summarized in one summary table. This approach supports the determination of temporal mass-in-place trends.

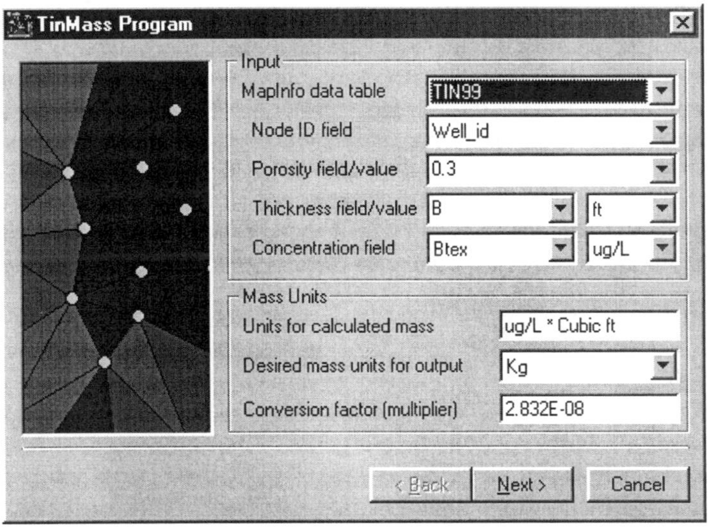

FIGURE 1. Example input parameter window for TINMASS.

Practical Use and Limitations. Mass-in-place estimates computed by TINMASS can be used to estimate site-specific natural attenuation rates at sites where sufficient historical data are available, and the existing monitoring network has defined the vertical and horizontal limits of the plume. The natural attenuation rate can be approximated as the slope of an exponential regression on a log-linear plot of mass vs. time. When using TINMASS for this purpose it is important to use a consistent set of monitoring locations across time. Even if the number of available data points increased during the relevant period, the same set of points should be used in order to limit the extent to which spatial/temporal variability will bias the rate estimate.

Use of TINMASS for estimating natural attenuation rates is best suited to sites where the plume is receding or has reached a steady-state condition. The method is less suitable for sites where a significant portion of the contamination occurs outside the existing monitoring network. At such sites it may not be possible to prove that temporal mass reductions were caused by attenuation (and not due to plume migration beyond the TIN network).

TINMASS can estimate mass-in-place of a constituent in a variably thick aquifer or within a discrete interval of an aquifer (depending on the layout of the monitoring network). In order to estimate mass-in-place in three dimensions, it is sometimes appropriate to estimate mass-in-place for separate layers (monitoring depth zones) and sum the results.

Because concentrations typically decrease in an exponential manner away from the source, the linear interpolation used in this procedure has a tendency to overestimate mass-in-place. The potential for overestimation decreases as the number of monitoring points in the TIN network increases. It should be noted that TINMASS only examines mass trends for the dissolved phase. Where present, non-aqueous phase liquids (NAPLs) will usually account for a dominant portion of the total mass, but estimation of NAPL mass is subject to a very high level of uncertainty. If undetected, NAPL presence may cause misinterpretation of mass removed versus mass-in-place trends. Despite the uncertainty of the mass estimate for a given sampling event, the uncertainty is relatively constant over time when using TINMASS to assess mass-in-place trends for dissolved plumes (assuming that the network is constant through time). Consequently, the natural attenuation rate estimate for dissolved plume mass is not significantly affected by errors associated with each individual mass estimate.

EXAMPLE APPLICATIONS

Example A. Natural attenuation is treating a portion of the dissolved chlorinated solvent plume beneath the Sanitary Landfill (SLF) at the U.S. Department of Energy Savannah River Site (WSRC, 1997). Much of the trichloroethene (TCE) in the plume has been reductively dechlorinated to *cis*-1,2-dichloroethene (DCE) and vinyl chloride (VC). An abundance of dissolved organic carbon is the likely reductant supporting dechlorination. The landfill was capped in 1997, and a horizontal well biosparging system is now being operated to treat the groundwater contamination. A modeling analysis was recently completed to evaluate the performance of natural attenuation for eliminating the residual contamination that escapes the biosparging treatment zone. Estimation of site-specific natural biodegradation rates was the first step in the natural attenuation modeling process. TINMASS was used to specify a TIN network within the plume, and to quantify the dissolved mass for TCE, DCE, and VC over time (see Figure 2). A uniform porosity of 20% was specified as an input parameter. The unit thickness for each point was specified as the difference between the water level and a datum elevation. Although the monitoring network consisted of multi-level wells screened at three depths, all the wells were pooled together for this analysis since the aquifer formation and contaminant distribution are relatively homogeneous

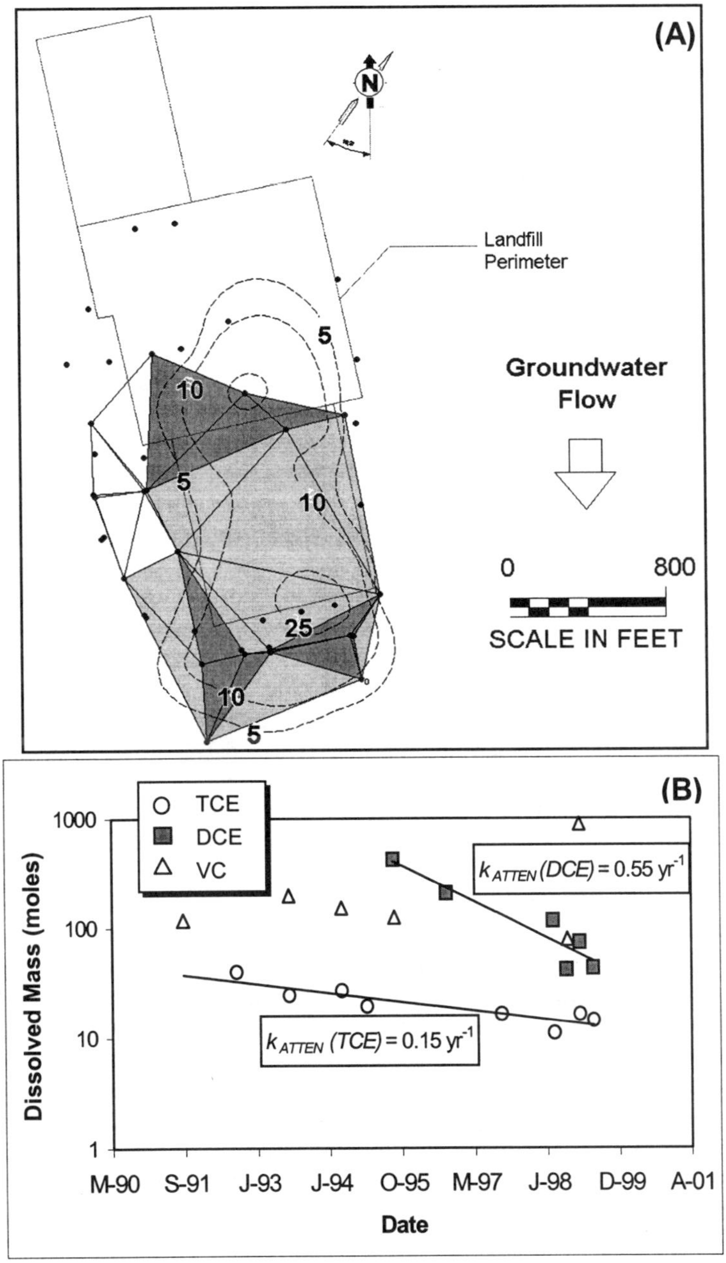

FIGURE 2. (A) Configuration of 1993 TCE plume (μg/L) and TIN network used to estimate dissolved chloroethene mass at the Savannah River Site SLF. Shading in TIN triangles is user-specified, and represents different mass per volume quantities (values not shown). (B) Time series plots of dissolved mass for TCE, DCE, and VC.

over the interval (approximately 18 meters).

As illustrated in Figure 2, the TIN network covered reasonably well the configuration of the TCE plume at the SLF. TINMASS was used to estimate dissolved masses for TCE, DCE, and VC. The TINMASS estimates (kilograms) were converted to a moles basis to allow a mass balance assessment of the sequential decay of TCE over time (under anaerobic conditions, one mole of TCE decays to one mole of DCE, and one mole of DCE decays to one mole of VC). The data were plotted on a log-linear graph, and decay coefficients were approximated as the slope of an exponential regression of the data (see Figure 2). From this analysis, the estimated first-order biodegradation rate is 0.15 yr^{-1} TCE, and 0.55 yr^{-1} for DCE. Fewer data were available for the DCE rate estimate since laboratory analysis for this constituent did not commence until 1995. Note that these rates apply to the attenuation of the whole plume, without regard for spatial variability in redox conditions.

Additional TINMASS analyses were performed to estimate biodegradation rates for aerobic and anaerobic zones (data not shown). In general, the anaerobic regions coincide with the footprint of the landfill perimeter, and aerobic zones occur outside the landfill perimeter. Mass estimates from TINMASS suggest that the DCE biodegradation rate is 0.44 yr^{-1} in the anaerobic zone, and 0.66 yr^{-1} in the aerobic zone. The mass of VC appeared to show a net increase directly underneath the landfill cap during the period analyzed, and the rate of VC accumulation in the anaerobic zone was consistent with the rate of DCE mass decay. In the aerobic zone downgradient of the landfill, the VC biodegradation rate estimated from TINMASS is 0.29 yr^{-1}.

The data in Figure 2B suggest that TCE and DCE are attenuating at the SLF. Sufficient data are available to suggest that the TCE has degraded as a first order trend. There is more scatter in the DCE trend, which likely reflects spatial variability in the processes and conditions affecting DCE degradation.

Example B. During the 1980s, a jet fuel (JP-4) release introduced hydrocarbon contamination into a shallow aquifer beneath a confidential site in the Northeastern U.S. Sampling performed during the early 1990s detected benzene, toluene, ethylbenzene, and xylene (BTEX), and naphthalene in the site groundwater. More recent sampling events have confirmed that benzene and naphthalene are the only two contaminants present at concentrations above clean-up levels. Other straight-chain and branched hydrocarbons typical of JP-4 are also present in the groundwater. These have not been differentiated, and are characterized only as total petroleum hydrocarbon (TPH). The residual light non-aqueous phase liquid (LNAPL) that was present in the 1990s likely still remains in certain locations.

In addition to declining constituent concentrations, evidence of natural attenuation includes the accumulation of biodegradation end-products (e.g., H_2S, CH_4, and Fe^{II}). TINMASS was used to estimate mass removed and attenuation rates for the various contaminants. The mass of dissolved H_2S, CH_4, and Fe^{II} was also estimated with TINMASS, and these data were used with stoichiometric relationships to estimate the potential amount of biodegradation associated with each biodegradation end-product. Biodegradation rates were also estimated by the method of Buscheck and Alcantar (1995) for the purpose of comparison with the rate estimates from the mass-in-place calculations. The results of these analyses are summarized in Table 1.

TABLE 1. Summary of dissolved mass-in-place and associated natural attenuation rate estimates for groundwater constituents at JP-4 site.

	Attenuation Rate by TINMASS (yr^{-1})	Degradation Rate by Buscheck and Alcantar (yr^{-1})	Mass Removed 1996 – 2000 (kg)	Mass Produced by 2000 (kg)	Possible Incurred Biodegradation Based on Stoichiometry[1] (kg)
BTEX	0.40	3.8	20	--	203
Benz	0.40	3.0	6	--	209
Naph	0.26	2.4	49	--	215
TPH	0.88	0.7	1050	--	184
Fe^{II}	--	--	--	184	--
H_2S	--	--	--	120	--
CH_4	--	--	--	98	--

1. Stoichiometric relationships presented by Wiedemeier et al. (1998) were used to calculate the amount of biodegradation possible – assuming no microbial growth - for the observed masses of accumulated biodegradation end-products. Octane was used as a surrogate for TPH.

As shown in Table 1, application of TINMASS allowed an estimate of the dissolved mass of BTEX, naphthalene, and TPH degraded during the period of 1996 to 2000. This analysis indicates that TPH represents the greatest fraction removed during this period. By comparison, the masses of BTEX and naphthalene removed were significantly smaller. This finding is not surprising given that BTEX and naphthalene together represent less than 5% by weight of typical virgin JP-4.

The attenuation rates estimated from the TINMASS calculations for BTEX, benzene, and naphthalene are within the range normally reported for biodegradation of these compounds in aquifers. However, these estimates likely underestimate the actual rate of intrinsic biodegradation since this did not account for contributions from source dissolution. During the period of 1996 to 2000, additional BTEX and naphthalene likely were introduced into the groundwater by dissolution from the LNAPL source. If source dissolution had not occurred, the apparent plume attenuation (and mass removal) would have been more rapid. Contributions from source dissolution likely explain, in part, the differences in the rate estimates between the TINMASS method and the method of Buscheck and Alcantar. The Buscheck and Alcantar method focuses on biodegradation in the plume, while the TINMASS method applies to both source and plume. As shown in Table 1, rates estimated with the Buscheck and Alcantar method are nearly an order of magnitude faster than attenuation rate estimates derived from the TINMASS calculations (with the exception of TPH). Others have noted, however, that the method by Buscheck and Alcantar is prone to overestimating the rate of biodegradation (McNab and Dooher, 1998). As such, we conclude that the attenuation rate estimated by TINMASS method is a conservative approximation to the biodegradation rate.

The last column in Table 1 provides an estimate of the mass of each constituent that could have been degraded if all the H_2S, CH_4, and Fe^{II} observed was due to the biodegradation of that constituent. These data suggest that at least a portion of the observed mass removed from 1996 to 2000 can be attributed to biodegradation under ferrogenic, sulfidogenic, and methanogenic conditions. Comparison of the mass removal estimates for BTEX, naphthalene, and TPH with their calculated potential incurred biodegradation provides insight as to the principal electron donors driving the accumulation of H_2S, CH_4, and Fe^{II}. This comparison suggests that the biodegradation of benzene, BTEX, and naphthalene alone during 1996 to 2000 does not account for all the H_2S, CH_4, and Fe^{II} that has accumulated. It follows that biodegradation of the TPH has contributed to a significant portion of the H_2S, CH_4, and Fe^{II} that have accumulated.

SUMMARY

The software programs BIOSCREEN (Newell et al., 1996) and BIOCHLOR (Aziz et al., 2000) provide a variety of tools for analyzing natural attenuation, including an ability to estimate contaminant mass-in-place and mass biodegraded. Although TINMASS is not an alternative to these powerful tools, it offers a unique capability to estimate dissolved contaminant mass across a variably contaminated domain. Unlike BIOSCREEN and BIOCHLOR, TINMASS is not limited to symmetrical and homogeneous model domains, and does not employ transport equations or flow parameters. Instead, TINMASS focuses primarily on the trends in empirical data. TINMASS allows users to estimate dissolved mass across formations of variable porosity, variable thickness, and variable contaminant distribution. It also provides a simple framework for integrating historical mass-in-place trends across multiple monitoring points.

In this paper, mass-in-place calculations were used to estimate site-specific attenuation rates and assess mass balances between biodegradation parent compounds and daughter products. These applications suggest that TINMASS is a useful tool for natural attenuation analysis. Additional evaluation at other sites is needed to further examine the use of this method for estimating site-specific natural attenuation rates.

ACKNOWLEDGMENTS AND DISCLAIMER

The U.S. EPA Robert S. Kerr Environmental Research Laboratory and the Westinghouse Savannah River Company (WSRC) funded portions of this work. This paper has not been subjected to EPA peer and administrative review. Therefore, no official endorsement may be inferred. Mention of commercial products in this paper does not represent endorsement by EPA or WSRC.

REFERENCES

Aziz, C.E., C.J. Newell, J.R. Gonzales, P. Haas, T. P. Clement, and Y. Sun. 2000. *BIOCHLOR Natural Attenuation Decision Support System, User's Manual,*

Version 1.0, U.S. Environmental Protection Agency Report, EPA/600/R-00/008, National Risk Management Research Laboratory, Ada, OK.

Buscheck, T.E., and C.M. Alcantar. 1995. "Regression Techniques and Analytical Solutions to Demonstrate Intrinsic Bioremediation." In R.E. Hinchee, J.T. Wilson, and D.C. Downey (Eds.), *Intrinsic Bioremediation*, pp. 109-116. Battelle Press, Columbus, OH.

Dupont R.R., K. Gorder, D.L. Sorensen, M.W. Kemblowski, and P. Haas. 1997. "Case Study: Eielson Air Force Base, Alaska." In *Proceedings of the Symposium on Natural Attenuation of Chlorinated Organics in Ground Water.* U.S. Environmental Protection Agency Report, EPA/540/R-97/504, Office of Research and Development, Washington, D.C.

Istok, J. 1989. *Groundwater Modeling by the Finite Element Method.* Water Resources Monograph No. 13, American Geophysical Union, Washington, D.C.

McNab, W.W., and B.P. Dooher. 1998. "A Critique of a Steady-State Analytical Method for Estimating Contaminant Degradation Rates." *Ground Water, 36*(6):983-987.

Newell, C.J., R.K. McLeod, and J.R. Gonzales. 1996. *BIOSCREEN Natural Attenuation Decision Support System, User's Manual, Version 1.3*, U.S. Environmental Protection Agency Report, EPA/600/R-96/087, National Risk Management Research Laboratory, Ada, OK.

U.S. Environmental Protection Agency. 1998. *Seminars on Monitored Natural Attenuation for Ground Water.* Report EPA/625/K-98/001.

U.S. Environmental Protection Agency. 1999. *Use of Monitored Natural Attenuation at Superfund, RCRA Corrective Action, and Underground Storage Tank Sites.* Office of Solid Waste and Emergency Response, Directive 9200.4-17P, Washington, D.C.

Watson, D.F. 1982. "ACORN: Automated Contouring of Raw Data." *Computers and GeoSciences. 8*(1):97-101.

Wiedemeier, T.H., M.A. Swanson, D.E. Montoux, E.K. Gordon, J.T. Wilson, B.H. Wilson, D.H. Kampbell, P.E. Haas, R.N. Miller, J.E. Hansen, and F.H. Chapelle. 1998. *Technical Protocol for Evaluating Natural Attenuation of Chlorinated Solvents in Ground Water.* U.S. Environmental Protection Agency Report, EPA/600/R-98/128, National Risk Management Research Laboratory, Ada, OK.

WSRC. 1997. *Intrinsic Bioremediation of Landfills Interim Report.* Westinghouse Savannah River Company Report WSRC-RP-97-17, Rev. No.: 0.

QUANTITATIVE VALIDATION OF A MODEL OF CHLORINATED ETHENE NATURAL ATTENUATION

Harold C. Young, Charles A. Bleckmann, Junqi Huang, Daniel E. Reynolds, and Mark N. Goltz
(Air Force Institute of Technology, Wright-Patterson AFB, OH, USA)

ABSTRACT: The implementation of natural attenuation as a contaminated groundwater remediation strategy requires a detailed understanding of how physical, chemical, and biological processes impact contamination at a particular site. One way to gain this understanding is through contaminant fate and transport modeling. In this study, a deterministic model that included relevant fate and transport processes was applied to a chlorinated ethene-contaminated field site, at which spatial and temporal data had been collected. Parameters used for model input were obtained from the literature, experimental data, and by calibrating the model using concentration data from 1993. The model was then run in a predictive mode, and simulation results were compared to field data from 1999. Model performance was measured by comparison of observed and simulated concentration contour plots and evaluation of goodness-of-fit statistics. Over the six years the model was run in a predictive mode, the model was found to predict contaminant concentrations reasonably well for the three contaminants that were monitored.

INTRODUCTION

The contamination of groundwater by chlorinated solvents is a well-recognized problem. The extent of contamination across the United States is such that the EPA has declared trichloroethylene (TCE) and its degradation products, cis-1,2-dichloroethylene (DCE) and vinyl chloride (VC), to be priority pollutants (Bloom et al., 2000). This paper focuses on management of dissolved phase chlorinated ethene contamination through monitored natural attenuation (MNA).

Natural attenuation encompasses the "physical, chemical, or biological processes that, under favorable conditions, act without human intervention to reduce the mass, toxicity, mobility, volume, or concentration of contaminants in soil and groundwater" (USEPA, 1999). Recently, natural attenuation has come under criticism for being used in situations where its effectiveness has not been adequately demonstrated. According to a report from the National Research Council (NRC), "natural attenuation should be accepted as a formal remedy for contamination only when the processes are documented to be working and are sustainable" (NRC, 2000). One way to demonstrate that natural attenuation may be occurring in the subsurface is through fate and transport modeling. Several computer models have been presented that simulate the natural attenuation of TCE (Clement et al., 2000; Feng, 2000), but few have been validated using data from a contaminated site. Model validation could result in increased credibility for natural attenuation as a remediation strategy for chlorinated solvent contamination, with a subsequent rise in implementation and reduction in overall costs. To demonstrate a model validation process, the contaminant fate and transport model BR3D was applied to Site LF-03 at F.E. Warren Air Force Base (AFB), Cheyenne, Wyoming. After calibration, the model was run in a predictive mode. Results were then evaluated against field data through comparison of concentration contour plots and statistical analyses. It is suggested that this validation process could prove useful when applying models that simulate natural attenuation of chlorinated ethenes at contaminated sites.

MODEL APPLICATION

Site History: Landfill 03 (LF-03) was operational on F.E. Warren AFB from approximately the mid-1950s to the mid-1960s for the disposal of industrial and residential wastes generated by the base (Parsons Engineering Science, 1999). In the mid-1980s it was determined that leachate from LF-03 could possibly pose a threat to human health and the environment. It is hypothesized that TCE, used as a metal cleaner at base shops, had been disposed in LF-03 according to the standard industry practices of the day. A focused remedial investigation (RI) was started for LF-03 in 1995, and a treatability study (TS) to evaluate the use of MNA was conducted by Parsons Engineering Science and the USEPA National Risk Management Research Laboratory (NRMRL) in 1999.

In the TS, the computer code MT3D was used in conjunction with MODFLOW (a groundwater flow model) to simulate the fate and transport of TCE. A 220-cell by 125-cell by three layer model domain was used, with each cell measuring 20 feet by 20 feet (6.1m by 6.1m). MODFLOW and MT3D were calibrated using data collected from 27 monitoring wells. Simulations were then run to ascertain the effectiveness of different remedial alternatives.

While the TS proved informative to decision-makers at F.E. Warren AFB, it had several shortcomings. First, TCE was the only contaminant simulated, even though DCE and VC were also detected at the site. Second, the study did not involve any effort, outside of model calibration, to validate the application of the model to the site. To better understand the processes taking place at Site LF-03, the computer model BR3D (Huang and Goltz, 1998) was applied to simulate the fate and transport of TCE, DCE, and VC at the site. BR3D is a three-dimensional transport model that, unlike the MT3D code used in the TS, is capable of simulating the sequential decay of chloroethenes in an aquifer.

Groundwater Flow Model: For the purposes of this study, the Visual MODFLOW (Version 2.8.2.0) groundwater flow model used in the TS is assumed to be valid (Parsons Engineering Science, 1999).

Contaminant Fate and Transport Model: Key assumptions made in the TS were incorporated into the application of BR3D to the site: TCE infiltration into groundwater started in 1960, DCE and VC are present solely due to the reduction of TCE, chlorinated ethene attenuation is due to first-order reductive dechlorination, and sorption may be modeled as a linear equilibrium process.

These assumptions were incorporated into BR3D to create a site-specific model that simulated 3D advection and dispersion, linear equilibrium sorption, and first-order biodegradation. Steady state groundwater flow and constant porosity (both spatially and temporally) were also assumed, such that the important physicochemical and biological processes could be incorporated into the general contaminant transport equation (Charbeneau, 2000):

$$R\frac{dC}{dt} = -\underline{v} \bullet \nabla C + \nabla \bullet (\underline{\underline{D}} \bullet \nabla C) - kC \quad (1)$$

where
C = concentration of contaminant in aqueous phase (TCE, DCE, or VC) $[M/L^3]$
t = time [T]
$\underline{v}$ = average linear velocity vector [L/T]
$\underline{\underline{D}}$ = dispersion coefficient matrix $[L^2/T]$
k = contaminant biological decay first order rate constant $[T^{-1}]$
R = retardation factor of contaminant due to sorption [-]

The value of R for each chlorinated ethene (Table 1) was calculated using literature values (Howard and Meylan, 1997) and site-specific information (Parsons Engineering Science, 1999). The value of each rate constant (k_{TCE}, k_{DCE}, and k_{VC}) was determined through model calibration, as described below.

TABLE 1. Model retardation coefficients.

R_{TCE}	1.33
R_{DCE}	1.28
R_{VC}	1.22

Model Calibration. The first order decay rates for the chlorinated ethenes were calibrated using concentrations of TCE, DCE, and VC observed in 1993. It was decided to perform model calibration using the log transform of the observed concentrations, as this would minimize the influence of large values observed at select observation points. The root mean squared error (RMSE) of the transformed data was then calculated:

$$RMSE = \sqrt{\overline{(\log(Q_O) - \log(Q_S))^2}} \quad (2)$$

where
Q_O = Observed Values [g/L]
Q_S = Corresponding Simulated Values [g/L]
Note: The overline denotes the average of the indicated quantities.

Chlorinated ethene decay rate constants were estimated using a forward, trial-and-error process. Using the sum of the three chlorinated ethene RMSEs as the objective function, the decay rates for TCE, DCE, and VC were varied until k_{TCE}, k_{DCE}, and k_{VC} each converged on values that resulted in a minimized objective function. After a simulation was run using the fit decay rates, a set of goodness-of-fit statistics was calculated. These statistics included unconditional bias, non-dimensional bias (B'^2) and the Nash-Sutcliffe coefficient of efficiency.

Unconditional bias (B) is a measure of the model's tendency to over- or under-estimate contaminant concentration, and is defined as

$$B = m_S - m_O \tag{3}$$

where
m_S = average simulated value
m_O = average observed value

Non-dimensional bias measures the relative magnitude of bias, and is given by

$$B'^2 = \frac{B^2}{s_O^2} \tag{4}$$

where
s_O = standard deviation of the observed values

The coefficient of efficiency (E) is an estimate of how closely a plot of ($\log(Q_O)$, $\log(Q_S)$) points fall along the 1:1 line, or line-of-perfect-fit, and is defined as

$$E = 1 - \frac{\overline{(\log(Q_O) - \log(Q_S))^2}}{s_O^2} \tag{5}$$

A maximum value of E = 1 suggests perfect model performance in simulating observed data, while a negative value of E indicates that the model "introduces more ambiguity than that introduced by simply using the mean value of the observation as an estimator" (Imam et al., 1999). Using E as a gauge of model performance is relatively new to contaminant transport modeling. Accordingly, there is no consensus of what value of E indicates an acceptable fate and transport model, especially when the data have been log transformed. However, as this model did not require a major calibration (only three parameters varied), it may be possible to follow the example given by Frankenberger et al. (1999) for the validation of a hydrology model. If this example is applicable, values in the range of E = 0.60 can be considered good, while values of $E \leq 0.5$ tend to indicate poor model performance.

Model Validation. After BR3D was calibrated to 1993 data, a predictive simulation was run from 1960 to 1999, with results compared to the 1999 observed concentrations of TCE, DCE, and VC. Concentration contour plots (generated by Surfer v5.0.3) were compared and RMSE, bias (B) and coefficient of efficiency (E) were reported for each contaminant.

RESULTS AND DISCUSSION

Calibration. The calibrated decay rates are listed in Table 2. The fit values for first-order decay are reasonable, in that they fall within the range of values reported in the literature (Suarez and Rifai, 1999).

TABLE 2. Calibrated model parameters.

	First Order Rate Constant, k (d^{-1})
TCE	2.3×10^{-4}
DCE	1.8×10^{-4}
VC	0

Contaminants with greater number of chlorine substituents had greater rates of decay ($k_{TCE} > k_{DCE} > k_{VC}$), as would be expected in an anaerobic aquifer where reductive dehalogenation is taking place.

After a simulation was run using the fitted decay rates, a set of goodness-of-fit statistics was calculated (Table 3). The model bias (B) of each contaminant reveals that concentrations for all three are being under-estimated, with DCE having the greatest magnitude of bias (B'^2).

TABLE 3. Calibration goodness-of-fit statistics.

	RMSE (log(g/L))	E (-)	B (log(g/L))	B'^2 (-)
TCE	0.55	0.71	-0.27	0.069
DCE	0.74	0.51	-0.42	0.16
VC	0.21	0.60	-0.030	0.008

Fits to TCE and VC data are considered good (E > 0.60), while the value of E for DCE data (E = 0.51) tends to indicate poor model performance.

Validation. Figures 1, 2, and 3 compare the simulated and observed chlorinated ethene concentrations for 1999, while Table 4 lists the goodness-of-fit statistics for the predictive simulation. Inspection of the 4×10^{-5} g/L contour lines in Figure 1 indicates that simulated TCE concentration matches well to observed data, while an inspection of Figure 2 shows that DCE is considerably underestimated in the simulation. In contrast, the model overestimates VC concentrations (see Figure 3). It should be noted, however, that the simulated VC concentrations are

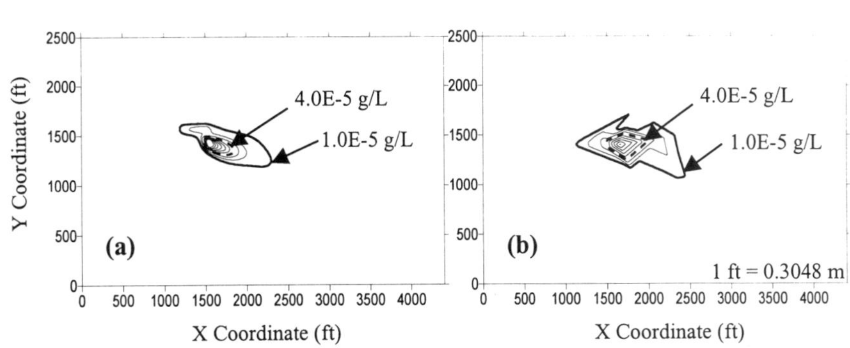

FIGURE 1. Simulated (a) and observed (b) TCE concentrations for 1999.

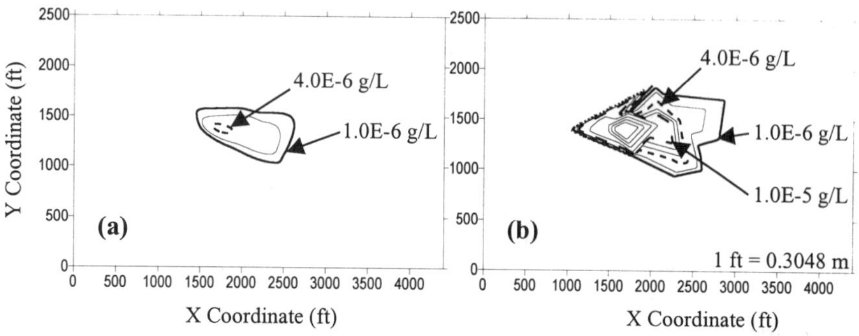

FIGURE 2. Simulated (a) and observed (b) DCE concentrations for 1999.

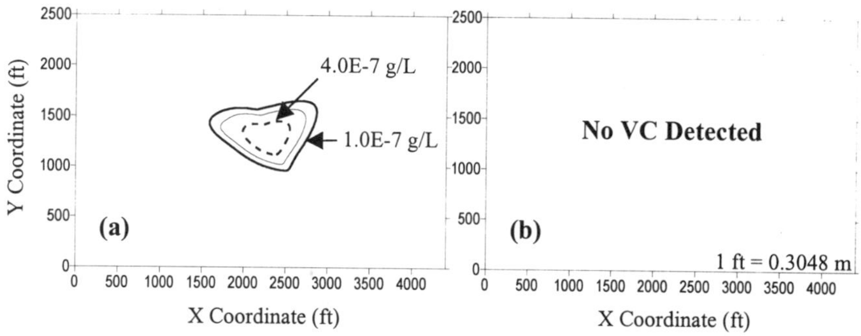

FIGURE 3. Simulated (a) and observed (b) VC concentrations for 1999.

nearly at the VC detection limit of 1×10^{-7} g/L, such that the overestimation is relatively small.

As VC was not observed at Site LF-03 in 1999, the observed variance of VC concentrations equals zero, preventing calculation of the Nash-Sutcliffe coefficient of efficiency (E) and the non-dimensional bias (B'^2) for VC. Regardless, two interesting observations can be made. First, the magnitude of non-dimensional bias (B'^2) for TCE decreased from 0.069 in 1993 to 0.009 in 1999. It is hypothesized that the TCE loading rate in the first few time steps may have been insufficient, resulting in the underestimation of TCE concentration later in the simulation. Due to the mathematical relation between E and B'^2, this reduction in the magnitude of bias results in an increase in E, and the unusual observation that the model prediction of the observed TCE data from 1999 is statistically better (E=0.77) than the model calibration to the observed TCE data from 1993 (E = 0.71). The relatively high value of E for TCE suggests

TABLE 4. Validation goodness-of fit statistics

	RMSE (log(g/L))	E (-)	B (log(g/L))	B'^2 (-)
TCE	0.50	0.77	-0.09	0.009
DCE	0.77	0.47	-0.39	0.14
VC	0.24	-	0.13	-

that the TCE concentration fit is good. Second, the value of E for DCE is relatively low (below 0.50), indicating that the fit of simulated DCE to the observed values is relatively poor (Frankenberger et al., 1999). Lastly, the positive value of VC bias (B) indicates that VC is being over-predicted. As no VC was observed at Site LF-03 in 1999, any model prediction over $1x10^{-7}$ g/L (the detection limit for VC) would result in over-prediction.

CONCLUSIONS

Predictive Simulation and Comparative Analysis. While chlorinated ethene advection at F.E. Warren AFB Site LF-03 is relatively slow, and only six additional years were simulated, we found that BR3D provided good predictions of TCE concentrations and fair predictions of DCE concentrations. Due to a lack of quantifiable data, no conclusions were drawn regarding VC predictions. From these observations regarding model performance, we may conclude that chlorinated ethene fate and transport at the F.E. Warren site could be simulated by an advection/dispersion model that assumes linear equilibrium sorption and first-order decay.

REFERENCES

Bloom, Y., R. Aravena, D. Hunkeler, E. Edwards, and S.K. Frape. 2000. "Carbon Isotope Fractionation during Microbial Dechlorination of Trichloroethene, *cis*-1,2-Dichloroethene, and Vinyl Chloride: Implications for Assessment of Natural Attenuation." *Environmental Science and Technology. 34*(13):2768-2772.

Charbeneau, R.J. 2000. *Groundwater Hydraulics and Pollutant Transport.* Prentice-Hall: New Jersey.

Clement, T.P., C.D. Johnson, Y. Sun, G.M. Klecka, and C. Bartlett. 2000. "Natural attenuation of chlorinated ethene compounds: model development and field-scale application at the Dover site." *Journal of Contaminant Hydrology. 42*:113-140.

Feng, P.P. 2000. "Modeling the Effect of Nonlinear and Rate-Limited Sorption on the Natural Attenuation of Chlorinated Ethenes." MS thesis, AFIT/GEE/ENV/00M-04. School of Engineering and Management, Air Force Institute of Technology (AU), Wright-Patterson AFB OH.

Frankenberger, J.R., E.S. Brooks, M.T. Walter, M.F. Walter, and T.S. Steenhuis. 1999. "A GIS-based variable source area hydrology model." *Hydrological Processes. 13*: 805-822.

Howard, P.H., and W.M. Meylan (Eds.). 1997. *Handbook of Physical Properties of Organic Chemicals*. Lewis Publishers: Boca Raton, FL.

Huang, J., and M.N. Goltz. 1998. "A Model of *In Situ* Bioremediation that Includes the Effect of Rate-Limited Sorption and Bioavailability." *Proceedings of the 1998 Conference on Hazardous Waste Research*, Snow Bird, Utah, May 19-21.

Imam, B., S. Sorooshian, T. Mayr, M. Schaap, H. Wosten, and B. Scholes. 1999. *IGBP-DIS Working Paper #22: Comparison of Pedotransfer Function to Compute Water Holding Capacity Using the van Genuchten Model in Inorganic Soils*. May 1999, Report to IGBP-DIS Soil Data Tasks. National Geophysical Data Center: Colorado.

National Research Council. 2000. *Natural Attenuation for Groundwater Remediation*. National Academy Press: Washington, D.C.

Parsons Engineering Science, Inc. 1999. *Treatability Study in Support of Monitored Natural Attenuation at Landfill 3 (Site LF-03), F.E. Warren Air Force Base, Cheyenne, Wyoming*. Prepared for the U.S. Air Force, Center for Environmental Excellence.

Suarez, M.P. and H.S. Rifai. 2000. "Biodegradation Rates for Fuel Hydrocarbons and Chlorinated Solvents in Groundwater." *Bioremediation Journal*. *3*(4): 337-362.

United States Environmental Protection Agency (USEPA). 1999. *Use of Monitored Natural Attenuation at Superfund, RCRA Corrective Action, and Underground Storage Tank Sites*. Office of Solid Waste and Emergency Response Directive 9200.4-17P.

COMPUTER MODELING OF NATURAL ATTENUATION OF ORGANIC CONTAMINANTS AND CONCOMITANT GEOCHEMICAL PROCESSES IN SANDY AQUIFER MATERIALS

Per Aagaard, ***Zuoping Zheng*** and Gijs Breedveld (University of Oslo, Norway)

ABSTRACT: Natural attenuation of organic contaminants and concomitant geochemical processes were investigated by applying PHREEQC program coupling equilibrium chemistry of inorganic constituents, kinetically mediated degradation of organic contaminants with different electron acceptors, and advection-dispersive transport as well. One substrate, toluene, several electron acceptors (dissolved O_2, NO_3^-, Fe(III), SO_4^{2-}), and heterotrophic microbial populations were included in the simulation. Toluene degradation and biomass growth were modeled using the dual-Monod kinetics subjected to basic assumptions, where model parameters were based on literature data. The model run for the batch reaction case resulted in a gradual decrease of toluene concentration coupled with consumption of different electron acceptors, leading to changes of pH and pe; for the one-dimensional reactive transport case, an initial increase, followed by an almost linear decrease of toluene concentration in the effluent was observed, due to biodegradation and microbial growth during transport. Sensitivity analyses for one-dimensional reactive transport in aquifers indicate that the simulation results are quite sensitive to changes in the microbial parameters, particularly the initial biomass concentration and the maximum specific substrate utilization rate. In addition, the physicochemical characteristics including pore-water velocity, initial solution composition and mineral assemblage have also significant influences on the biodegradation processes.

INTRODUCTION

Geochemical modeling is a useful tool to integrate physical, chemical and biological processes occurring in groundwater contaminated with organic contaminants. This effort has in recent years widely been based on quantitative representations of either the kinetics of contaminant degradation or the physicochemical processes (Hunter et al., 1998; Murphy and Ginn, 2000). In general, the physical processes considered in modeling include advection, diffusion and dispersion, which are primarily based on the structure and properties of the groundwater flow system in porous media. The chemical processes include ion exchange, mineral dissolution and precipitation due to changes in pore water chemistry caused by biological activities. In contrast, the biological processes of growth/decay are characteristics of the bacterial population. The coupling of these processes in a dynamic contaminant system, however, is not well understood. Actually, very few existing models take coupling of these processes into account. For example, McNab and Narasimhan (1994) present a model capable of handling geochemical interactions between organic and inorganic species.

The present study aims at improving our understanding of the intrinsic attenuation of organic contaminants coupled with different electron acceptors in an aquifer system, with the purpose of identifying the most influential and key limiting factors of biodegradation on the basis of PHREEQC simulations.

MODEL FORMULATION

PHREEQC is a computer program developed at the U. S. Geological Survey, incorporating equilibrium chemistry and kinetically mediated biodegradation of soluble aromatic contaminants coupled with sequential electron acceptors (Parkhurst and Appelo, 1999).

Microbial Reactions. Microbially mediated degradation of organic contaminants are commonly considered to be kinetically controlled rather than being in thermodynamic equilibrium (McNab and Narasimhan, 1994). Biodegradation reactions involving dissolved aromatic hydrocarbons in the groundwater occur at specific rates, which are a function of the prevailing environmental conditions and the availability of electrons as well as acceptors (Wiedemeier et al., 1999). Currently five microbial groups carrying out a range of processes are included within PHREEQC. The processes to be considered can be roughly classified: aerobic respiration, denitrification, iron reduction, sulfate reduction and methanogenesis. This is not intended as a complete list of microbial metabolism in aquifers, but to represent the major important microbial pathways. Although potentially important in some aquifer systems, not much is known about the biodegradation of organic contaminants under Mn(IV)-reducing conditions. In addition, the amount of reducable Fe(III) solids outweights Mn (IV) solids on a mass basis. Therefore, Mn reduction was not considered as a terminal electron acceptor in this modeling. The switching from one pathway to another pathway can be governed by the estimations of threshold concentration of the terminal electron acceptors in aquifers (Chapelle et al., 1995). Aerobic respiration, dentrification, iron and sulfate reduction, and methanogenesis are all modelled based on Monod kinetics. Bacterial decay and growth are also modelled using Monod kinetics as described by Brusseau et al.(1999). Microbial substrate removal is related to biomass growth through the biomass yield coefficient. All microbial groups modelled are subject to growth and decay. In this paper, toluene was selected as the representative organic contaminant because experimental and field studies have shown that toluene is biodegradable under both aerobic and anaerobic conditions (Barker and Wilson , 1997).

The organic substrate (hydrocarbon) is consumed to produce microbial biomass. The equations used to describe biomass growth are as follows.

$$\frac{dS}{dt} = -M_t k\left(\frac{S}{K_s + S}\right)\left(F\frac{C_i}{K_i + (F * C_i)}\right)$$

$$\frac{dM_t}{dt} = M_t kY\left(\frac{S}{K_s + S}\right)\left(F\frac{C_i}{K_i + (F * C_i)}\right) - bM_t$$

where S is the hydrocarbon substrate concentration (mol/L); C_i is the electron acceptor concentration (mol/L); M_t is the total microbial concentration (mol/L); k is the maximum hydrocarbon utilization rate per unit mass microorganisms (mol/g cells/s) ; Y is the microbial yield coefficient (g cells/ mol hydrocarbon); K_s is the hydrocarbon half saturation constant (mol/L) ; K_i is the electron acceptor half saturation constant (mol/L); b is the microbial decay rate (/s); F is the mass ratio of electron acceptor considered to organic compound consumed.

Basic Assumptions. (1). Different electron acceptors are utilized during the consumption of organic substrate. When high-energy electron acceptor concentration decreases below threshold values, the electron acceptor switches from this acceptor to another electron acceptor. (2). Microbial population responsible for biodegradation is assumed to be heterotrophic bacteria as a function of the organic substrate, the electron acceptors and biomass concentrations. No bioavailability is considered in the simulations

Geochemical Parameters. It is essential to accommodate relationships between microorganisms and hydrogeochemical processes in the PHREEQC program, as the model is not designed to model microbiology alone; rather it is designed to simulate the impact of microbial activity (microbial biodegradation of organic contaminants) on the aqueous geochemistry. Microbial activity impacts the chemistry in several ways: (1) Geochemical speciation changes due to the production and/removal of dissolved oxygen, nitrate, ferric iron, sulfate, carbon dioxide, and toluene; (2) pH and alkalinity change due to the production and/or removal of the chemical species; (3) Redox potential (pe) changes due to the removal of microbial terminal electron acceptors.

Changes in the bulk chemistry resulting from microbial activity within PHREEQC are utilized as input data by the geochemical speciation section of PHREEQC resulting in a new estimate of the current pH. This then feeds back into the microbial growth equation through the pH so ensuring that microbial activity and chemistry are linked. Redox reactions between the terminal electron acceptors oxygen, nitrate, ferric iron, sulfate, carbon dioxide and electron donors toluene, are modelled via the microbial growth equations. The redox potential is then determined from the most oxidizing redox couple, using standard redox functions. These microbially mediated changes to both the modelled pH and redox determine speciation and precipitation/dissolution reactions of minerals.

RESULTS AND DISCUSSIONS

Two cases,including batch reaction systems and one-dimensional reactive transport, were simulated to show how biodegradation of organic contaminants was coupled to different terminal electron acceptors, and how these processes affected aqueous geochemistry as a function of reaction progress and transport.

Batch Experiment Simulations. In this case, toluene was selected as the representative contaminant with a concentration of 0.5 mM. The initial composition of the solution and equilibrium mineral phases were defined and given in

Table 1. The microbial parameters used here were taken from the literature (Essaid et al., 1995). The threshold concentrations for different electron acceptors are from Chapelle et al.(1995). The total simulation time was 10 days with a time step of 0.01 day. Figure 1 shows the simulated concentration profiles of toluene, electron acceptors, microbial popultion, pH and pe as a function of time. This simulation managed to reproduce the major features of the biodegradation processes and chemical changes as a function of time (Bouwer, 1993). When oxygen was depleted, the nitrate consumption commenced. Microbial reduction of Ferric iron was characterized by gradual decrease of Fe(III) concentration. This is in contrast to lab and field observations where an increase of Fe(II) concentration is observed. Sulfate reduction commenced after depletion of Fe(III). During these degradation reactions, the growth of the microorganism occurred. As a result of changes in electron acceptors, pe and pH change correspondingly. pH was stable under aerobic and denitrification conditions but shifted to high values under other anaerobic conditions. Changes in pe values indicated clearly biodegradation of toluene coupled with different electron acceptors. In general, the simulated results are in good agreement with the evolution of groundwater geochemistry impacted by degradation processes of organic contaminants coupled with different electron acceptors reported in the literature (Bouwer, 1993).

TABLE 1. Parameters used for the model simulation

Initial Composition and equilibrium mineral phases	Values
pH	7.0
Oxygen (mmol/L)	0.25
NO_3^- (mmol/L)	0.25
SO_4^{2-} (mmol/L)	0.25
HCO_3^- (mmol/L)	0.82
Cl^- (mmol/L)	1.71
Na^+ (mmol/L)	1.71
Ca^{2+} (mmol/L)	0.58
Mg^{2+} (mmol/L)	0.11
Toluene (mmol/L)	0.50
Equilibrium mineral phases: Magnetite and $Fe(OH)_3$(a)	
One-dimensional transport parameters	
Length (meter)	2.5
Residence time (day)	1.0
Dispersivity (m)	0.04
Diffusion coefficient (m^2/s)	0
Single time-step (day)	0.1
Boundary conditions	flux
Initial biomass (g/L)	$1.05*e^{-6}$

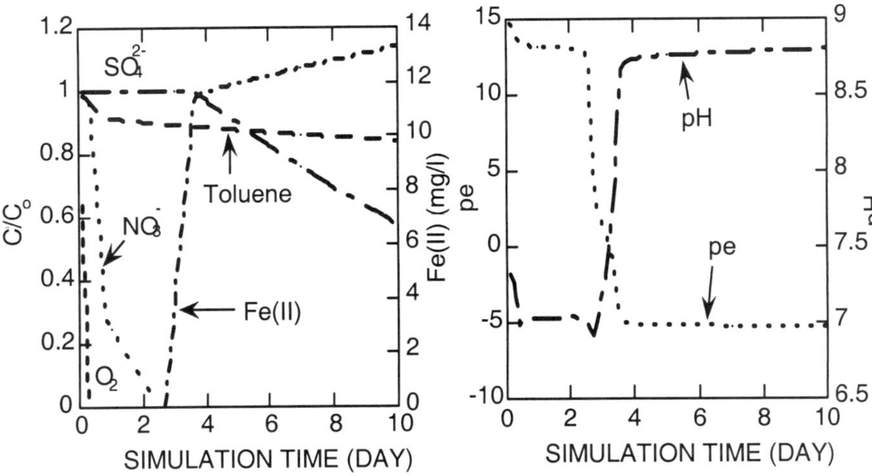

FIGURE 1. Changes of electron acceptors and toluene geochemical parameters including pH and pe as a function of time for the batch reaction.

One-Dimensional Reactive Transport. This case consists of a 2.5 meter column, discretized to 10 cells of 0.25 meter each. The injection concentration of toluene is 0.5mmol/L, but the initial concentration was zero elsewhere in the column. The initial electron acceptors are the same as those used in the batch reaction case (Table 1). The flow and transport parameters used in the simulation were given in Table 1. The total simulation time were 300 days. Figure 2 illustrates the breakthrough curves of toluene and sequential electron acceptors with pore volume. Aqueous parameter profiles of alkalinity, pH, CO_2, and CH_4 with pore volume are shown in Figure 3. The toluene concentrations gradually decline, which is in agreement with consumption of different electron acceptors. Oxygen was consumed just after 0.2 days and plume enters anaerobic conditions. It also shows that biodegradation of toluene coupled to the iron reduction takes more time than other processes, because Fe(III) concentration, controlled by both equilibrium iron mineral phases and aqueous properties, was higher than other electron acceptor concentration. In addition to biodegradation and changes in the terminal electron acceptors, aqueous chemistry parameters correspondingly change. In general, increases of pH and alkalinity, but decrease of CO_2 were observed with pore volume. Interestingly, alkalinity was constant under aerobic conditions, but continued to increase under anaerobic conditions. The appearance of higher alkalinity under sulfate-reducing conditions is probably due to higher solution pH and reducing environment. On the other hand, an increase of CH_4 clearly indicated that redox conditions have changed from aerobic to highly anaerobic conditions in the solution.

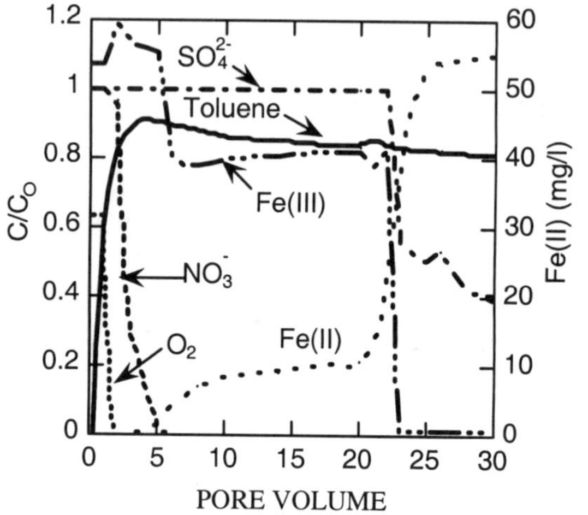

FIGURE 2. Breakthrough curves of toluene and terminal different electron acceptors for one-dimensional reactive transport.

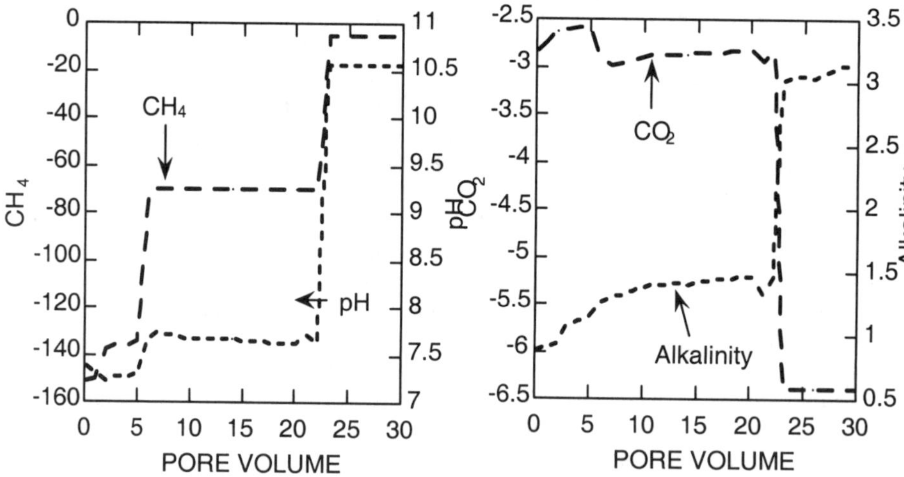

FIGURE 3. Changes of main geochemical parameters in the one-dimensional reactive transport.

SENSITIVITY ANALYSIS

A sensitivity analysis was performed to investigate the sensitivity of the most uncertain input parameters and to determine the key controlling factors of natural attenuation of aromatic compounds coupled to different electron acceptors in the aquifer. The sensitivity analysis with respect to microbial parameters and transport parameters was evaluated by systematically halving and doubling each of the parameters. This procedure is useful to rank the significance of the parameters according to their impact on chosen model properties. The one-dimensional simulation presented in Figure 2 served as the reference scenario for this sensitivity analysis. Results of the sensitivity analysis were compared, following Bosma (1994), based on: (1) the maximal relative concentration break-through; (2) the observed decline in the effluent after breakthrough relative to the reference case. Table 2 summarizes the results of sensitivity runs with respect to different parameters tested.

TABLE 2. Sensitivity analysis for the uncertain parameters

Parameter	Maximal relative concentration		Decline	
	Halving	Doubling	Havling	Doubling
Initial Biomass	0.95	0.80	0.08	-0.19
Yield Coefficient	0.90	0.90	0.03	-0.09
Decay Coefficient	0.90	0.90	0.01	0
Half Saturation Coefficient (K_s)	0.90	0.90	0.01	0
Half Saturation Coefficient (K_i)	0.93	0.87	0.06	-0.07
Maximal Utilization Coefficient (qm)	0.95	0.78	0.1	-0.43
Residence Time	0.95	0.79	0.09	-0.33

Negative values represent a decrease; positive values an increase relative to the reference run

CONCLUSIONS

This study represents an attempt to include biodegradation schemes into PHREEQC program, incorporating equilibrium chemistry and kinetically mediated reactions, to the bioremediation problem involved in natural attenuation of organic contaminants coupled to sequential electron acceptors in aquifer system.

The two cases, representing idealized batch reaction and one-dimensional reactive transport, showed that natural attenuation of organic contaminants and combined biomass growth can be modelled with the sequential electron acceptors provided that reasonable threshold values are given for various electron acceptors in aquifers. Sensitivity analyses with respect to uncertainty of input parameters revealed that the available electron acceptors, Monod kinetic parameters, in particular initial biomass, maximum utilization rate, and residence time are the key controlling factors in the modeling processes.

REFERENCES

Barker, J.F., and J.T. Wilson . 1997. "Natural biological attenuation of aromatic hydrocarbons under anaerobic conditions". In C.H.Ward, J.A.Cherry, and M.R.Scalf (Eds.), *Subsurface Restoration*, pp. 289-300. Ann Arbor Press.

Bouwer, E.J. 1993. "Bioremediation of subsurface contaminants". In R. Mitchell (Ed.), *Environmental microbiology*.pp. 287-318. Wiley-Liss, New York.

Brusseau, M.L., L.H. Xie, and L. Li. 1999. "Biodegradation during contaminant transport in porous media: 1. Mathematical analysis of controlling factors". *J. Contam. Hydrol.* 37: 269-293.

Chapelle, F.H., P.B. McMahon, N.M. Dubrovsky, R.F. Fujii, E.T.Oaksford, and D.A. Vroblesky . 1995. "Deducing the distribution of terminal electron-accepting processes in hydrologically diverse groundwater systems". *Water Resour. Res.* 31: 359-371.

Essaid, H.I., B.A. Bekins , E.M. Godsy, and E. Warren .1995. "Simulation of aerobic and anaerobic biodegradation processes at a crude oil spill site". *Water Resour. Res.* 31: 3309-3327.

Hunter , K.S., Y. Wang , and P. Van Cappellen . 1998. "Kinetic modeling of microbially driven redox chemistry of subsurface environments: coupling transport, microbial metabolism and geochemistry". *J. Hydrol.* 209: 53-88.

McNab, W.W., and T.N. Narasimhan . 1994. "Modeling reactive transport of organic compounds in groundwater using a partial redox disequilibrium approach". *Water Resour. Res.* 30: 2619-2635.

Murphy , E.M., and T.R. Ginn .2000. "Modeling microbial processes in porous media". *Hydrogeol. J.* 8: 142-158.

Parkhurst , D.L., and C.A.J. Appelo .1999. "User′s guide to PHREEQC (version2) -a computer program for speciation, batch-reaction, one-dimensional transport, and inverse geochemical calculations". *Water resources Investigation Report* 99-4259. U. S. Geological Survey.

Wiedemeier, T.H., H.S. Rifai , C.J. Newell, and J.T. Wilson .1999. *Natural attenuation of fuels and chlorinated solvents in the subsurface*. John Wiley & Sons, New York.

RECOMMENDATIONS FOR IMPROVED DISSOLVED OXYGEN SAMPLING

Brian Scott Aiken (Parsons Engineering Science, Phoenix, AZ)
Dr. Laurie LaPat-Polasko (Parsons Engineering Science, Phoenix, AZ)

ABSTRACT: Dissolved oxygen (DO) is a key parameter in the characterization of contaminated groundwater, particularly groundwater contaminated with volatile organic carbon (VOC) compounds. DO is generally considered to be a routine environmental measurement. However, after reviewing data from dozens of sites, it is apparent that the accuracy of DO measurements is alarmingly poor in a majority of the sites. The most common error is false positives due to the contamination by atmospheric oxygen. This has significant consequences because sites that have anaerobic groundwater may be characterized as aerobic. In the case of VOC contamination, this one measurement can lead to an inaccurate characterization of the site and possibly to the selection of an inappropriate remedial strategy. In some cases, VOC sites that would be good candidates for natural attenuation by reductive dechlorination may not be considered viable, even if there are other indicators of reductive dechlorination, such as the presence of dechlorination by-products. This paper discusses some of the common causes of poor DO groundwater data, provides data from three example sites, and suggests methods to ensure accurate DO data.

BACKGROUND ON ENVIRONMENTAL SAMPLING PROCEDURES

Research continues to identify limitations of accepted sampling techniques. For example, Gibs and Imbrigiolotta (1990) reported that the standard practice of purging three casing volumes gave stable volatile organic compound (VOC) measurements only 55% of the time. The study also concluded that commonly accepted field parameters, such as specific conductance and dissolved oxygen (DO), are not reliable indicators of stable VOC readings. Gosselin et al. (1994) discusses the inaccuracies that can occur when high capacity wells are used for environmental monitoring. These inaccuracies can affect many parameters, including DO. Powell and Puls (1997) discuss several reasons why "standard" sampling techniques give inaccurate results. Among the concerns cited in this paper are: 1) Wells with long well screens can produce groundwater that is "diluted" from mixing across vertical strata; 2) High pump rates can induce flow vectors through highly permeable zones; and 3) Inserting a sample pump or device can cause mixing within the well casing, which can lead to a change in water quality parameters. Powell and Puls recommend low-flow purging techniques as an appropriate sampling procedure in most cases. Barcelona et al. (1994) also discusses the merits of low flow sampling. It should be noted that many of the concerns related to DO measurements presented in this paper are addressed using low flow techniques.

It is particularly difficult to obtain representative DO data because groundwater can be easily contaminated by atmospheric oxygen. Complacency by many samplers exacerbates the problem. Reviewing data from numerous sites showed that there is an alarming consistency in the poor quality of DO data, primarily false positives. Often, a quick review of the available data reveals inconsistencies that are indicative of false high DO in groundwater. Examples of possible contra indicators are:

- Negative redox
- High total organic carbon
- Reductive dechlorination byproducts
- Absence of nitrate
- Samples collected from deep and/or semi-confined aquifers
- Indicators of alternate electron acceptor processes

FIELD OBSERVATIONS AND DATA REVIEW

Data for this paper were collected over a five-year period by observing and conducting field sampling and after reviewing data from approximately 50 VOC contaminated sites. As this was not part of a funded comprehensive study, this paper addresses only circumstances that were encountered during routine site analyses. The intention here is not to be exhaustive, but to heighten awareness about the potential for inaccurate DO results. Each site must be evaluated individually as to the accuracy and effectiveness of the sampling techniques.

Field Observations. Inaccurate DO concentrations were observed in the following circumstances. In all of these cases, alternative methods were used to measure DO and showed that the standard method often led to false positive measurements.

- Well purging by bailing. DO is often measured as a parameter, even in wells that have been purged and or sampled by bailing.
- Excessive well drawdown during pumping. Excessive well drawdown creates a cone of depression, which can allow groundwater from upper intervals or from the capillary fringe to enter the pump. This water may have different water quality characteristics than the bulk aquifer. The capillary fringe and upper interval water in particular is more likely to be exposed to atmospheric oxygen. In some extreme cases, the groundwater level was drawn down to the point that the pump began cavitating, drawing in both air and water.
- Inappropriate pump level. Many standard sampling procedures have the pump set just below the groundwater surface where atmospheric oxygen can partition into the water.
- Slow recovery wells. Slow recovering wells may be pumped to the point of extreme drawdown or pumped dry and allowed to refill before sampling. In either case, the water within the well casing is exposed to atmospheric oxygen. In several cases, the groundwater sampling pumps were placed within the screened interval of very deep wells. While waiting for stable readings, the DO concentrations started strongly

anaerobic but increased and became aerobic as the water level dropped, sometimes as much as 100 ft (35 m) during purging. It is standard practice in these circumstances to record the final stable DO reading, even though early reading confirmed the presence of strongly anaerobic water.

- Inappropriate equipment use. Some DO meters require constant stirring to obtain accurate results. In several cases, DO probes were lowered into a well, but were not continually moved, leading to inaccurate results. Another practice that was observed when using DO meters was placing the probe very near the groundwater surface where air exchange with the atmosphere has occurred.

RESULTS FROM EXAMPLE SITES

Data from three sites are presented. Each site was inaccurately characterized as aerobic when in fact the bulk groundwater is anaerobic.

Site #1. The first site is a landfill in Phoenix, Arizona. Groundwater upgradient of the landfill had been characterized as aerobic, with an anaerobic zone resulting from total organic carbon (TOC) released from the landfill. Chlorinated solvents had been released, and concentrations of both dichloroethene (DCE) and vinyl chloride (VC) near the source area were greater than 1,000 µg/L. The site appeared to be a good candidate for mixed environment natural attenuation (as defined by Wiedemeier et al. 1999), with highly chlorinated solvents reduced near the source and subsequent oxidation of the reductive dechlorination by-products. Since DO was an important factor in the characterization of this site, a review of DO data and sampling methods was performed during the subsequent sampling events. Table 1 lists DO concentrations that were collected in 1995 and 1999. Sampling procedures were not changed during the 1999 sampling events; however, the sampling team was alerted to the possibility of inaccurate DO data.

TABLE 1. Dissolved Oxygen Results (in mg/L) in selected wells at a Phoenix, AZ landfill site.

Well ID (EW-)	1*	4	5	9	15	18	NE*	PZ1	PZ10
Sample Date									
06/95	2.4	3.0	0.2	0.4	0.1	0.3	2.2	0.3	1.0
06/99	0.4	0.2		0.3	0.4	0.3	3.1	0.4	2.3
11/99	1.1	0.4	0.9	0.2	0.4	0.3		0.3	

* The 1999 sample logs were marked to show that these wells were purged dry prior to collecting samples.

Well EW-4 was formerly thought to be aerobic (1995 data) and was shown to be anaerobic (1999 data). Purging the wells dry prior to sampling could cause the aerobic groundwater in EW-1 and EW-NE. The aerobic groundwater in well PZ-10 is also suspect because:

1) low and negative redox readings, and

2) Well EW-PZ10 is down gradient of EW-6 and up gradient of EW-5, both of which contain anaerobic groundwater. All three wells are in a semi-confined portion of the aquifer.

While the data at this site are not compelling, there is sufficient reason to question the assumption that the bulk aquifer conditions are aerobic.

Site #2. The second site is a Superfund site in Phoenix, Arizona that is contaminated with VOCs and hydrocarbons. This site was generally accepted to be aerobic. Trichloroethene (TCE) and trichloroethane (TCA) were the primary contaminants, with significant concentrations of *cis*-1,2-DCE and 1,1-DCE. The *cis*-1,2-DCE was thought to have been produced primarily in the source areas or in anaerobic micro-niches. Figure 1 shows the DO concentrations for wells across the site, which appear to show predominantly aerobic conditions. Note that the database contained DO results that were as high as 24 mg/L, which is unlikely since the groundwater temperature (in Phoenix, Arizona) is about 70°F (21°C), yet these results were not flagged.

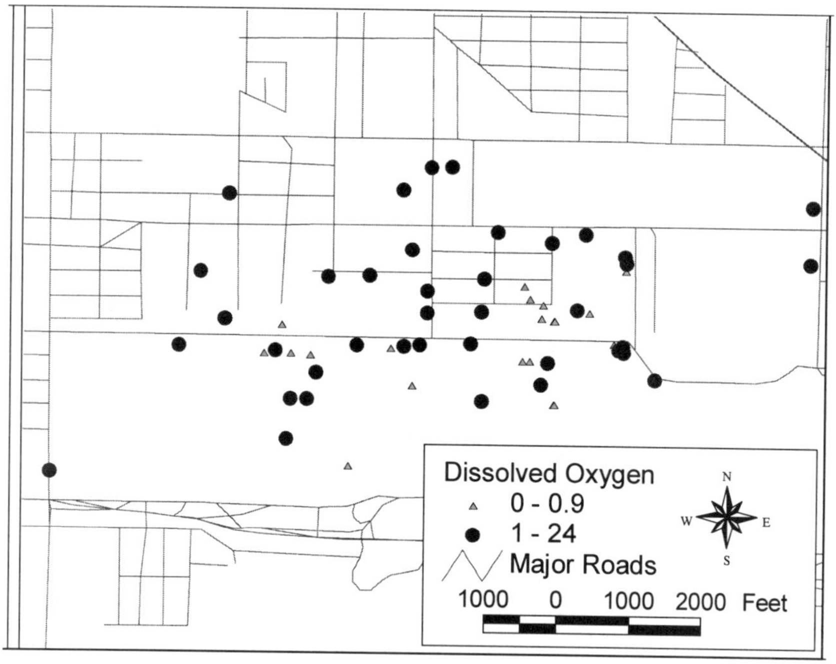

FIGURE 1. Dissolved oxygen across the site.

From the wells in Figure 1, 28 wells were selected for DO analysis using an in-situ colorimetric sampler (Figure 2). The analysis shows that only six of the 28 wells contained aerobic water, and were confined to an isolated area of the site (Figure 3). Two of the six aerobic wells are suspect because both are screened in the deep bedrock with overlying alluvial groundwater, and both had significant drawdown during purging (in-situ sampling was performed after purging).

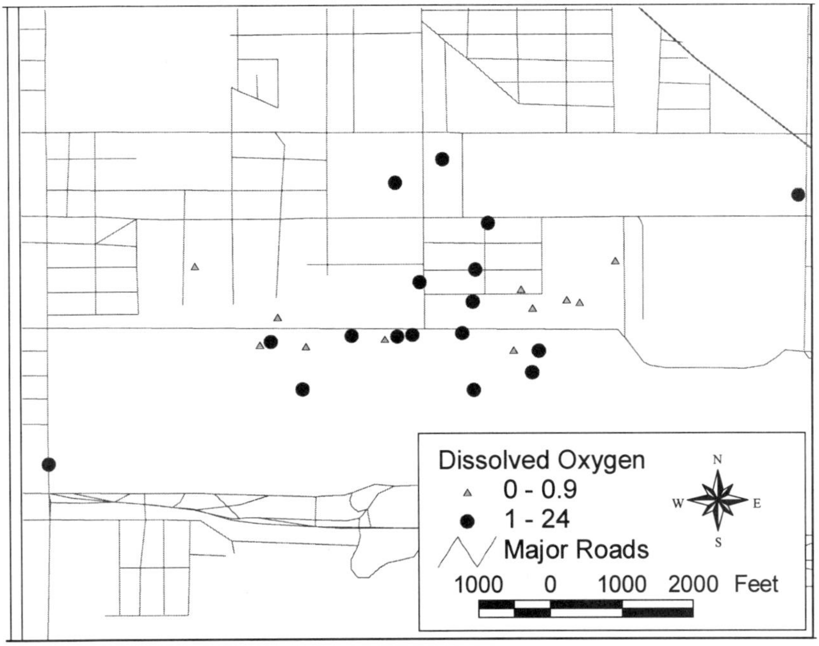

FIGURE 2. Wells selected for dissolved oxygen sampling.

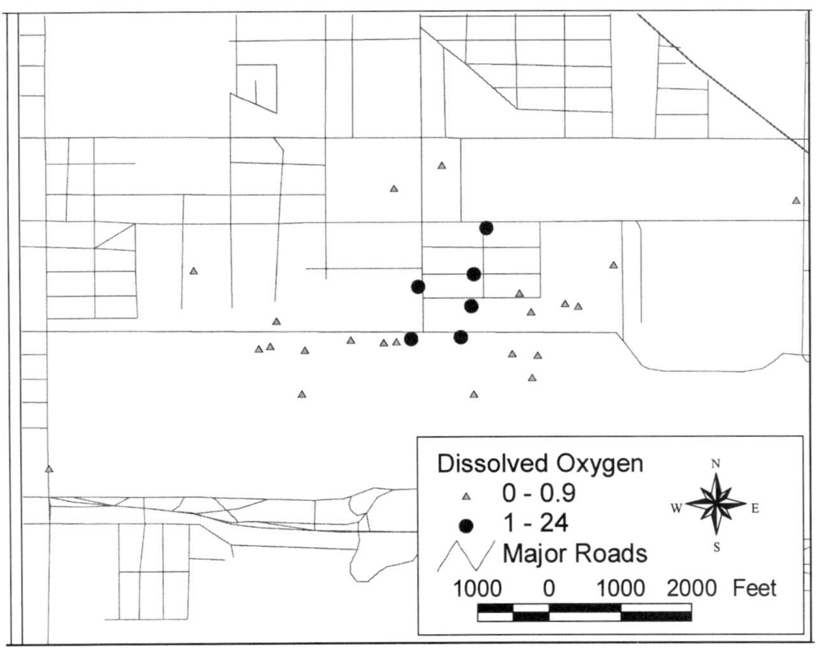

FIGURE 3. Dissolved oxygen in selected wells after re-sampling to eliminate false positives.

Site #3. The third site is a VOC contaminated site in Memphis, Tennessee. The site was contaminated with perchloroethene (PCE), and had significant concentrations of the reductive dechlorination byproducts TCE and *cis*-1,2-DCE. The DO concentration at the site indicated strongly aerobic conditions. The site was downgradient of a fertilizer manufacturer, and as a result had elevated levels of ammonia. As a result of the high ammonia concentrations, this site was being considered by a United Stated Geological Survey research team to study aerobic cometabolism of TCE by ammonia mono-oxygenase producing bacteria. An anaerobic zone near the source was thought to be responsible for the dechlorination by-products TCE and *cis*-1,2-DCE.

DO measurements were made at this site as part of a comprehensive sampling plan to evaluate natural attenuation. DO was measured using several different techniques. Table 2 summarizes the sampling results.

TABLE 2. Comparison of different dissolved oxygen sampling techniques

	Dissolved Oxygen [mg/L]			
Well Number[a]	Average value (prior to natural attenuation sampling)	In-Situ Probe[c]	Winkler Titration	In-situ Colorimetric Kit[d]
MW-24	NM	0.39	0.4	<0.05
MW-17	NM	0.55	0.8	<0.05
MW-27	2.3	0.80	0.6	0.1
MW-20	6.2	0.32/0.40	0.6	0.1
T4-3M	NM	0.57	1.2	0.3
MW-E2D	4.4	0.81/0.51	2.2	0.2
MW-E5D	4.6	0.45/0.42	2.2	0.10
MW-48	NM	0.37	<0.2	<0.05
T3-2M	NM	0.41	1.2	0.2
Overall site Average[b]	4.9	NM	NM	NM

NM - Not Measured

[a] - Wells selected for natural attenuation sampling.

[b] - Includes data from 14 wells: min. value = 1.8: total # of samples = 90.

[c] - Before/after purging

[d] - Kit was a CHEMetrics K-7501 with an in-situ sampling tool (A-0139).

The natural attenuation sampling showed that the site was anaerobic, not aerobic. This site was used as a case study to compare methods of DO analysis. Sampling with an in-situ DO probe provided accurate results, and was not affected by sampling before or after purging. The wells were purged using a peristaltic pump and Winkler titrations were performed using a LaMotte 5860 kit. The apparent aerobic results using this kit underscore how easy it is to obtain false positive results, particularly under field conditions. The colorimetric kits are very sensitive to low DO concentrations and have a detection limit of about 0.05 mg/L. These tests showed that the groundwater at depth was under highly anaerobic conditions. Although the data are not shown here, DO results were also obtained during dissolved gas (hydrogen) analysis, and the results were mixed, with some wells aerobic and some anaerobic.

The site had originally been sampled by bailing the wells, which is the most likely source of initially high DO concentrations. The aerobic conditions had been confirmed using a DO probe, but measurements were taken near the groundwater surface (about 10 ft or 3 m below ground surface), whereas the contamination was in a lower interval about 60 ft (20 m) below ground surface. As a result of the DO and comprehensive natural attenuation sampling, this site was characterized as being under primarily sulfate reducing conditions, and is currently being monitored for natural attenuation of VOCs.

DISSOLVED OXYGEN SAMPLING GUIDELINES

Data Goals and Objectives. The goal of DO sampling is generally to establish whether the aquifer is under aerobic or anaerobic conditions. A DO concentration of 1 mg/L is a suitable threshold for aerobic water. A conservative approach is to assume that the groundwater is anaerobic (< 1 mg/L DO), and view the sampling plan as a tool to disprove the existence of aerobic water. While these guidelines cover most cases, it is not possible to foresee every well configuration; therefore, it is important to flag DO results in questionable circumstances.

Sample Collection Location. If possible, samples should be collected from:

- within the screened interval of the well, and if possible,
- near the center of the screened interval, unless a specific portion of the screened interval has been identified, through depth specific sampling, to be more appropriate.
- at least three ft (1 m) below the water surface - this allows a concentration gradient to develop between the sample water and water exposed to the atmosphere.
- at least three ft (1 m) from the bottom of the well to avoid suspending silt.

Sample Collection Technique. Any of the three methods listed can provide accurate DO readings. The in-situ sampler and colorimetric kit provides the most sensitivity (down to 0.05 mg/L DO) and greatest protection against false positives.

- Collect the sample in-situ using a CHEMetrics K-7501 colorimetric kit and an A-0139 in-situ sampler (this is not meant as an endorsement, but at the time of publication, the authors are not aware of another manufacturer with a similar device). With this sampler, groundwater can be sampled at any depth. The sampler must be carefully lowered into the well to limit air-water mixing at the water surface. Since the sample is collected in-situ at depth, there is minimal potential for contamination from atmospheric oxygen. DO may be collected before well purging.

- Collect the sample in-situ using a DO probe. DO probes are generally limited to wells that are less than 200 ft (60 m) in depth. As is the case with the colorimetric sampler, the sample is collected in-situ at depth and there is minimal potential for contamination from atmospheric oxygen.
- Collect the sample from the water stream during well purging. This is the most common method since there are flow-through cells available with multiple probes including DO, redox, conductivity, temperature and pH. This method is useful in establishing stable well readings during purging. Care should be taken to ensure there is no air entrainment during purging as a result of a loose fitting. This is particularly important when using a peristaltic or other negative pressure pump. This method can lead to inaccurate results if there is significant drawdown in the well.

Well Purging. The preferred purging method is micropurging. Micropurging involves pumping at a flow rate that is comparable to the recharge rate of the well. A well that cannot be pumped without causing drawdown should be sampled before purging. In the case of a dedicated pump, reducing the flow rate to minimize drawdown is helpful; however, the depth of the dedicated pump is often not known with sufficient accuracy to ensure that the pump is not near or at the groundwater surface. In most cases, the in-situ sampler described above can be lowered through a sounding tube or by loosening the well cap.

CONCLUSIONS

Sampling for DO can often lead to false positive results, resulting in inappropriate site characterization. Three example sites are present where DO data were critically evaluated. In each case, there was sufficient evidence to question the widely held view that the bulk aquifer was aerobic. At the third site, a field demonstration of the sampling procedures (CHEMetrics colorimetric kit) was required to convince the agencies that the site was strongly anaerobic, and not aerobic, as previously believed.

Data were reviewed from dozens of sites, and in the majority of the cases, high DO values could be questioned based on a review of other physical and chemical parameters. Some of the most common indicators of false positive DO data are low redox, reductive dechlorination by-products, and high TOC. Since it was not possible to observe sampling techniques in these cases, the causes of the possible errors is not known.

At the sites where sampling procedures were observed, the cause of false positives was usually a result of wells that were difficult to sample accurately, or lack of adequate sampling procedures. Although skilled samplers were present in most cases, there is an apparent complacency about collecting DO data, which can contribute to the potential for false positive readings. Sampling using the guidelines described can greatly minimize false positive DO results.

REFERENCES

Barcelona, M.J., Wehrmann, H.A. and M.D. Varljen. 1994. "Reproducible Well-Purging Procedures and VOC Stabilization Criteria for Ground-Water Sampling." *Groundwater. 32*(1): 12-22.

Gibs, J. and T.E. Imbrigiotta. 1990. "Well-Purging Criteria for Sampling Purgeable Organic Compounds". *Groundwater. 28*(1): 68-78.

Gosselin, D.C., Ayers, J.F. and Y.K. Zhang. 1994. "Modeling Concentration Variations in High-Capacity Wells: Implications for Groundwater Sampling." *Water Resources Bulletin. 30*(4): 613-622.

Powell, R.M. and R.W. Puls. 1997. "Hitting the Bull's-Eye in Groundwater Sampling." *Pollution Engineering*. June *29*(6): 51-54.

Weidemeier, T.H., Rifai, H.S., Newell, C.J. and J.T. Wilson. 1999 *Natural Attenuation of Fuels and Chlorinated Solvents in the Subsurface.* J. Wiley and Sons: New York, NY

THEORY AND PERFORMANCE OF A FIELD-PORTABLE DISSOLVED HYDROGEN ANALYZER

Patrick J. Evans (Camp Dresser & McKee Inc., Bellevue, Washington)

ABSTRACT: A field-portable dissolved hydrogen (DH) analyzer (patent pending) that uses a solid-state sensor was developed. This DH analyzer uses an automated gas-liquid equilibration step in order to transfer DH to a carrier gas. Gas-phase hydrogen is detected using a solid-state sensor. A series of catalysts and adsorbents are used to remove interfering gases prior to sensor exposure. The DH analyzer was tested at Ft. Lewis, near Tacoma, Washington, where groundwater contains trichloroethene and varying concentrations of DH. Correlation between the DH analyzer and the standard bubble strip/reduction gas analyzer method were observed.

INTRODUCTION

Dissolved hydrogen (DH) is an important parameter in many fields of engineering and science. In aquifers it can indicate the dominant terminal electron accepting processes (TEAPs) being utilized by microorganisms to obtain energy for growth (Chapelle et al., 1997). It has also been shown to indicate the optimal conditions for reductive dechlorination of organic chemicals (Yang and McCarty, 1998). DH has been proposed as a process control variable in fermentation and anaerobic wastewater treatment processes (Cleland et al., 1984; Strong and Cord-Ruwisch, 1995). DH measurement has found application in monitoring methanogenic activity in the rumen (Robinson et al., 1981). DH is important with respect to hydrogen embrittlement of metals and high-pressure water boiler operation in the nuclear industry (Wilson, 1982). DH in oil can also indicate potential for electrical transformer failure.

DH concentration is dependent on dominant TEAPs under steady-state conditions. A DH concentration between 0.2 to 0.8 nanomolar (nM) indicates an iron-reducing TEAP, between 1 to 4 nM indicates a sulfate-reducing TEAP, and between 5 to 30 nM indicates a methanogenic TEAP (Chapelle et al., 1997). The presence or absence of specific TEAPs in an aquifer can indicate patterns of contaminant transformation. For example, trichloroethene (TCE) can be reductively dechlorinated completely to ethene via vinyl chloride (VC) under methanogenic conditions, but only to cis-dichloroethene (cDCE) under iron-reducing conditions. Additionally, iron-reducing microorganisms have been shown to be capable of oxidatively degrading cDCE to carbon dioxide.

DH has been proposed and can be used for monitoring and control of various biological systems because it is rapidly turned over. While DH in theory is an attractive parameter, DH measurement in practice has proven to be fraught with difficulty and expense.

Among other challenges, the concentrations that must be detected in groundwater range from 0.1 nM to 30 nM. Nearly every analytical method that is capable of hydrogen detection cannot detect DH at these low levels. The one exception is the bubble strip method used in combination with a reduction gas analyzer (Chapelle et al., 1997). This method is difficult to use and the instrumentation is expensive and intended for laboratory use.

Preparation of equilibrated gas samples using the bubble strip method followed by gas analysis in a fixed-base laboratory using a reduction gas analyzer is another approach. Problems associated with adapting sophisticated laboratory instruments for use in the field can be avoided. However, many artifacts can arise during DH measurement. For example, use of direct current centrifugal groundwater sampling pumps can result in hydrogen generation. Field measurement is the only means that provides for timely quality assurance and quality control of DH analysis. Furthermore, field sampling and analysis eliminates any concerns about sample stability.

There is a need for a field-portable DH analyzer that is easy to use and relatively inexpensive. The DH analyzer should be capable of operating either on 12 VDC or a battery. It must be capable of detecting DH in the range of 0.2 nM to 30 nM. Sampling and analysis time should be less than 60 minutes and preferably 30 minutes or less. The theory and performance of such an analyzer is presented.

METHODS

To evaluate the DH analyzer, DH concentrations in water were measured using the bubble strip method in combination with a reduction gas analyzer (Chapelle et al., 1997). The reduction gas analyzer consisted of a Trace Analytical reduction gas detector model RGD2 connected to the discharge of a carbosieve chromatography column. Carbon monoxide was measured on the same instrument. At Ft. Lewis, a Trace Analytical RGA3 was used. Oxygen was measured using a Hewlett Packard 5890 gas chromatograph with thermal conductivity detection.

Groundwater sampling and analysis was conducted at Ft. Lewis near Tacoma, Washington. The study area was the East Gate disposal yard (EGDY), where TCE is the primary contaminant of concern. Sampling and analysis was done concurrently with an ongoing demonstration of reductive anaerobic biological *in situ* treatment technology (RABITT), which involved introduction of a mixture of volatile fatty acids into groundwater to stimulate reductive dechlorination of TCE.

RESULTS AND DISCUSSION

Solid-state sensors are known to be capable of detecting numerous chemicals. One type of sensor is the metal oxide semiconductor (MOS) illustrated in Figure 1. Application of the MOS design to hydrogen can involve the use of palladium as the metal that is referred to as the gate. The semiconductor shown in Figure 1 is p-doped silicon. Between the semiconductor and the gate is a thin silicon dioxide oxide or insulator. Molecular hydrogen is absorbed by the palladium gate and dissociates into protons and electrons. This dissociation results in a change of the electrical conductivity properties of the silicon semiconductor thus giving an electrical response to hydrogen. This solid-state sensor is the final detection element of the DH analyzer.

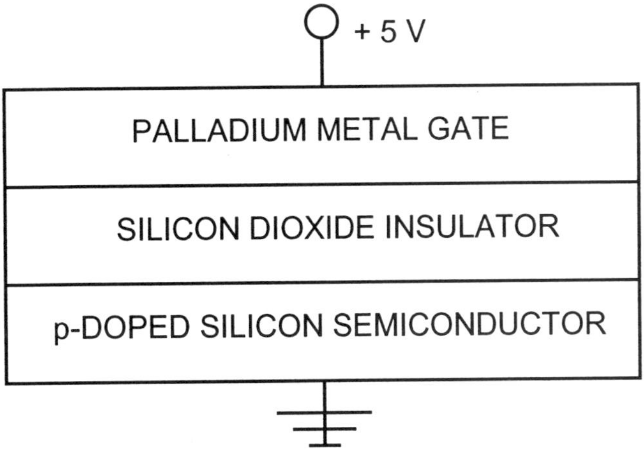

FIGURE 1. Schematic representation of a metal oxide semiconductor (MOS) solid-state sensor.

An example of the response of such a sensor is shown in Table 1. For this test, the specific MOS solid-state sensor was a Shottky diode. Test 1 shows the sensor response to hydrogen at different concentrations in nitrogen. The response of 49 mV/min to 0.1 ppm was an acceptable sensitivity since 0.1 ppm hydrogen is equivalent to 0.08 nM DH at equilibrium based on Henry's law.

TABLE 1. Sensor response (mV/min) to hydrogen in nitrogen in the presence of oxygen and carbon monoxide.

		Gaseous hydrogen concentration (ppm)		
Test	**Conditions**	**0.1**	**1**	**10**
1	Control	49	250	2,200
2	0.5% O_2	2.9	38	550
3	0.1% CO	-6.3	35	88

The data in Table 1 also demonstrate the effect of oxygen and carbon monoxide on sensor response to hydrogen. Oxygen and carbon monoxide severely inhibited the response to hydrogen. The effect of oxygen was due to the catalytic effect of palladium, which catalyzed the reaction of hydrogen with oxygen to form water. In doing so, less dissociated hydrogen remained in the MOS sensor and a lower response was observed. These results indicate that acceptable hydrogen sensitivity was dependent on the absence of oxygen. Even though DH is typically measured in anaerobic aquifers, trace levels of oxygen can be present and significantly affect hydrogen sensitivity. Carbon monoxide can also be present in anaerobic groundwater samples and must also be removed prior to exposure of the sensor.

Removal of oxygen and carbon monoxide was a challenge since most methods for their removal resulted in hydrogen generation or consumption. For example, a standard method for generation of anaerobic gas atmospheres is based on the use of sodium borohydride that produces hydrogen upon contact with water. Generated hydrogen then reacts with oxygen on a platinum or palladium catalyst. One suitable method for oxygen removal involved the use of quinones or other organic reagents, as opposed to inorganic reagents. Table 2 shows results for the removal of oxygen from nitrogen containing low concentrations of hydrogen. The control contained no oxygen scavenger. The organic-based oxygen scavenger demonstrated that oxygen was removed without hydrogen generation or consumption. However, carbon monoxide was generated as a side reaction. A metal-oxide reagent was used to adsorb carbon monoxide successfully.

TABLE 2. Removal of oxygen and carbon monoxide from carrier gas.

		Concentration					
		Oxygen (%)		Hydrogen (ppm)		Carbon monoxide (area)	
Test	Conditions	Initial	Final	Initial	Final	Initial	Final
1	Control	2.6	2.9	1.9	1.9	320	340
2	O2 scavenger	2	0	0.23	0.25	410	5,900
3	O2 & CO scavengers	2	0	0.18	0.21	170	0

Sensing of hydrogen described above is based on hydrogen being in the gas phase. MOS solid-state sensors do not work when immersed in water. Thus, DH must be transferred from groundwater to a carrier gas. Gas-liquid mass transfer is described by the following equation:

$$N = K_L a(C^* - C)$$

Where N is the rate of flux of hydrogen across the gas-liquid interface, K_L is the mass transfer coefficient (liquid phase mass transfer controlling), a is the surface area of the gas-liquid interface, and C^*-C is the driving force. C^* is the theoretical gas phase hydrogen concentration in equilibrium with ambient groundwater DH. C is the actual gas phase hydrogen concentration. K_L can be increased with energy such as greater gas and liquid velocities. a can be increased by dispersal or other methods that involve differing levels of energy input. C^* and C cannot be changed unless the temperature or the pressure of the system is changed. Thus the rate of DH transfer is proportional to the product K_La. The bubble strip method is inherently a low K_La system. One method that can easily increase K_La is based on use of a spray nozzle. Groundwater is dispersed into fine droplets greatly increasing a and thus K_La. Initial results with such a device demonstrated equilibration in less time than the bubble strip method.

Figure 2 is a process flow diagram of the DH analyzer (patent pending). First, a gas-liquid equilibrator (GLE) is used to transfer DH from groundwater to the carrier gas that is being circulated by a blower. The GLE contains a spray nozzle that disperses flowing groundwater into fine droplets that have a high surface area for contact between carrier gas and groundwater, resulting in large hydrogen mass transfer rates. A peristaltic or a bladder pump is used to pump groundwater into the GLE. A direct-current centrifugal pump cannot be used because it will generate DH. Henry's law indicates that about 0.8 nM DH will equilibrate with 1.0 ppm of gaseous hydrogen. Once the carrier gas is equilibrated with the groundwater, a series of automatic valves redirect the carrier gas flow so it is in contact with a series of catalysts and adsorbents. These catalysts and adsorbents serve to remove interfering gases from the carrier gas. Interfering gases can include oxygen, carbon monoxide, hydrogen sulfide, water, and methane. Once the carrier gas is treated, the flow pattern is redirected once again to expose the sensor. The sensor outputs an electric signal in response to the hydrogen concentration.

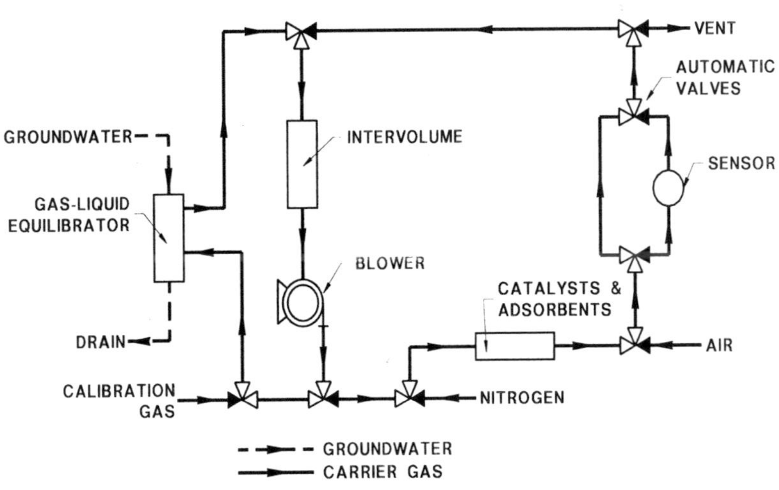

FIGURE 2. Dissolved hydrogen analyzer process flow diagram (patent pending).

Figure 3 is a photograph of the DH analyzer and Figure 4 presents results comparing the DH analyzer to the bubble strip/reduction gas analyzer method at Ft. Lewis. The line drawn through the data is what would be expected for a perfect correlation between the DH analyzer and the bubble strip method. The actual correlation coefficient (r^2) for the data was 0.91. The data show the best correlation between 0.2 nM and 3 nM. At the low end of the range a bubble strip DH of 0.1 nM gave 0.04 nM on the DH analyzer. At the high end of the range, a bubble strip of 24 nM gave a 200 nM on the DH analyzer. This high result may be attributable to methane interference with the sensor.

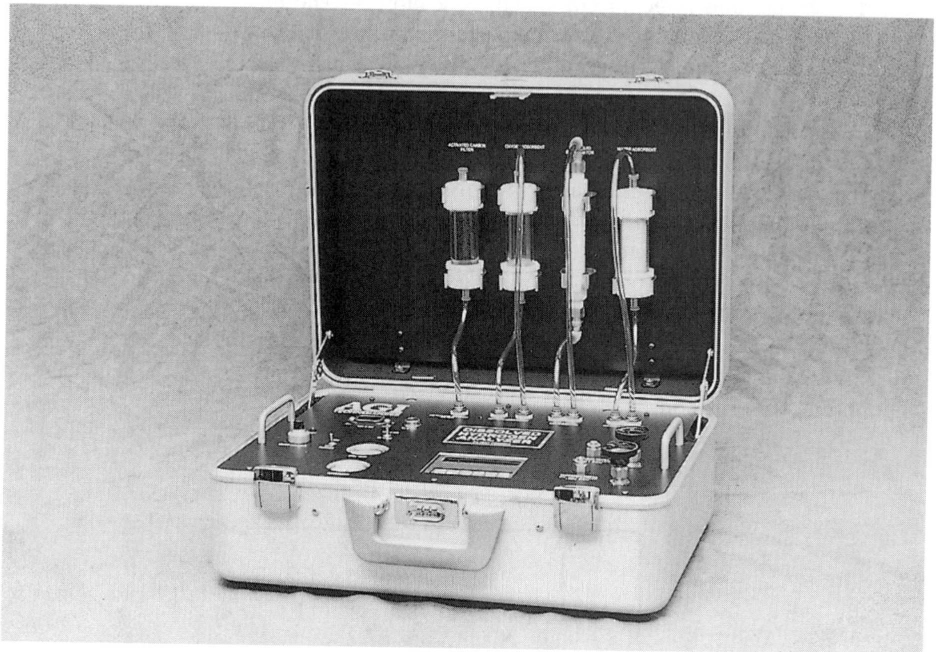

FIGURE 3. DH analyzer photograph (patent pending).

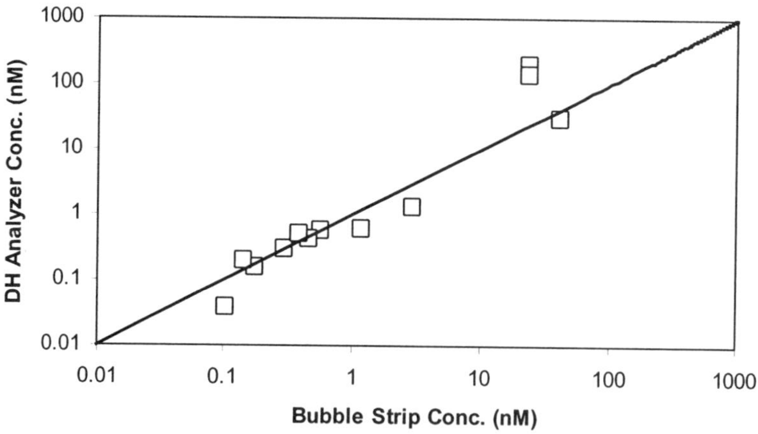

FIGURE 4. Correlation between the DH analyzer and the bubble strip/reduction gas analyzer methods at Ft. Lewis.

CONCLUSIONS

DH is a useful parameter having applications in many fields. It can indicate patterns of organic contaminant biotransformation under monitored natural attenuation or engineered bioremediation scenarios. A field-portable DH analyzer was developed and shown to be capable of measuring DH concentrations as low as 0.2 nM. The DH analyzer is field-portable and is based on use of a solid-state sensor. Correlation between the DH analyzer and the standard bubble strip/reduction gas analyzer method was observed during operation at a TCE-contaminated site. Further work will involve improvements to the mechanical system to increase reliability and optimization to minimize response to interfering gases. These refinements will be implemented during demonstration and validation of this analytical technology at four additional sites over the next year.

Commercialization of the DH analyzer is planned once these refinements have been incorporated and a suitable manufacturing and sales partner has been identified. The cost of the DH analyzer is expected to be between $5,000 and $10,000. This cost compares favorably to the reduction gas analyzer that costs about $25,000.

ACKNOWLEDGMENTS

This project was supported by contracts with the U.S. Navy under the Environmental Security Technology Certification Program and the U.S. Air Force under the Small Business Innovative Research program. The author wishes to thank Carmen LeBron of the Naval Facilities, Engineering Service Center, Steve Cox of the U.S. Geological Survey, Richard Wilson of Ft. Lewis, C.C. Liu of Case Western Reserve University, Lt. Lisa Ackert of the U.S. Air Force, and Lisa Nickens, Mary Trute, and Naht Rammachat of CDM.

REFERENCES

Chapelle, F.H., D.A. Vroblesky, J.C. Woodward, and D.R. Lovley. 1997. "Practical Considerations for Measuring Hydrogen Concentrations in Groundwater." *Environ. Sci. Technol. 31*: 2873-2877.

Cleland, N., E.G. Horsten, H. Elwinng, S.-O. Enfors, and I. Lundstrom. 1984. "Measurement of Hydrogen Evolution by Oxygen-Limited *E. coli* by Means of Hydrogen Sensitive Pd-MOS Sensor." *Appl. Microbiol. Biotechnol. 20*:268-270.

Robinson, J.A., R.F. Strayer, and J.M. Tiedje. 1981. "Method for Measuring Dissolved Hydrogen in Anaerobic Ecosystems: Application to the Rumen." *Appl. Environ. Microbiol. 41*:545-548.

Strong, G.E., and R. Cord-Ruwisch. 1995. "An *In Situ* Dissolved-Hydrogen Probe for Monitoring Anaerobic Digesters under Overload Conditions." *Biotechnol. Bioeng. 45*:63-68.

Wilson, C.R. 1982. "Hydrogen-Evolution Monitoring as a Measure of Steam-Generator Corrosion." Electric Power Research Institute Report NP-2650.

Yang, Y, and P.L. McCarty. 1998. "Competition for Hydrogen within a Chlorinated Solvent Dehalogenating Anaerobic Mixed Culture." *Environ. Sci. Technol. 32*: 3591-3597.

A NOVEL METHOD FOR INVESTIGATING BIODEGRADATION PROCESSES AT THE FRINGE OF A CONTAMINANT PLUME IN POROUS MEDIA

Wei Huang, David N Lerner, Colin C Smith, and Steven F. Thornton
(Groundwater Protection and Restoration Group, University of Sheffield, Sheffield, UK)

ABSTRACT: A novel non-invasive imaging technique was developed to investigate the biodegradation processes occurring at the fringe of a simulated contaminant plume. In conditions where the supply of electron acceptors is limited, biodegradation may be expected to be focused at the plume fringe and transverse dispersion will play a very important role. The technique permitted investigation of the mixing process of dissolved oxygen with the plume in a two dimensional porous media at laboratory scale. A small transparent aquifer (160 mm length, 130 mm width and 3 mm thickness) was made up of quartz plates and quartz sand (60 -100 micrometers). Potassium acetate was used as a substrate and was continuously injected into the aquifer to form a plume. Ruthenium (II)-dichlorotris(1,10-phenanthroline) ($Ru(phen)_3Cl_2$), a water soluble fluorescent dye which can be quenched by dissolved oxygen, was used as an indicator of dissolved oxygen concentration in the porous media. Under UV light (450 nm), $Ru(phen)_3^{2+}$ can be excited and emits fluorescence (590-600 nm). The intensity of the fluorescence is dependent on the ambient concentration of oxygen. The oxygen distribution within the model aquifer was interpreted from images recorded by a CCD (Charged Coupled Device) camera. The laboratory experiments show that at the core of the plume, oxygen was exhausted by the biodegradation processes and at the fringe, oxygen was supplied only by transverse mixing.

INTRODUCTION

Contamination of groundwater by organic chemicals is a widespread environmental problem. Due to the potentially high costs of remediation, the scientific prediction of natural attenuation processes is of very practical importance. Natural attenuation is usually limited by the supply of electron acceptors (Cirpka et al. 1999; Oya and Valocchi 1998). In most cases, the contaminant content of the plume is substantial, while the availability of electron acceptors (O_2, NO_3, SO_4, Fe(III)) is limited. Biodegradation is predominantly concentrated at the fringe of the plume, and significantly controlled by transverse dispersion processes. Electron acceptors may be replenished by transverse mixing, and the toxicity of contaminants can be reduced by dilution processes arising from transverse dispersion.

Therefore, the rate of biodegradation in a stable plume depends to a large extent on the mixing processes at the fringe. Information on contaminant mixing processes in the subsurface becomes crucial in order to predict natural attenuation.

However, the transverse mixing zone at the fringe can be very thin. Sampling in this thin zone is very difficult and will inevitably disturb the plume. Measurements based on a non-invasive imaging technique therefore become very attractive. Huang et al. (2001) described a non-invasive imaging technique employing fluorescent dye tracers and demonstrated its ability to accurately simulate transverse dispersion processes with a conservative tracer. Here we extend this methodology to the more complex process of biodegradation within a plume. The ruthenium(II) complex is employed as an indicator of oxygen concentration (Amao and Okura 2000; Carraway et al. 1991; Garcia-Fresnadillo et al. 1999; Li and Walt 1995). Under UV light, the fluorescent intensity of the ruthenium(II) complex is a function of the concentration of oxygen (Carraway et al. 1991). By combining the imaging technique with these fluorescent indicators, the mixing of oxygen and plume can be directly measured without any disturbance.

Objective. This paper reports a novel approach to the simulation and quantification of biodegradation processes in the laboratory. It allows a direct and non-invasive observation of oxygen mixing and consumption at the fringe of a contaminant plume.

MATERIALS AND METHODS

Materials

Mineral medium. The mineral medium employed consisted of ultra high quality water containing :

0.1% K_2HPO_4,
0.1% KH_2PO_4,
0.1% NH_4Cl,
0.05% $MgSO_4$,
and saturated $CaCl_2$ and $FeSO_4$ solution, 5-8 drops/1000mL
(all chemicals supplied by Sigma-Aldrich).

Fluorescent indicator. Ruthenium(II)-dichlorotris(1,10-phenanthroline) (supplied by Aldrich), a water soluble fluorescent dye which can be quenched by dissolved oxygen, was used at a working concentration of 1.0×10^{-4} M as an indicator of dissolved oxygen in the porous media. $Ru(phen)_3Cl_2$ is a very stable poly aromatic hydrocarbon. There was no evidence that bacteria could degrade it in the experiments undertaken. Under UV (450 nm) light, $Ru(phen)_3^{2+}$ can be excited and fluoresces at 590-600 nm. The quenching process is described by the Stern-Volmer equation (Garcia-Fresnadillo et al. 1999):

$$I_0/I = 1 + K_{SV}[O_2] \tag{1}$$

Here I_0 and I are the luminescence intensities in the absence and in the presence of quencher (O_2), respectively, K_{SV} is the Stern-Volmer constant and is equal to the product $\tau_0 k_q$ (the emission lifetime under O_2-free conditions times the bimolecular quenching rate constant).

Carbon source. Potassium acetate (KAc) (supplied by Sigma Aldrich)

Bacteria. Bacteria were isolated from field soil and were enriched in the mineral medium and 1000 mg/L KAc.

Matrix. The matrix employed in the solute transport experiments was made up of high quality quartz sand (60-100 μm). Quartz was used in preference to ordinary glass since it transmits UV with little loss.

Microbe culture in matrix. The solute transport experiments required a microbe population to be present on the matrix. This was achieved by mixing the quartz sand with the mineral medium and packing it in a glass column. The bacteria (isolated from the field soil) and KAc (as the sole carbon source) were continuously introduced into the column, while oxygen was bubbled from the column base. After 24 hours, the supply of KAc was cut off. Samples of liquid were then taken at regular intervals until the concentration of KAc (measured by Ion Chromatograph DIONEX DX-120) decreased to zero. At this stage the quartz sand was ready to be incorporated into the solute transport experiment without introducing additional unwanted KAc.

Methods

Batch experiment. Initially batch experiments were undertaken to investigate the relationship between oxygen consumption and acetate concentration during biodegradation. The apparatus is depicted in Fig 1. The bacteria, 500 mL mineral medium, fluorescent indicator 1.0×10^{-4} M, and KAc (100 mg/L) were mixed in a brown bottle. The concentration of bacteria was measured as Abs=0.084 (measured by spectrophotometer at 600 nm). Every 20 minutes, samples were taken out of the bottle and filtered by 0.2 μm membranes and then measured by Ion Chromatography for acetate concentration. Dissolved oxygen was continuously monitored by an oxygen probe (OX20 portable oxygen meter, Walden).

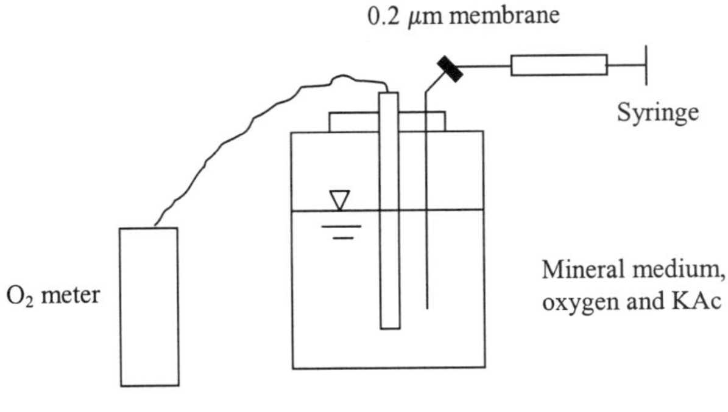

FIGURE 1. Batch experiment

Solute transport experiments. The experimental setup is indicated in Fig. 2. An aquifer has water flow induced by a pump. A tracer and substrate (KAc) were introduced, illuminated by UV light, and observed by a CCD (Charged Couple

Device) camera, with images captured on computer. All experiments were conducted at a temperature of 16 °C. These components are described below.

Aquifer. The quartz sand with cultured microbes was packed into a quartz box of dimensions 130×160×3 mm to form a micro-aquifer. The thickness of the box relative to its width and length means that it can effectively be regarded as a 2-dimensional aquifer. Each end of the aquifer terminates in a chamber, separated from the aquifer by uniform filters. The aquifer box was partially filled with mineral medium, then the quartz sand/water mixture from enrichment procedure was slowly poured into the box. At the same time, the emplaced beads were stirred to release gas bubbles. The mass of quartz sands was measured in order to allow the average porosity n of the aquifer to be determined ($n = 0.49$).

Water flow. The aquifer water consisted of the mineral medium, $Ru(phen)_3Cl_2$ at the working concentration, and 7.4 mg/L dissolved oxygen.

Pump. The effluent from the aquifer was removed by a Fluid Metering, Inc. QG6 ceramic valveless dispenser and metering pump. This permits a constant flow rate to be applied to the model. The flow rate is 0.14 mL/min, equivalent to a pore water velocity of 1.25×10^{-6} m/s.

Plume source. KAc at a concentration of 1000 mg/L was continuously injected into the aquifer through a 6 mm internal diameter pipe by a syringe pump (6.8 μL/min). The pipe was inserted through the face of the aquifer model and passed through the full 3 mm thickness of the aquifer. Within the aquifer, the pipe had a 2.5 mm wide slot through which tracer was injected resulting in a relatively uniform distribution of tracer through the aquifer thickness.

UV light. Two 20W BLB (Black Light Blue) UV tubes were used as light source to excite the tracer. The distance between UV light and quartz box was 176 mm.

Camera. The process of mass transport was recorded by a Hitachi color KP-D581 CCD camera. The resolution of the camera is 811(H)×508(V) pixels and the minimum object illumination is 0.02 lux. An UV filter was installed in the front of the camera to block the UV light. The distance between the CCD camera and the box was 596 mm.

Data processing. A Data Translation, Inc. DT3131 Frame Grabber and software were installed in a computer PC to support image capture. GLOBAL LAB IMAGE (version 2.5) software was used to process image data.

Oxygen calibration. Different concentrations of dissolved oxygen within the mineral medium (including $Ru(phen)_3Cl_2$ at 1.0×10^{-4} M) were passed through the whole aquifer and imaged as described above. The fluorescence intensity was used to calibrate the indicator as a measure of oxygen concentration.

RESULTS AND DISCUSSION

KAc Degradation Analysis. In the batch experiments, both the acetate and oxygen concentrations decayed with time. The relationship between the KAc and oxygen consumption is shown in Fig. 3.

This indicates an experimental ratio of O_2/Ac=0.3922. It is therefore inferred that the decrease of oxygen only results from degradation of Ac and that dissolved oxygen is the only electron acceptor for Ac degradation.

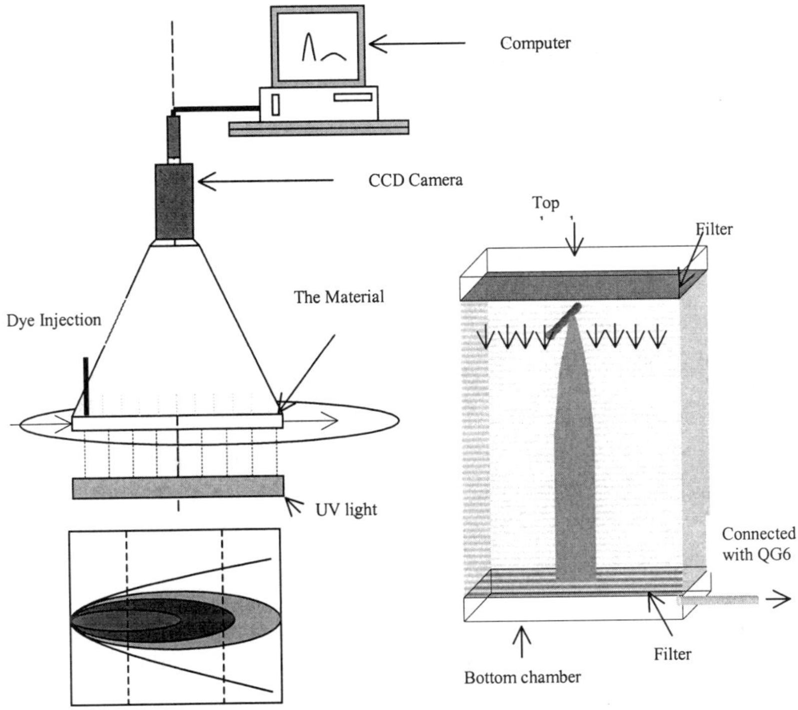

FIGURE 2. Schematic of 2-D aquifer and CCD-camera measurement system

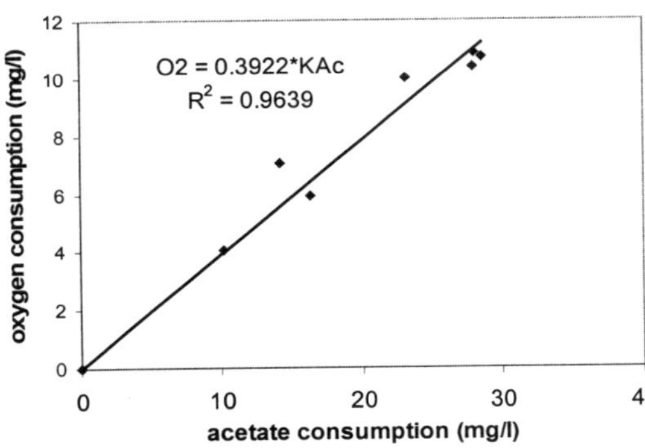

FIGURE 3. The relationship between the acetate and oxygen

Oxygen Calibration Analysis. Oxygen free water shows the strongest fluorescence intensity. The presence of oxygen reduces the intensity. Quenching of the fluorescence is induced by molecular collision and it is a dynamic process. The quenching effect depends on temperature and oxygen concentration. Fig. 4 indicates the effect of different concentrations of oxygen in porous media. The lines/banding seen in the matrix in Fig 4(b), are due to placement effects of the quartz sand. Various techniques for minimising this effect are being investigated.

Since all experiments were conducted at 16 °C, the intensity of $Ru(phen)_3^{2+}$ was controlled by the oxygen concentration only and the experimentally derived relationship is plotted in Fig. 5. It can be seen that the data fit the Stern-Volmer equation (1), with K_{sv}=0.598.

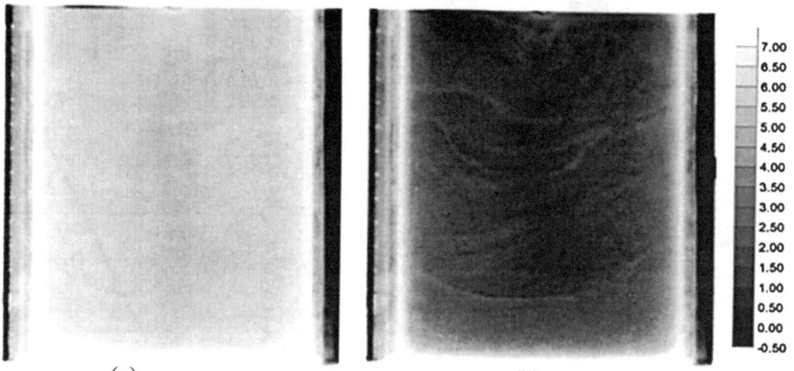

(a) (b) DO (mg/L

FIGURE 4. Images of aquifer saturated with different concentrations of oxygen. (a) DO (Dissolved oxygen) = 6.40 ppm. (b)DO = 0.00 ppm

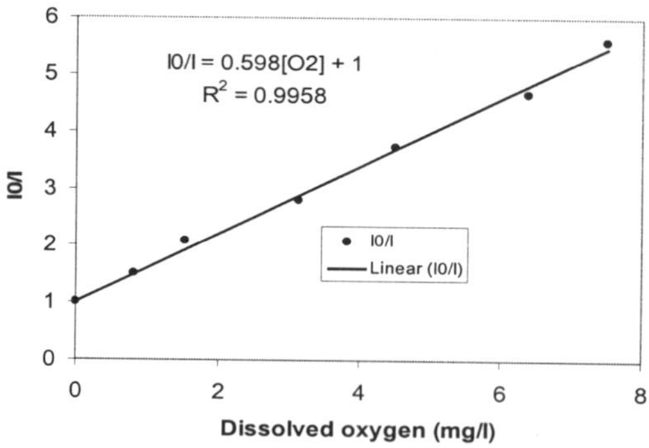

FIGURE 5. Calibration of dissolved oxygen in quartz sands, I_0 is the intensity without oxygen, I is the intensity with oxygen present

Solute Transport in Porous Media. Fig. 6 shows the oxygen maps in the porous media at different times due to the KAc plume. After injecting KAc for 744 minutes (Fig. 6(a)), the plume had not yet reached steady state. It may be observed that the core of the plume is much brighter than the background, indicating that the oxygen was nearly exhausted. After 1440 minutes (Fig. 6(b)), the plume has reached a steady state. The core of the plume is an oxygen free zone (Fig. 7), and biodegradation should not happen here. However at the fringe, oxygen is mixed into the plume by transverse dispersion, and biodegradation may occur in this zone.

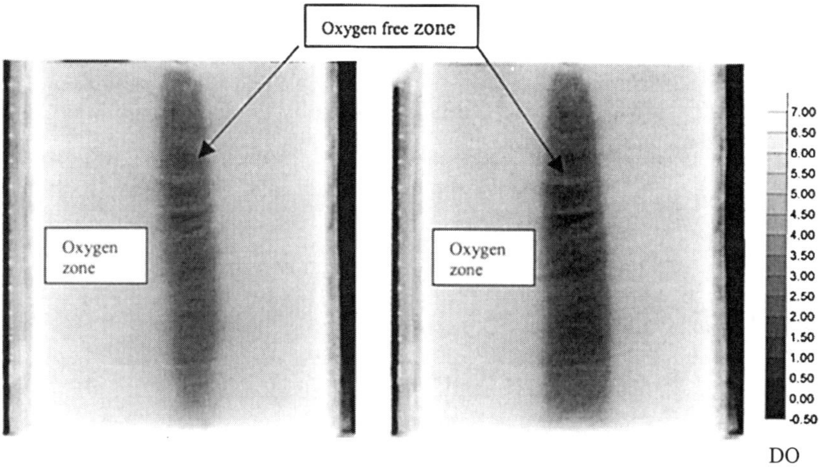

FIGURE 6. Distribution of oxygen. (a) t = 744 min, unsteady state, (b) min, steady state

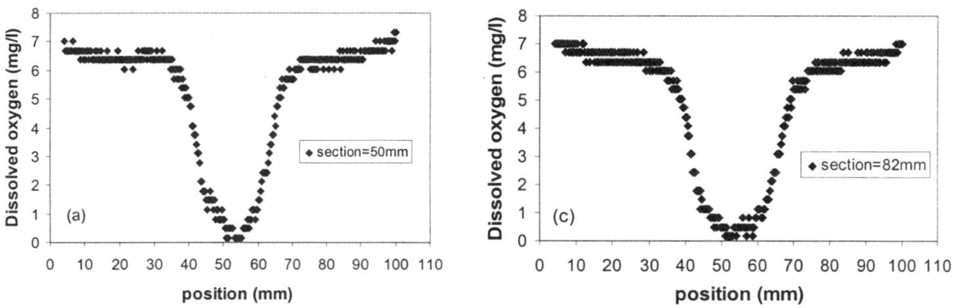

(a) (b)

FIGURE 7. Oxygen distribution at different sections in steady state. The distance between section and source is (a) 50 mm (b) 82 mm

CONCLUSIONS

1. Non-invasive quantification of oxygen consumption and mixing during biodegradation processes within a simulated 2-D contaminant plume has been successfully demonstrated in a model aquifer;
2. Experimental observation confirms that dissolved oxygen is mainly provided by transverse mixing at fringe of the plume.

To the knowledge of the authors this is the first time that electron acceptor fringe mixing has been demonstrated by experimental observation.

FURTHER WORK

Work is planned to extend the techniques described in this paper to include additional indicators of degradation processes, and to compare the experimental results with theoretical and numerical models of biodegradation and fringe mixing processes.

Acknowledgements This research is sponsored by the UK Environment Agency and an ORS award from Universities UK.

REFERENCES

Amao, Y. and Okura, I., 2000. "Optical oxygen sensing properties of tris(4,7 '-diphenyl-1,10 '-phenanthroline) ruthenium (II)-polyacrylic acid complex thin film". *Polym. J.* *32*(5): 452-455.

Carraway, E.R., Demas, J.N., Degraff, B.A. and Bacon, J.R., 1991. "Photophysics and Photochemistry of Oxygen Sensors Based On Luminescent Transition-Metal Complexes". *Anal. Chem.* *63*(4): 337-342.

Cirpka, O.A., Frind, E.O. and Helmig, R., 1999. "Numerical simulation of biodegradation controlled by transverse mixing". *J. Contam. Hydrol.* *40*(2): 159-182.

Garcia-Fresnadillo, D., Marazuela, M.D., Moreno-Bondi, M.C. and Orellana, G., 1999. "Luminescent nafion membranes dyed with ruthenium(II) complexes as sensing materials for dissolved oxygen". *Langmuir.* *15*(19): 6451-6459.

Huang W., Smith C.S., Lerner D.N., Thornton S.F. and Oram A. 2001. "Solute transport in porous media: evaluation of a non-invasive imaging technique". *Wat. Res.* submitted.

Li, L. and Walt, D.R., 1995. "Dual-Analyte fiberoptic sensor for the simultaneous and continuous measurement of glucose and oxygen". *Anal. Chem.* *67*(20): 3746-3752.

Oya, S. and Valocchi, A.J., 1998. "Transport and biodegradation of solutes in stratified aquifers under enhanced in situ bioremediation conditions". *Wat. Resour. Res.* *34*(12): 3323-3334

SURGICAL SITE CLOSURE – INTEGRATING NATURAL ATTENUATION AND FOCUSED SOURCE TREATMENT

Steven P. Sittler, P.G. (APT, Limited – Mishawaka, Indiana)

ABSTRACT: "Surgical site closure" is an innovative remedial strategy designed to intelligently integrate natural attenuation, risk-based cleanup goals, and focused source removal/treatment to cost-effectively remediate hydrocarbon-impacted sites. The first step in a surgical site closure is to develop a thorough understanding of site-specific groundwater geochemistry, via an evaluation of the presence and relative contribution of intrinsic bioremediation processes. The primary intent of this initial step is to demonstrate the stability or regression of a dissolved-phase groundwater plume, eliminating the need for off-site or plume-wide remedial efforts and/or plume containment. Application of risk assessment principles, including a detailed evaluation of potential receptors and associated exposure pathways and possible development of site-specific, risk-based cleanup objectives, is the second key component in the surgical site closure process. In many cases, actual potential receptors are limited, and/or potential exposure pathways are incomplete, resulting in a significant reduction in risk and facilitating a much less aggressive closure approach. The third step in the surgical site closure process involves focused source removal (e.g., soil excavation), where necessary, accompanied by enhanced aerobic biodegradation to degrade the remaining dissolved-phase groundwater impacts to concentrations at or below closure levels. "Surgical site closures" have been completed at nine sites, and are in progress at 10 additional sites in Indiana for a major oil company. Groundwater geochemistry evaluations demonstrated plume stability and eliminated the need for remediation of off-site groundwater impacts, but also demonstrated that most of the dissolved-phase plumes were predominantly anaerobic, supporting the use of enhanced aerobic biodegradation to accelerate remedial efforts. Limited, focused soil excavation then successfully eliminated or significantly reduced the source area(s) for groundwater impacts at all of the sites. Oxygen release compound (ORC®) slurry injections were subsequently completed in and near the source areas and resulted in reduced target constituent concentrations in groundwater at nearly all of the sites. Necessary data were collected during remedial activities to permit use of Indiana's proposed Risk-Integrated System of Closure (RISC) policy, as appropriate, now that it has become effective.

"SURGICAL" SITE CLOSURE APPROACH

The "surgical" site closure approach is best applied at industrial sites where released materials are amenable to biodegradation, and where long-term, natural attenuation or monitoring-only type strategies are not suitable for reasons of property transfer or potential off site liability. "Surgical" site closures are typically implemented in a phased approach, as described in the following paragraphs.

Step 1 - Intrinsic Bioremediation Evaluation. The first step in the "surgical" site closure approach is to perform an intrinsic bioremediation evaluation to estimate the total degradational capability (or "assimilative capacity") of the formation. Biogeochemical data must be obtained from multiple monitoring wells to evaluate the presence and relative contribution of various intrinsic bioremediation processes to the overall assimilative capacity. Typical industry protocol (e.g., Wiedemeier, et al, 1995) recommends that samples be obtained from non-impacted upgradient and downgradient wells, as well as from several impacted wells within the plume. The wells included in the evaluation should be aligned as closely as possible in the direction of groundwater flow.

Sample collection should be performed using standard, low-flow purging techniques to minimize disruption to the formation, enabling an accurate calculation of the assimilative capacity. Key indicator parameters, including pH, specific conductance, temperature, dissolved oxygen (DO) and oxidation/reduction potential (ORP), must be monitored during purging until the values are stable, indicating that the pump is withdrawing formation water.

When stable or equilibrium conditions are achieved, the stabilized indicator parameter values are recorded, and laboratory-supplied sample containers are filled while the pump remains in low-flow mode. Per industry protocol (Wiedemeier, et al, 1995), each sample should be analyzed for dissolved hydrocarbons (typically BTEX, but also possibly MTBE, 1,2,4- or 1,3,5-trimethylbenzene, or other similar regulated compounds), nitrate, nitrite, total iron, ferrous iron, sulfate, sulfide, ammonia, alkalinity, and permanent gasses (oxygen, carbon dioxide, and methane).

Indicator parameters including pH, temperature, specific conductance, ORP, alkalinity, carbon dioxide, and ammonia are used to evaluate the biogeochemical conditions at each site, specifically with regard to suitability for biodegradational processes. Dissolved oxygen, nitrate, ferrous iron, sulfate, and methane concentrations in the clean upgradient and downgradient wells are then compared to similar data from the impacted wells to determine the utilization rates (for dissolved oxygen, nitrate, and sulfate) or production rates (for ferrous iron and methane) of each individual electron acceptor. The individual utilization/production rates for each individual electron acceptor are then multiplied by stoichiometric relationships and summed to determine the total assimilative capacity for the water-bearing unit.

This assimilative capacity is compared to the highest total BTEX concentration in groundwater at the site. If the assimilative capacity exceeds the highest total BTEX concentration, the dissolved-phase groundwater plume is stable (i.e. not continuing to migrate downgradient). Plume stability results when the input of dissolved hydrocarbons from residual sources is balanced by losses through biodegradation, integrated over the extent of the plume (McNab & Dooher, 1998). When plume stability is demonstrated (which is frequently the case for most dissolved-phase hydrocarbon plumes), off-site remediation or containment technologies are usually unnecessary, limiting the scope of remedial efforts to on-site impacts only. If the plume is not stable, steps can be taken to eliminate source areas (i.e. free product, heavily-impacted soil) to ultimately achieve stability without the necessity for site-wide remedial efforts.

Step 2 - Risk Assessment. The next step in the "surgical approach" to site closure is to use risk-based corrective action (RBCA) programs (where available) to develop realistic, achievable site-specific cleanup objectives. Although the specific protocols vary from state to state, RBCA programs generally involve some variation of a tiered approach to cleanups. The exact requirements to achieve closure under each tier varies by state. Most RBCA programs also consider the current and projected future land use (i.e. residential, commercial, industrial, etc.) and some require evaluation of ecological risks, as well as potential risks to human health. Use of RBCA also typically requires a somewhat more detailed and thorough site characterization than currently required by some state regulatory programs, so that a valid exposure assessment can be prepared.

Of critical importance is a detailed evaluation of potential receptors and associated exposure pathways. In many cases, actual potential receptors are limited, and/or potential exposure pathways are incomplete, resulting in a significant reduction in potential risk. If potential exposure pathways are present, existing available data can be compared to default cleanup objectives to determine whether additional remedial efforts are needed. These default values are developed using conservative risk assumptions, but are usually significantly higher than other types of default objectives such as the federal maximum contaminant levels (MCLs) for drinking water. Since risk assessments using default values typically involve little or no additional expense, the benefits of reducing the volume of soil or groundwater requiring remediation can be significant. If target constituent concentrations in soil or groundwater greatly exceed Tier I cleanup objectives, or are prevalent across a large area, additional data collection and calculations can be performed to generate site-specific risk-based cleanup objectives, via a Tier II or Tier III risk assessment.

Although risk-based cleanup objectives are not essential to success for the "surgical" site closure approach, they can frequently decrease the remedial timeframe and, consequently, overall remedial costs. An evaluation of the development of applicable risk-based cleanup objectives, where available, should always be integrated into a site-specific closure strategy.

Step 3 - Focused Source Removal/Treatment. The final step in the "surgical site closure" approach is to perform limited, focused, source-area removal or treatment to reduce target constituent soil and groundwater concentrations where necessary. In the absence of site-specific, risk-based cleanup objectives, the optimum volume of soil or groundwater to be removed or treated is determined through a cost-benefit analysis. Long-term monitoring costs if the material is left in place or untreated are compared to removal/treatment costs (and the associated reduced monitoring schedule) to determine the scope of the active remedial effort.

Source removal options may include free product removal (a minimum requirement for virtually any closure strategy), soil excavation and off-site disposal or treatment, and/or limited groundwater removal/treatment (usually via mobile, vacuum-based units). Treatment strategies may include short-term air injection and/or soil vapor extraction (again, using a mobile unit) or injection of chemicals such as hydrogen peroxide or ORC® to oxidize contaminants or increase subsurface dissolved oxygen concentrations, stimulating aerobic biodegradation. The intent of source removal/treatment is to maximize mass removal in those areas where intrinsic bioremediation alone would require unacceptably long timeframes to achieve closure goals.

CASE HISTORIES - "SURGICAL" SITE CLOSURES

"Surgical" site closures are currently underway at approximately 20 hydrocarbon release sites in Indiana for a major oil company. All of the sites are underlain by unconsolidated sediments, ranging from low-permeability silty clays with sand stringers to sand-and-gravel formations. Released hydrocarbon products primarily consisted of unleaded gasoline and diesel fuel, and the facilities included service stations and bulk storage terminals.

Intrinsic Bioremediation Evaluations. Intrinsic bioremediation evaluations were performed at 11 of the sites with off-site groundwater impacts. The primary purpose of the evaluations was to demonstrate plume stability and eliminate the need for aggressive remediation of off-site groundwater impacts, allowing the responsible party to concentrate remedial dollars on focused, on-site, source-area cleanups. Table 1 and Figure 2 summarize the assimilative capacity calculations for the 11 sites. The maximum total BTEX concentration in groundwater at each site (prior to initiation of remedial efforts) is also included in Table 1.

Table 1 - Summary of Assimilative Capacity Calculations

Site Location	Diss. Oxygen	Nitrate	Ferrous Iron	Sulfate	Methane	Total	Max. Total BTEX (mg/L)	Plume Stable?
	Assimilative Capacity (mg/L)							
Mishawaka, IN	1.71	None	None	0.76	None	2.47	0.98	YES
Anderson, IN	1.07	0.28	0.06	None	None	1.41	0.71	YES
Ft. Wayne, IN	0.69	None	1.26	9.18	11.86	22.99	29.9	NO
Evansville, IN	0.252	None	0.30	6.16	5.97	12.682	1.681	YES
Kokomo, IN (West)	0.102	0.296	0.34	23.45	5.78	29.968	6.59	YES
Muncie, IN	0.355	None	0.49	7.27	3.85	11.97	6.72	YES
South Bend, IN	None	None	0.493	None	0.28	0.773	0.47	YES
South Bend, IN	None	None	0.232	5.78	1.90	7.912	4.29	YES
Kokomo, IN (East)	1.31	1.02	0.41	9.9	3.17	15.81	6.53	YES
Kouts, IN	None	None	1.66	41.14	3.86	46.66	13	YES
New Castle, IN	0.253	None	0.224	5.21	4.17	9.857	1.67	YES
Averages	**0.52**	**0.15**	**0.50**	**9.90**	**3.71**	**14.77**		

	Diss. Oxygen	Nitrate	Iron	Sulfate	Methane
	Contribution to Total Assimilative Capacity (%)				
Study Site Average	**3.5%**	**1.0%**	**3.4%**	**67.0%**	**25.1%**

NOTES:
None = No contribution to assimilative capacity for this electron acceptor

Figure 2 - Summary of Assimilative Capacities

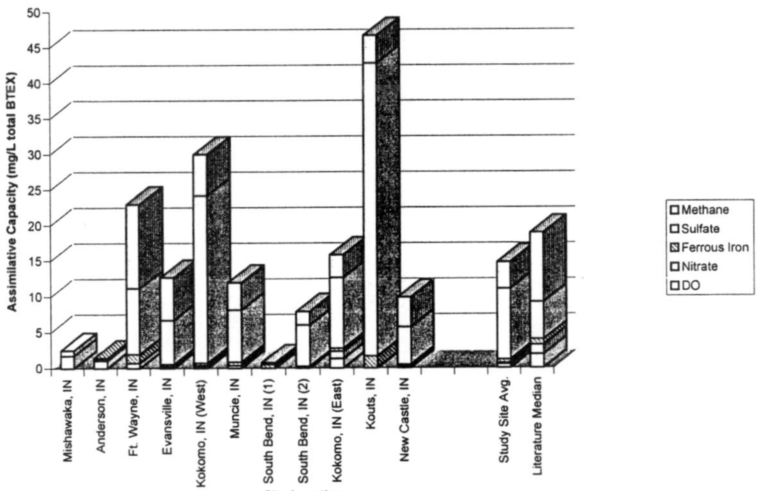

As the data indicate, the dissolved-phase hydrocarbon plumes appear to be stable at all but one of the sites. However, at all but two sites the majority of the available assimilative capacity (greater than 90% of the total) was due to anaerobic processes, which are several orders of magnitude slower than aerobic processes. Consequently, for most of the sites, focused source removal and/or treatment was warranted.

Soil Excavation/Treatment. Following completion of the intrinsic bioremediation evaluations, source removal/treatment activities were initiated. At seven of the sites, heavily-impacted soils in source areas (e.g., UST basins, dispenser islands) were excavated and transported off-site for treatment in a permitted biotreatment cell (biocell). The biocell was chosen because in addition to being cost-competitive compared to a conventional landfill, at the end of the treatment process, the client will receive a certificate that the soil has been remediated to applicable cleanup objectives, eliminating future liability for the soil to the client. The volume of soil excavated at each site ranged from approximately 50 to 800 tons, with an average of approximately 375 tons. Soil removal activities were successful in eliminating the source area(s) for groundwater impacts at all of the sites. Confirmatory soil sampling performed after excavation indicated that residual target constituent concentrations in native soils were below applicable default cleanup objectives at most of the sites, effectively "closing" the sites with respect to soil impacts.

Oxygen-Release Compound (ORC®) Injection/Application. To accelerate degradation of dissolved-phase hydrocarbon constituents in groundwater, focused, source-area groundwater treatment was performed at 18 sites by injecting or applying an ORC® slurry to increase subsurface dissolved oxygen concentrations and enhance aerobic biodegradation. ORC® is a patented formulation of magnesium peroxide that slowly releases oxygen when hydrated. The amount of ORC® needed at each site was determined using proprietary software developed by the manufacturer, Regenesis Bioremediation Products (Regenesis). An injection point spacing of 10 feet was chosen for most sites, based on previous experience, and manufacturer's recommendations. This spacing was adjusted in the field during actual injection, as needed, in order to maximize the potential for a uniform distribution of ORC® throughout the impacted portion of the saturated unit. At four of the sites, ORC® was mixed into the upper two to three feet of the saturated zone at the base of the excavated area, using a trackhoe excavator, rather than incurring additional costs for injection.

Groundwater analytical data were collected from all of the sites prior to ORC® injection/application, and quarterly monitoring was performed for one to two years afterward. A representative sampling of benzene and total BTEX data from key wells at these sites, including data obtained prior to ORC® injection/application is presented in Tables 2 and 3, respectively. DO data were also obtained for the same wells prior to ORC® injection/application and for several months afterward, and the presence or absence of a DO increase is noted in Tables 2 and 3.

TABLE 2 - SUMMARY OF BENZENE DATA
SURGICAL SITE CLOSURES WITH ORC APPLICATION

SITE	BENZENE (MG/L)					DO INCREASE OBSERVED?
	PRE-ORC	AFTER ORC				
		3 MO.	6 MO.	12 MO.	24 MO.	
High Permeability						
SOUTH BEND (NW)	0.09	0.043	0.013	ND	ND	YES
	0.001	ND	0.012	ND	0.001	YES
SOUTH BEND (W)	0.53	0.64	0.463	0.46	0.146	NO
ANDERSON	0.003	ND	ND	ND	NFA	YES
	ND	0.005	ND	---	NFA	YES
ELKHART	0.011	0.006	0.015	0.010	---	YES
MISHAWAKA	0.013	ND	ND	ND	NFA	YES
KOUTS	2.1	0.107	0.156	0.094	---	NO
	3.0	4.1	7.3	1.64	---	NO
Medium Permeability						
FORT WAYNE	14.0	0.006	ND	0.006	---	YES
MARION	0.008	0.007	0.006	0.001	NFA	NO
Low Permeability						
ELWOOD	0.076	0.013	0.030	0.003	NFA	YES
INDIANAPOLIS (NW)	0.44	0.053	0.025	0.008	0.022	YES
	3.7	8.2	6.9	4.3	1.75	NO
INDIANAPOLIS (E)	0.33	0.29	0.24	0.367	0.206	NO
	0.19	0.045	0.36	0.99	0.549	NO
INDIANAPOLIS (W)	0.014	ND	0.023	0.023	NFA	NO
NEW CASTLE	0.091	0.089	0.18	0.079	0.047	NO
	0.27	0.69	0.14	ND	ND	NO
	0.007	0.002	0.056	ND	ND	NO
	0.18	0.24	0.41	0.23	0.22	NO
CUMBERLAND	0.281	0.002	0.137	0.07	---	YES
	9.8*	2*	2.5*	2.0*	---	YES
MUNCIE (Far W)	0.006	0.016	0.009	ND	NFA	NO
MUNCIE (Near W)	0.052	0.001	ND	---	---	NO
	0.328	0.227	0.276	---	---	NO
CONNERSVILLE	0.019	ND	ND	---	---	NO
	0.009	0.006	ND	---	---	NO
	0.012	0.017	0.076	---	---	NO
	ND	0.002	0.002	---	---	NO

NOTES:
* Numbers are MTBE concentrations
NFA - No Further Action required by the IDEM

TABLE 3 - SUMMARY OF TOTAL BTEX DATA
SURGICAL SITE CLOSURES WITH ORC APPLICATION

SITE	TOTAL BTEX (MG/L)					DO INCREASE OBSERVED?
	PRE-ORC	AFTER ORC				
		3 MO.	6 MO.	12 MO.	24 MO.	
High Permeability						
SOUTH BEND (NW)	2.62	0.386	0.151	ND	ND	YES
	0.106	0.030	0.355	0.104	0.103	YES
SOUTH BEND (W)	26.13	21.65	18.63	11.646	---	NO
ANDERSON	0.185	0.066	0.153	0.028	NFA	YES
	2.759	0.708	0.65	---	NFA	YES
ELKHART	2.956	0.643	1.793	0.469	---	YES
MISHAWAKA	0.318	ND	1.356	0.017	NFA	YES
KOUTS	10.23	0.303	0.387	0.24	---	NO
	8.10	10.82	16.86	6.495	---	NO
Medium Permeability						
FORT WAYNE	41.9	0.031	ND	0.023	---	YES
MARION	0.008	0.007	0.006	0.001	NFA	NO
Low Permeability						
ELWOOD	0.093	0.013	0.030	0.003	NFA	YES
INDIANAPOLIS (NW)	0.524	0.062	0.033	0.002	0.016	YES
	4.68	9.26	7.49	10.22	7.31	NO
INDIANAPOLIS (E)	1.32	0.77	0.68	0.61	0.655	NO
	0.285	0.045	0.796	0.764	0.638	NO
INDIANAPOLIS (W)	1.084	1.01	0.597	0.359	NFA	NO
NEW CASTLE	0.091	0.091	0.191	0.017	0.05	NO
	0.294	0.724	0.153	ND	ND	NO
	0.127	0.004	0.137	ND	ND	NO
	0.188	0.252	0.427	0.071	0.182	NO
CUMBERLAND	8.539	0.007	11.388	1.92	---	YES
	9.8*	2*	2.5*	2.0*	---	YES
MUNCIE (Far W)	0.007	0.019	0.012	ND	NFA	NO
MUNCIE (Near W)	4.44	0.731	ND	---	---	NO
	0.806	0.413	0.52	---	---	NO
CONNERSVILLE	0.041	0.022	ND	---	---	NO
	0.068	0.031	ND	---	---	NO
	0.133	0.186	0.789	---	---	NO
	0.009	ND	0.002	---	---	NO

NOTES:
* Numbers are MTBE concentrations
NFA - No Further Action required by the IDEM

As Tables 2 and 3 indicate, reductions in benzene and total BTEX concentrations in groundwater were observed at most of the sites after soil excavation and ORC® injection/application. In several cases, benzene and total BTEX concentrations in groundwater were reduced to below MCLs in less than three months, probably primarily as a result of source removal. Regulatory closure has been granted at seven of the sites, with pending closures at several others. Monitoring data were also used to aid in determining whether additional ORC® injection at some of the more resistant sites was warranted. At three of the sites with high initial BTEX concentrations, additional ORC® injection was performed in February 2001 to further reduce BTEX concentrations in groundwater. Ongoing groundwater monitoring is continuing at these sites.

Risk Assessment. Because these sites were all located in the State of Indiana, which does not yet have a risk-based corrective action program in place, risk assessments were not performed as part of the closure strategy. However, the Indiana Department of Environmental Management (IDEM) has recently developed a program-wide policy known as the Risk-Integrated System of Closure (RISC), which was implemented in February 2001. Consequently, as a proactive approach toward possible future use of this policy at the surgical site closure sites, necessary background, hydrogeological, and analytical data (e.g., BTEX concentrations in soil, hydraulic conductivity estimates, surrounding property uses) were collected during the soil removal and ORC® slurry injection activities. Now that the RISC policy is available for use, each site with BTEX concentrations exceeding default cleanup criteria is being evaluated to determine whether application of the RISC guidance can cost-effectively accelerate site closure. Following these evaluations, the RISC policy guidance will be used where appropriate to develop site specific cleanup goals and aid in expediting site closure.

SUMMARY

Cost analyses performed during development of the closure strategies for these sites indicated that if traditional remedial approaches were used (i.e. active off-site remediation and/or containment, combined with active, site-wide remedial efforts), total costs to reach closure at most sites would have ranged from at least \$100,000 to potentially more than \$1 million. Actual implementation costs for the "surgical" site closure approach at the 19 sites ranged from approximately \$25,000 to \$75,000, with additional monitoring costs of \$5,000 to \$10,000 per year, per site for one to two years following implementation.

REFERENCES

Chappelle, Francis H., 1999, *Bioremediation of Petroleum-Hydrocarbon Contaminated Ground Water: The Perspectives of History and Hydrology*, Ground Water, Vol. 37, No. 1, January-February 1999

McNab, Walt W. Jr., and Brendan P. Dooher, 1998, *Uncertainty Analyses of Fuel Hydrocarbon Biodegradation Signatures in Ground Water by Probabilistic Modeling*, Ground Water, Vol. 36, No. 4, July-August 1998

Wiedemeier, Todd, Wilson, John T., Kampbell, Donald H., Miller, Ross N., and Jerry E. Hansen, 1995, *Technical Protocol for Implementing Intrinsic Remediation with Long-Term Monitoring for Natural Attenuation of Fuel Contamination Dissolved in Groundwater*, Air Force Center for Environmental Excellence, Technology Transfer Division, Brooks AFB, San Antonio, Texas

ASSESSING THE POTENTIAL FOR BIOREMEDIATION

Antony Jones (Komex H_2O Science INC., Huntington Beach, California, USA)
Diane Lorton, Jeremy Mason, Steve Smith (King's College London, UK)
Caroline Fenwick, Anne Glover (Aberdeen University, Aberdeen, UK)
Rolf Henkler (ICI Paints Ltd.)
Bruce Anderson (PlanReal AG, Victoria, Australia)

ABSTRACT: In this field-based study the potential for using natural attenuation at three benzene, toluene, ethylbenzene and xylene (BTEX)-contaminated sites has been investigated using a combination of microbial and molecular analysis of the hydrocarbon degrading microbial community, toxicity assessment and chemical analysis. Biodegradative activity has been quantified using radiolabeled mineralization assays on sediment and groundwater samples. Groundwater toxicity has been assessed using a *lux*-marked microbial biosensor. Chemical analysis of sediment and groundwater has quantified contamination and environmental parameters. At a paint-manufacturing site, toxicity of the groundwater pre- and post-air sparging averaged 64% and 17% inhibition respectively, indicating volatile organic carbon (VOC) toxicity. ^{14}C-toluene biodegradative activity varied considerably (0.001 to 1.503 μg toluene h^{-1} sample $unit^{-1}$) between different well samples with activity increasing with total VOC, indicating an adapted VOC degrading microbial consortia in high VOC environments. At a wood-preservative-manufacturing site, significantly lower toxicity constraints were found (2-8%), and ^{14}C-toluene biodegradative activity was significantly higher than at the previous site. At the third site, another paint-manufacturing site, toxicity was low (8% inhibition); however, ^{14}C-toluene biodegradative activity was undetectable. Further microbiological and chemical analysis identified both low nutrient concentrations and low microbial density as the constraining factor for bioremediation. In this paper, we demonstrate how microbial biodegradative activity, biosensor toxicity, and chemical analysis have been combined to assess the potential for natural attenuation.

INTRODUCTION

Natural attenuation is rapidly acquiring credibility as an option in the remediation of contaminated land (Stapleton and Sayler, 1998). It is an attractive, cost-effective alternative to more intrusive and expensive remediation strategies (National Research Council, 1993). Natural attenuation relies on the potential ability of indigenous microorganisms to transform or degrade contaminants and the suitability of the site environment to support microbial biodegradation. The potential of the indigenous microorganisms to degrade the contaminants *in situ* must be shown when evaluating natural attenuation (Madsen et al., 1991). Three lines of evidence have been used to investigate the potential for natural attenuation. Site characterization and the monitoring of contaminants provides geochemical parameters and an evaluation of plume length and migration. Biosensor toxicity analysis demonstrates whether the indigenous microbial community may be inhibited by toxic constraints. Quantification of biodegradative activity demonstrates the potential of the microbial community to degrade the contaminants given the appropriate geochemical parameters, and toxicity. Natural attenuation has been assessed in the context of toxicity, biodegradative activity and geochemical parameters and appropriate remedial strategies consequently adopted at three sites in the United Kingdom, Germany and Brazil.

OBJECTIVES

The objectives of this study were to develop an inert reticulated vitreous carbon (RVC) matrix for representative sampling of biomass from the groundwater at contaminated sites. To then field test the RVC to enable an assessment of *in situ* biodegradative activity using radiolabeled mineralization assays, ultimately, combining the FROGs with environmentally relevant biosensors and chemical analysis to enable assessment of both the potential and ongoing success of natural attenuation.

MATERIALS AND METHODS

Quantification of Groundwater Biodegradative Activity. Sediment samples and FROGs (Filters for the Recovery Of Groundwater samples) were used for the quantification of biodegradative activity. FROGs were prepared using a modified method of Henkler (2000), by cutting RVC from sheets (1m x 50cm x 5cm) using a mycological borer (size 6). Each RVC, measuring 5cm by 1cm diameter, was pre-wetted in decreasing concentrations of ethanol solution (96%, 50%, 25%, 10%, 1% v/v) for 24h at each concentration. After pre-wetting, each RVC was thoroughly washed twice in distilled water to remove any residual ethanol solution, prior to being fixed to cryocane aluminium holders. At each sampling point, 20 reticulated vitreous carbon units were secured to a cassette of 5 aluminium holders, autoclaved at 121°C, 15 psi for 21 mins, and transported to site in sterile sealed containers (Jones et al., 2000). FROGs were installed in monitoring wells for a period of three weeks. FROG holders were suspended on covered wire 1m below the surface of the groundwater and secured to the borehole cover. On removal from the sampling points, ten FROGs were frozen at –70°C and the remaining FROG'S were maintained at 4°C.

Four of the FROGS maintained at 4°C were aseptically crushed in 50ml Falcon tubes, centrifuged at 5 500 rev min^{-1} for 5mins at 4°C and made up to a final volume of 10mL with either well water or Hutner's medium (Cohen-Bazire et al., 1957). Samples were vortexed for 2mins at 2 400 rev min^{-1} and 1ml of FROG suspension was used to determine biodegradative activity with toluene-(ring-UL)-^{14}C (Sigma-Aldrich), using a modified method of Jones et al., (1999).

Total heterotrophic bacteria and toluene-degrading bacteria were enumerated using plate counts for FROG and sediment samples. Total heterotrophs were enumerated with R2A agar, and toluene-degrading bacteria with Hutner's agar supplemented with toluene. All plates were incubated at 25°C and counted after 48 and 96 hours.

Toxicity Assessment. Sediment and groundwater samples obtained from BTEX-contaminated sites were screened using a bioluminescence-based (*lux* gene-marked) bacterial biosensor to identify constraints to remediation (Sousa et al., 1998).

RESULTS AND DISCUSSION

Development and application of FROGs. Obtaining representative microbial communities from soil cores at contaminated sites may be both expensive and prone to variability. Groundwater samples tend to have too low a density of microorganisms and the majority of hydrocarbon-degrading organisms are known

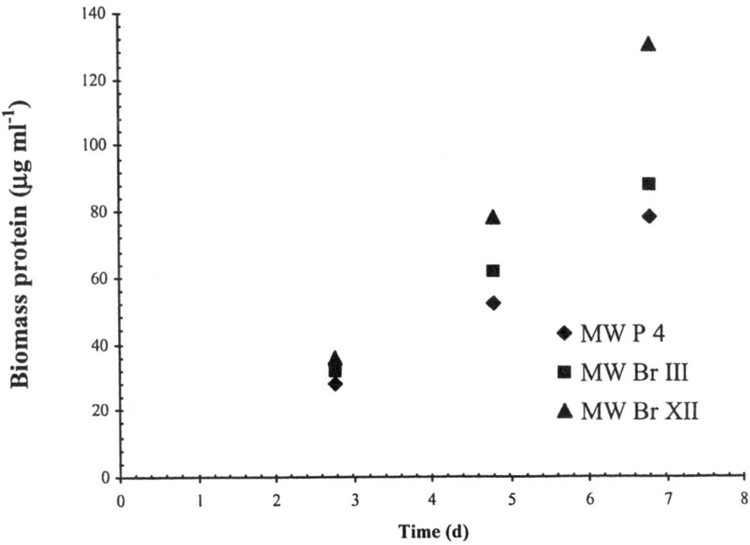

FIGURE 1 Biomass accumulation on FROGs from three monitoring wells (P4, Br III and Br XII) from a contaminated paint-manufacturing site.

to grow as biofilms. Consequently, FROGs have been developed to allow a cost-effective method of repetitive sampling of biofilm microorganisms *in situ* from the same monitoring well. FROGs consist of a porous inert carbon matrix that facilitates the growth of a microbial biofilm.

The generation of a microbial biofilm in three monitoring wells from site A over a period of seven days has been demonstrated (Figure 1). After a period of 21 days the amount of microorganisms stabilizes and an established microbial biofilm has developed (Henkler, 2000). Scanning electron microscopy has confirmed the formation of a biofilm (Henkler, 2000), which has been assayed for biodegradative activity and the presence of Class II and IIB dioxygenase genes.

TABLE 1 Summary of Assessment for the Potential for Natural Attenuation at three contaminated sites.

Site	Toxicity Assessment % Inhibition		Biodegradative Activity	Potential Constraint
	Pre N_2-sparge	Post N_2-sparge	µg toluene degraded $h^{-1}FROG^{-1}$	
A	68.6	16.7	0.516	VOC
B	7.9	2.3	4.19	None
C	5.8	8.2	0.012	Nutrient Limitation

Assessment of Potential for Bioremediation. Data from two of the three lines of evidence used to assess the potential for natural attenuation have been presented (Table 1). Biodegradative activity has been assessed with ^{14}C-UL-toluene in this study, as this compound is the main contaminant of concern at all three sites. Evidence of toluene biodegradation does not by itself demonstrate that other compounds such as benzene or naphthalene will be degraded. But an aerobic toluene-degrading microbial community does illustrate that the sediment or groundwater environment facilitates the aerobic degradation of aromatic compounds through the use of mono- or dioxygenase ring-cleaving enzymes. Furthermore, the presence of the *tod*C1 gene (a gene encoding part of the aerobic biodegradation pathway of toluene) in frozen FROG samples concurrent with toluene biodegradative activity has been demonstrated (data not shown). Benzene mineralization assays using FROG samples from Site B have demonstrated benzene biodegradative activity in groundwater (data not shown). In site A, a paint-manufacturing site, toxicity assessment has demonstrated VOCs as a toxic constraint on natural attenuation, whilst the biodegradative activity assessment has demonstrated the ability of the *in situ* microbial community to degrade toluene. A combination of soil-vapor extraction and monitored natural attenuation would be anticipated to remediate the site.

In site B, a wood-preservative manufacturing site, no significant toxic constraints have been shown and biodegradative activity was correspondingly high, indicating the potential success of using monitored natural attenuation only.

In site C, a paint-manufacturing site, initial biodegradative activity derived from FROGs samples was very low. A maximum toluene biodegradative activity of only 0.012µg toluene h^{-1} $FROG^{-1}$ was observed, by comparison with 4.2µg toluene h^{-1} $FROG^{-1}$ (Site B) and 0.52µg toluene h^{-1} $FROG^{-1}$ (Site A).

TABLE 2 FROGs bacterial density (Site C)

2 FROGs were crushed and resuspended in 15mL Falcon tubes in Hutner's medium, vortexed thoroughly and 100µL spread-plated onto R2A and Hutner's plus toluene plates, incubated for up to 96h at 25°C.

a) Total hetertotrophs
b) Toluene degraders

Sample	Plate Counts (cfu $FROG^{-1}$)	
	R2A medium (a)	Hutner's plus toluene (b)
3	9.35×10^6	8.00×10^2
7	9.67×10^6	0
10	1.09×10^4	0
13	4.07×10^6	0
14	5.68×10^5	0

TABLE 3 Sediment bacterial density (Site C)

Approximately 2g of sediment was accurately weighed into 15mL falcon tubes and resuspended in Hutner's medium, vortexed thoroughly and 100µL spread-plated onto R2A and Hutner's plus toluene plates.

a) Total hetertotrophs
b) Toluene degraders

Sample	Plate Counts (cfu g^{-1} sediment)	
	R2A medium (a)	Hutner's plus toluene (b)
1	5.93×10^4	0
2	3.59×10^4	0
3	4.01×10^6	4.70×10^5
4	1.49×10^6	3.20×10^2
5	2.22×10^3	1.22×10^4

Microbial counts were similarly low for both heterotrophic and toluene-degrading bacteria (Table 2), with no detectable toluene-degrading bacteria found on any of the FROGs samples except sample no.3. Sediment bacterial density was correspondingly low (Table 3). Chemical analysis of the groundwater and sediment demonstrated limited nutrient content in the soil (data not shown), whilst toxicity analysis demonstrated very low toxicity (5.8% and 8.2% for pre- and post-N_2 sparge, respectively). Low microbial density and nutrient deprivation would inhibit the colonization of FROGs and are most likely to be the constraining factors on the natural attenuation of hydrocarbons at the site. Bioaugmentation, nutrient addition, and biopiling on site would be required in conjunction with enhanced monitored natural attenuation

SUMMARY AND CONCLUSIONS

The potential for intrinsic bioremediation at a selection of sites contaminated by differing groups of aromatic hydrocarbons has been assessed using several methods. Differing constraining factors have been identified at the three sites investigated and appropriate remedial strategies adopted.

The combination of biodegradative activity quantification and biosensor toxicity determination provides the investigator with a valuable toolkit in determining the most appropriate site-specific bioremediation strategy.

FURTHER WORK

FROGs are currently being used to assess the potential for natural attenuation at a number of sites and are being developed to monitor changes in gene expression, abundance, and microbial diversity both between samples and temporally.

ACKNOWLEDGEMENTS

This work has been sponsored by the UK Engineering and Physical Science Research Council (EPSRC) with additional support by ICI Paints Ltd.

REFERENCES

Cohen-Bazire, G., W. R. Sistrom, and R. Y. Stanier. 1957. Kinetic studies of pigment synthesis by non-sulphur purple bacteria. J. Cell. Comp. Phys. **49:**25-68.

Henkler, C.H. 2000. Development of a solid surface system for the isolation of indigenous pollutant degrading microorganisms from contaminated water. *PhD Thesis*, King's College London, University of London, UK.

Jones, A. D. G., J. Mason, S. Smith, H.S. Wheater, A.B. Butler, J.N.B Bell, H. Gao, A.R.G. Shields, P.E. Hardisty, S. Wallace. 1999. "Hydro-biological controls on transport and remediation of organic pollutants." In A.Leeson and B. Alleman (Eds.), *The Fifth International In situ and On-Site Bioremediation Symposium. San Diego, California, April 1999.* Battelle, USA.

Jones, A. D. G., C. Fenwick, A. L. Glover, R. D. Henkler, and J. R. Mason. 2000. Assessment of Constraints to Bioremediation in Contaminated Sites. Goldschmidt 2000, Journal of Conference Abstracts, **5**(2), 599.

Madsen, E. L., J. L. Sinclair, and W. C. Ghiorse. 1991. *In situ* biodegradation: Microbiological patterns in a contaminated aquifer. Science **252:**830-833.

National Research Council. 1993. *In Situ Bioremediation: When Does it Work?* National Academy Press, Washington, DC.

Sousa, S., C. Duffy, H. Weitz, L. A. Glover, and E. Bar. 1998. Use of a *lux*-modified bacterial, biosensor to identify constraints to bioremediation of BTEX-contaminated sites. Environmental Toxicology and Chemistry. **17(6):**1039-1045.

Stapleton, R. D., and G. S. Sayler. 1998. Assessment of the microbiological potential for the natural attenuation of petroleum hydrocarbons in a shallow aquifer system. Microbial Ecology. **36**:349-361

HYDROGEN AS INDICATOR FOR REDOX CONDITIONS AND DECHLORINATION

Jeroen Ter Meer, Miriam van Eekert, Jan Gerritse and Huub Rijnaarts (TNO Environmental, Energy and Process Innovation, Apeldoorn, The Netherlands)

ABSTRACT: A data analysis of seventeen different sites with over 300 samples was carried out to find relationships between hydrogen, macro chemical parameters, redox potential and the degree of dechlorination. This improved the understanding of redox and dechlorination processes, as well as the applicability of methods to assess redox conditions and dechlorination in different field situations. Results show that at hydrogen concentrations above 0.5 nM, the conditions for dechlorination of chloroethenes can be considered as favorable in more than 85 % of the cases. A weak correlation exists between hydrogen and most macro chemical parameters. This is probably due to the fact that hydrogen reflects the exact redox conditions in one particular point in time and space, whereas the macro chemical parameters give an overall indication of redox processes that have taken place in the past and up-stream from the sampled area. In contrast to hydrogen, macro chemical parameters are persistent and can be transported in groundwater. Redox parameters (both hydrogen and macro chemical species) are not only governed by redox processes, but also strongly affected by other factors and processes (volatilization, carbonate speciation, etc.). This complicates making a reliable redox interpretation.

INTRODUCTION

Redox conditions in the subsurface are among the most important factors influencing natural attenuation in general, and dechlorination processes in particular. At contaminated sites, the identification of redox conditions is therefore becoming more and more a standard procedure. When using traditional methods based on the redox potential or macro chemical parameters, redox identification is not always easy. Eh values and macro chemical redox compounds do not always reflect the redox processes that actually take place. Consequently, the exact conditions cannot be defined in detail or the results lead to a wrong interpretation. For this reason alternative redox indicators have been explored, including the analysis of hydrogen concentrations in groundwater (Chapelle et al., 1997; Hoehler et al., 1998; Jakobsen et al., 1998; Lovley and Goodwin, 1988).

In the past several years, hydrogen based redox identification has frequently been applied in the field. Together with macro chemical analysis and redox potential measurements, hydrogen measurements often lead to an improved and more complete interpretation. Unfortunately, the different methods do not always agree. The consistency of the different methods must be assessed in order to make an unbiased redox interpretation.

Objective. The goal of the present research is to determine the value of the different redox and dechlorination parameters in different situations. The parameters (including Eh, oxygen, nitrate, iron, sulfate, methane and hydrogen) may behave totally different in situations where the contaminants, the geohydrology or the remediation activities are different. Therefore data analysis of seventeen different sites with over 300 samples was carried out to find relationships between hydrogen, macro chemical parameters, redox potential and the degree of dechlorination.

THEORETICAL BACKGROUND

Microorganisms play a crucial role in redox processes and the fate of hydrogen. Hydrogen is an intermediate in the decay of organic material in the subsurface. The formed hydrogen reacts with help of microorganisms with other soil and groundwater constituents. Hydrogen is an electron donor, and therefore transfers electrons to electron acceptors. These can be natural electron acceptors like oxygen, nitrate, iron(III), sulfate, and carbondioxide, or pollutants that are dechlorinated through this process, such as chloroethenes (figure 1). The chloroethenes include: tetrachloroethene (PCE), trichloroethene (TCE). Cis-1,2-dichloroethene (CIS) and vinylchloride (VC).

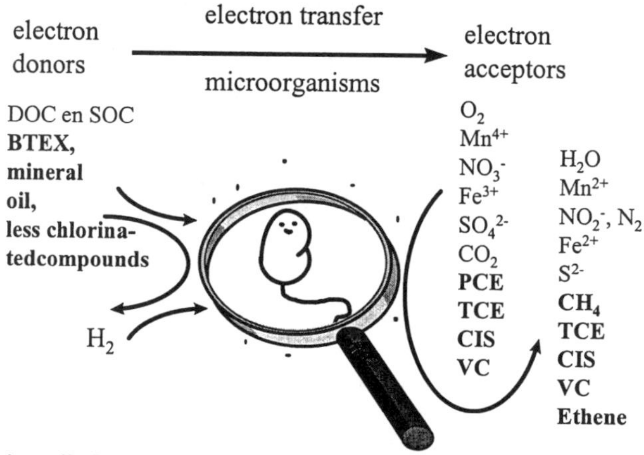

FIGURE 1. Role of hydrogen and microorganisms in the subsurface redox processes

The concentration level of hydrogen in groundwater is determined by a complex network of interrelated processes in which hydrogen is produced out of organic matter and consumed by reactions with electron acceptors. In most natural situations the production of hydrogen from organic matter is a relatively slow process compared to the consumption of hydrogen by electron accepting microorganisms (Hoehler et al., 1998; Jakobsen et al., 1998). Microorganisms consume hydrogen to a level in which the reaction just remains feasible. The reason that microorganisms act in this way is competition. Other (groups of) microorganisms that mediate in energetically less favorable electron accepting processes are not able to live on these low hydrogen levels.

It can be concluded that the higher the hydrogen concentrations are, the less efficient the redox processes. In this way hydrogen concentrations can be used as an indicator for identification of redox processes. The general order of redox processes in terms of energetic efficiency starts with oxygen and nitrate reduction as most efficient processes, the reactions become less efficient for iron (III) reduction, sulfate reduction and methanogenesis, respectively. This sequence does not always occur in practice because the efficiency of the processes is dependent on site-specific characteristics, such as pH, temperature, concentrations of several constituents and the (bio)availability of electron acceptors. Thermodynamic calculations may help to determine the energy of redox processes and to identify which redox processes are favorable and likely to take place. This is especially useful in situations where more processes take place simultaneously (redox transition zones).

DESCRIPTION OF THE DATASET

The dataset includes data from seventeen sites from The Netherlands. In contrast to many other countries, groundwater in large parts of The Netherlands is anaerobic, especially in the delta area. Aerobic conditions exist in the higher situated sandy areas. This combination of redox condition is of particular interest for this research. Locations have been selected in the aerobic as well as the anaerobic areas. An overview of the locations, including number of sampling points, the type of location, the contaminants present and the type of remediation that is carried out, is shown in table 1. Data include measurements of contaminants (mainly chlorinated solvents), redox parameters (hydrogen, oxygen, nitrate, iron, sulfate, methane, redox potential) and other relevant parameters (dissolved organic carbon, pH, temperature, etc.).

TABLE 1. Overview of the locations. Showing the number of samples, the type, the contaminants present and the current remedial action. (M)NA stands for (monitored) natural attenuation.

Location	No. of sampling points	Type	Contaminants	Remediation
C1	11	Petrochemical	CHC	MNA
C2	19	Railroad shunting yard	CHC/BTEX	MNA/ groundwater control
C5	9	Railroad shunting yard	CHC/BTEX	Mixing groundwater
C3	74	Car industry/ metal cleaning	CHC	MNA/ bioscreen
C4	59	Dry cleaning	CHC	Stimulated NA
C6	26	Dry cleaning	CHC	Stimulated NA
S1	28	Landfill	Mixed	MNA
S2	5	Landfill	Mixed	MNA
S3	6	Landfill	Mixed	MNA
S4	8	Landfill	Mixed	MNA
S5	7	Landfill	Mixed	MNA
S6	11	Landfill	Mixed	MNA
S7	6	Landfill	Mixed	MNA
S8	11	Landfill	Mixed	MNA
S9	10	Landfill	Mixed	MNA
S10	5	Landfill	Mixed	MNA
S11	15	Landfill	Mixed	MNA

RESULTS AND DISCUSSION

Relationships between the different redox and dechlorination parameters are studied in a bivariate way using scatter plots. The most interesting relationships are worked out and discussed below.

Nitrate. A relationship between hydrogen and nitrate can be observed (figure 2). High hydrogen concentrations correspond with low nitrate concentrations, and vice versa. This is in accordance to what can be expected: high nitrate concentrations indicate that nitrate reducing conditions exist, which is an energetically favorable redox process that involves low hydrogen concentrations. This behavior was observed for all locations.

Iron. No clear relationship between hydrogen and divalent iron concentrations was found. Although increased divalent iron concentrations generally indicate iron reducing or even stronger reduced conditions, high divalent iron concentrations did not correlate with high hydrogen concentrations. This is probably due to the fact that divalent iron may precipitate with sulfides or may be transported. In both cases it cannot be used as a reliable indicator for redox conditions.

Sulfate. Under sulfate reducing conditions decreased sulfate concentrations are expected together with relatively high hydrogen concentrations. However, sulfate and hydrogen concentrations do not show such a correlation. Sulfate concentrations are strongly determined by variations in background concentrations and transport phenomena. Consequently a decrease of the sulfate concentration resulting from redox processes cannot be found.

Methane. Methanogenic conditions are characterized by increased methane concentrations. Methanogenesis is energetically an unfavorable process and therefore associated with high hydrogen concentrations. From figure 2 can be concluded that such a relationship exists. Above concentrations of about 1 nM, methane concentrations are generally higher than 100 µg/L. Since methane is easily transported in groundwater it is not possible to indicate that at low hydrogen concentrations, methane concentrations are always low.

Redox Potential. Similar to hydrogen concentrations is the redox potential (Eh) a direct indicator of the intrinsic redox level. In contrast to the macro-chemical parameters they reflect the conditions at one particular point in space and time. Under reduced conditions it is expected that a low redox potential exists together with high hydrogen concentrations and vice versa. In figure 2, however, a direct relationship between both parameters is not observed. It is assumed that hydrogen and/or the Eh do not reflect the existing redox conditions properly. The failure of the Eh is most likely because it is generally accepted that in groundwater systems Eh measurements are not always a reliable redox indicator. This is due to the fact that they are not sensitive for all redox couples.

DOC. A relation between dissolved organic carbon (DOC) and hydrogen is expected, because hydrogen is produced from these compounds. The fact that no relationship could be found in figure 2 is in agreement with the hypothesis that the production of hydrogen from DOC is rate limiting. If this was not the case, high DOC levels would immediately lead to an increase of hydrogen concentrations. If within reasonable range DOC contents do no have an effect on the hydrogen concentration.

Dechlorination. High hydrogen concentrations are expected to favor dechlorination. In figure 2 a hydrogen threshold concentration can be found. Above a concentration of 0.5 nM, dechlorination can be considered as favorable in more than 85% of the cases. (The degree of dechlorination is calculated from the ratio between PCE and the degradation products TCE, CIS, VC and ethane, expressed as percentage). This suggests that dechlorination is effective when hydrogen concentrations are higher than 0.5 nM. This value is in the range of threshold values for iron reduction, sulfate reduction and methanogenesis. A competition for hydrogen between these processes is therefore expected. The process that is energetically most favorable under the given conditions is likely to win this competition.

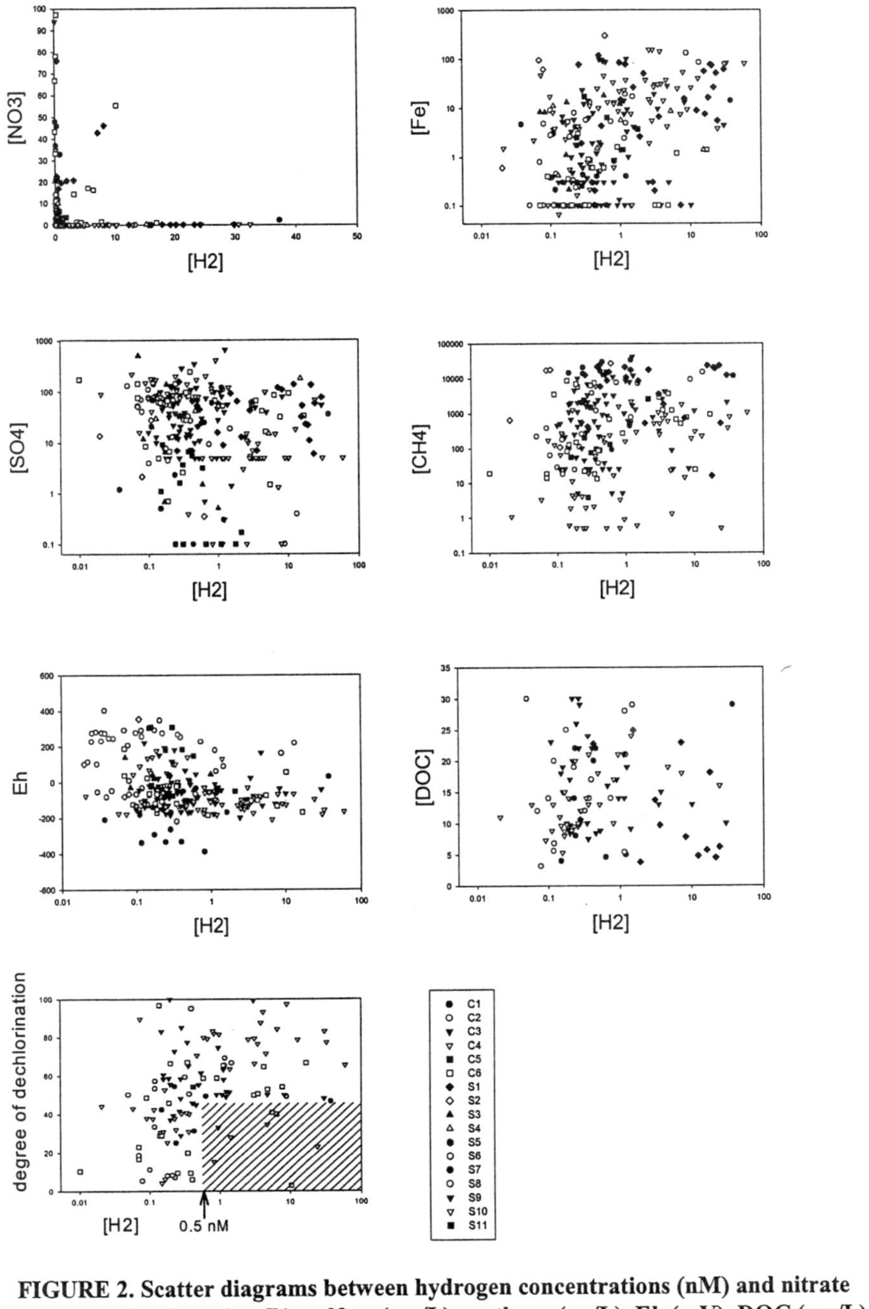

FIGURE 2. Scatter diagrams between hydrogen concentrations (nM) and nitrate (mg/L), divalent iron (mg/L), sulfate (mg/L), methane (μg/L), Eh (mV), DOC (mg/L) and the degree of dechlorination (%).

A factor analysis has been carried out to find underlying relationships in the data set and to understand the behavior of redox species. The maximum variance is determined and used to order data into multiple dimensions and interpret these as influential factors. These factors can be geochemical, biological or physical processes. In table 2 the results of the factor analysis is shown.

TABLE 2. Results of the factor analysis, showing the factor loadings.

Variable	Factor 1	Factor 2	Factor 3
Hydrogen	.32	.15	**-.60**
Temperature	-.29	**-.82**	.01
PH	.41	**.77**	.17
Redox Potential	**-.73**	.19	.39
Oxygen	-.43	.08	**-.65**
Nitrate	-.45	.53	.46
Iron (II)	**.87**	-.01	-.16
Sulfate	**-.69**	.25	-.37
Bicarbonate	**.66**	**.66**	.07
Methane	.31	**-.71**	.36
Degr dechlorination	**.69**	-.32	.01
DOC	**.90**	-.01	.06
Expl.Var	4.37	2.74	1.50
Prp.Totl	.36	.23	.12

The results show that three factors explain over 70% of the variance in the data set. The first factor explains 36%, the second 23% and the third 12%. For each factor relative strong factor loadings are found. Based on these factor loadings, the meaning of the three factors can be determined. The factors are explained by the following processes:

– **Factor 1: Redox**

The positive and negative factor loadings are interpreted to be related to redox processes. Parameters having a positive factor loading (such as divalent iron, DOC and dechlorination) increase upon decreasing redox conditions. The parameters with a negative factor loading (such as redox potential and sulfate) show a decrease.

– **Factor 2: Carbonate equilibrium**

Based on the variables with relative high factor loadings, the second factor is interpreted to be carbonate equilibrium. This equilibrium is associated with the bicarbonate concentration, the pH and the temperature. The methane concentration is indirectly influenced via the redox equilibrium with bicarbonate. Other parameters do not seem to be affected by this factor.

– **Factor 3: Gas-factor**

A third factor is associated with hydrogen and oxygen concentrations. Both are present a gas dissolved in the groundwater and can be volatilized. Other variables are not influenced by this factor.

In the first place, the behavior of geochemical parameters seems to be governed by redox processes. Additionally, carbonate equilibrium and volatilization seem to play an important role for some parameters. About 30% of the variance in the data set could not be explained and is probably the result of other processes, location specific factors or bias in measurements.

CONCLUSIONS

- The analysis of field data suggest that no strong correlations exist between hydrogen and most (macro-chemical) redox parameters. This is explained by differences in behavior. Hydrogen reflects the conditions at one particular point in space and time, whereas the concentrations of most other parameters are caused by processes that have taken place in the past and up-flow from the place of sampling.
- The best correlations were found between hydrogen and oxygen, methane and the degree of dechlorination.
- At hydrogen concentrations above 0.5 nM, the conditions for dechlorination of chloroethenes can be considered as favorable in more than 85 % of the cases. Hydrogen can therefore be used as indicator for dechlorination.
- There is no single best method to identify redox conditions. Macro-chemical parameters give a spatial and temporal average of the redox conditions, while hydrogen concentrations reflect the exact in-situ conditions. Depending on the type of information that is required, one of the two methods is preferred, or both.
- Factor analysis showed that the behavior of geochemical parameters is governed by redox processes (36%), carbonate speciation (23%) and volatilization (12%). 29% of the variance could not be explained. Interpretation of redox conditions based on geochemical parameters is strongly biased and complicated by other factors and processes.

REFERENCES

Chapelle, F.H., D.A. Vroblesky, J.C. Woodward and D.R. Lovley. 1997. "Practical considerations for measuring hydrogen concentrations in groundwater". *Environmental science and technology 31*: 2873-2877.

Hoehler, T.M., M.J. Alperin, D.B. Albert and C.S. Martens. 1998. "Thermodynamic control on hydrogen concentrations in anoxic sediments". *Geochimica et Cosmochimica Acta 62*: 1745-1756.

Jakobsen, R., H.-J. Albrechtsen, M. Rasmussen, H. Bay, P.L. Bjerg and T.H. Christensen, 1998. "H2-concentrations in a landfill leachate plume (Grindsted, Denmark): In situ energetics of therminal electron acceptor processes". *Environmental Science and Technology 32*: 2142-2148.

Lovley, D.R., and S. Goodwin. 1988. "Hydrogen as an indicator of the predominant terminal electron-accepting reactions in aquatic sediments". *Geochimica et Cosmochimica Acta 52*: 2993-3003.

HYDRAULICALLY FACILITATED REMEDIATION BY NATURAL ATTENUATION FOR GROUND WATER IMPACTED BY INORGANIC WOOD PRESERVATIVES

Jim Mueller (URS Corporation, Rolling Meadows, Illinois)
Michael Ritter (URS Corporation, Tampa, Florida)
Joanna Moreno (URS Corporation, Denver, Colorado)
Jeff Yang (URS Corporation, Cincinnati, Ohio)
Troy Hopper (URS Corporation, Houston, Texas)
Shannon Craig (Beazer East, Inc. Pittsburgh, Pennsylvania)
Jim Cook (Beazer East, Inc. Pittsburgh, Pennsylvania)

ABSTRACT: As a result of facility operations, soil and/or ground water at the Dantzler Lumber & Export Site in Jacksonville, Florida contain organic (pentachlorophenol) and inorganic (*e.g.*, boron and arsenic) constituents of interest (COI). Numerous ground water pumping configurations and remediation designs were modeled, including funnel-and-gate systems, a cut-off wall, and *in situ* permeable reactive barrier technology. Based on these results, we designed a hydraulically facilitated natural attenuation system for managing heavy metals in the dissolved phase. This design incorporated a ground water extraction system located along the longitudinal Site axis. The newly established hydraulic control forces ground water emanating from the source area to transverse the full extent of the Site. As such, the amount of time and space available for natural attenuation reactions was maximized. Optimization of the pumping was performed numerically with respect to the number, location, and pumping rates of the wells. Coupled with *in situ* source area reduction (air sparging), infiltration control (capping), and the processing of extracted ground water through an engineered wetland prior to final discharge, the hydraulically facilitated natural attenuation program represents a scientifically valid, effective, and cost-efficient remedial option that received regulatory approval for Site application.

INTRODUCTION

Arsenic is a naturally occurring ground water constituent, ranging in concentration from <1 to >1,000 μg/L in geothermal springs (Welch *et al.*, 2000). Inorganic arsenic in ground water is usually found as arsenite [As(III)] or arsenate [As(V)]. Under typical ground water conditions (*i.e.*, pH 6-8), the dominant arsenite species is neutral ($H_3AsO_3^0$; $pK_{a,1}$=9.20) and the negatively charged anions ($H_2AsO_4^-$ and $HAsO_4^{2-}$; $pK_{a,1}$=2.91 and $pK_{a,2}$=6.94) are the dominant arsenate species. Arsenic is also found naturally in coal, and as part of minerals such as arsenopyrite (Bhumbla and Keefer, 1994; Cullen and Reimer, 1989; Daus *et al.*, 1998). Arsenate and arsenite are subject to chemically and/or microbiologically mediated redox and methylation reactions. In theory, boron can also participate in these reactions, albeit to a lesser degree. In the oxidized state, dissolved arsenate (and perhaps boron) can form sulfide complexes with iron, nickel, and cobalt thereby precipitating out of solution.

Recently, the maximum contaminant level (MCL) for arsenic in ground water was reduced from 50 µg/L to 10 µg/L (U.S. EPA, 2001). The decision to reduce the MCL for arsenic was based largely on the report by Smith *et al.*, (1992) who found that the risk of developing cancer by drinking water containing 50 µg/L arsenic was one in 100. These findings are questionable given that arsenic concentrations exceeding 10 µg/L are widespread and common (Welch *et al.*, 2000). Nevertheless, this change in arsenic regulation has made it necessary to re-evaluate remedial strategies at a number of potentially impacted sites. A recent American Water Works Association study estimated the national compliance costs for a lower arsenic standard ranging from $330 million per year for a 20 µg/L MCL) to more than $4.1 billion per year for a 2 µg/L MCL (Frey *et al*, 1998). Thus, a safe, effective, and cost-efficient means of managing arsenic impacts is required.

Remediation by natural attenuation (RNA) incorporates a number of processes through which COI mass, volume, concentration, toxicity, and/or mobility are reduced to levels that do not pose a threat to human health or the environment, in a time frame that is reasonable compared to other remedial methods (U.S. EPA, 1999). Although it's recognized that RNA is potentially effective for managing sites impacted by inorganic COI, physical sorption and precipitation reactions represent the dominant RNA removal mechanisms for these constituents. As such, inorganic COI removal can be less rapid hence application of this remedial technology to sites impacted by heavy metals has been less popular than application to sites impacted by readily biodegradable organic COI (such as refined petroleum products).

MATERIALS AND METHODS

In determining whether RNA was a viable remedy for the Dantzler Site, the following investigations were performed:

- Review of Historical Water Quality Data: A review of historical water quality data was conducted to assess the migration and the distribution of concentrations of the COIs. Data from a number of monitoring points positioned along the longitudinal plume axis showed that the concentrations of the COIs are stable or decreasing as a function of space and time. However, off-site migration of arsenic and boron was evident. This was associated with the operation of two private, off-site irrigation wells.
- Laboratory Testing: Laboratory testing on aquifer solids was conducted to evaluate the ability of saturated zone solids to adsorb the COI and to determine site specific distribution coefficients. Testing confirmed that upon oxygenating the ground water to promote ferrous iron oxidation to ferric iron, water-soluble arsenic could become immobilized as an insoluble ferric arsenate. Albeit to a lesser extent, a similar reaction can occur between iron and boron, leading to the adsorption of boron on precipitated ferric hydrous-oxides.
- Pilot Testing: A ground water pilot test was conducted to demonstrate the use of air sparging to promote the chemical fixation of boron and arsenic, and the biological degradation of pentachlorophenol. Seven sparging points and

eleven monitoring wells were installed for the pilot test program. The effectiveness of the air sparging was documented by collecting ground water samples before, during, and after the testing to assess the changes in COI concentrations.

- Numeric Ground Water Modeling: A ground water extraction system was designed and optimized by numerical modeling in order to facilitate the RNA processes and keep the COI plume from migrating off-site. The United States Geological Survey (USGS) models MODFLOW and MODPATH were employed to simulate ground water flow conditions at the site and design and assess a ground water extraction system.

Site Description. Dantzler Lumber & Export Company operates a wood preserving facility in Jacksonville, Florida. Since 1982, Dantzler utilized chromated copper arsenate (CCA) to pressure-treat lumber; boron was used historically as a fire retardant. Prior to 1982, the Koppers Company, Inc. conducted wood treating operations at the Site. As a result of facility operations, soil and/or ground water at the 16-acre Site contain organic (pentachlorophenol) and inorganic (*e.g.*, boron and arsenic) COI. In 1987, a Consent Order was issued by the Florida Department of Environmental Protection to assess and remediate soil and ground water.

Site Hydrogeology. The surficial aquifer beneath the Site can be subdivided into four intervals based on variations in lithology, aquifer media characteristics, and depth (Healy, 1982; Krause and Randolph, 1989; Miller 1986). These aquifer intervals included the shallow (S), deep (D), bedrock interface (B) and bedrock (BR) sections. Monitoring wells installed to evaluate each of these intervals were previously completed with slotted well screens placed at depths of approximately 5, 14, 18, and 26 m below ground surface (bgs), respectively. Based on measurements at 33 S wells, 14 D well, 5 B wells, and three BR wells, ground water generally flows east/northeast beneath the site, with local variations due to the operation of on-site production wells and off-site, private irrigation wells. Flow at the B level (about 18 m bgs) is to the east and southeast. Horizontal hydraulic gradients vary little with depth, and range from 0.005 (D zone) to 0.0138 (BR zone), with a steeper gradient (0.022) in the east part of the site in the D zone. Small downward hydraulic gradients were observed in the nested wells. Slug test results performed in selected wells in all four zones give the following estimates for hydraulic conductivity: S zone = 0.4 to 3.4 m/day; D zone = 0.1 to 0.9 m/day; B zone = 0.27 m/day; BR zone = 0.06 m/day. An average conductivity of 0.9 m/day was used in modeling and design efforts. Vertical conductivities were not measured, and aquifer storage parameters were not evaluated. The 16-acre site is relatively flat, sloping from 18 m above mean sea level (amsl) in the southwest corner of the site to 11 m amsl in the northeastern corner. Over 76% of the site is paved, with surface water draining to the treating area sump (western part of site), the stormwater detention basin (eastern part of site), or to roadside ditches that flow to an unnamed tributary 457 m east of the site. The stormwater detention basin recharges ground water, except during larger storms when it

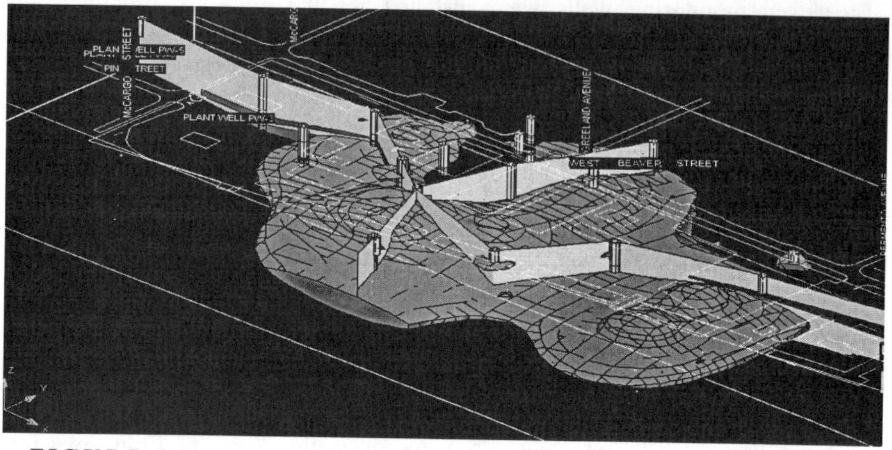

FIGURE 1. COI impacts prior to *in situ* remediation efforts.

overflows to the roadside ditch. Rainfall averages 130 cm annually. About 25 to 50 cm regionally percolates into the ground water.

COI Plume Characteristics. The concentrations of arsenic and boron in the shallow "S" monitoring wells exceeded the 1999 State of Florida ground water standards of 50 μg/L for arsenic and 630 μg/L for boron in an area that extends slightly north of Beaver Street and south of the property boundary (Figure 1). No off site impacts of penta were observed. Deep wells containing elevated COI were limited to the interior of the existing property.

RESULTS AND DISCUSSION

Air Sparging for Source Reduction. Shallow ground water at the site, particularly in the core of the arsenic plume, has a naturally low pH (4.5 to 6) and reduced redox conditions (-50 to –150 Mv). As such, metal arsenic exists in the reduced trivalent state (As^{+3}), which is soluble in water. Whether soluble As^{+3} is in the form of cations or complexed is not known. Pilot tests of air sparging were conducted to determine if the concomitant increase in the concentration of dissolved oxygen (DO) and oxidation-reduction potential of ground water could cause a decrease in the concentration of arsenic (and perhaps boron) within the effective radius of the sparging system.

A decrease in the amount of arsenite was observed during the pilot test, but absolute changes in oxidation conditions were marginal. It was speculated that arsenic precipitated in the form of mineral scordite ($FeAsO_4 \bullet 2H_2O$). In our opinion, however, it was more likely that the arsenic concentration was reduced as the consequence of co-precipitation of arsenate with iron, which was also converted through air sparge from soluble Fe^{2+} to insoluble Fe^{3+}. There was no discernible change in boron concentration.

Using change in DO concentration of ground water, air pressure in observation wells of unsaturated interval, and detection of helium used as a tracer gas, the radius of influence was determined to be approximately 40 feet under given operating pressures of the pilot tests. The short duration of the air sparge operation was considered responsible for the lack of significant and sustained change of environmental conditions. Nonetheless, *in situ* aeration was shown to have a positive affect on reducing the concentration of soluble arsenic.

Numerical Modeling. A three-dimensional numerical model (MODFLOW) was prepared and calibrated to the Site. Residual distribution contour maps for the S, D, and B Zones were prepared by subtracting the simulated water levels from the measured water levels. The residual distribution maps for the steady-state calibration simulation for the S and D zones generally exhibited low residuals for the middle portion of the site where the plumes and the simulated remediation system are located (Figure 1). These two zones (S and D) are where the bulk of the arsenic and boron reside, and thus were the focus of the conceptual remedial designs.

Conceptual designs numerous remedial alternatives were modeled:

- Ground water collection trench and sump pump;
- Ground water collection trench and a sheetpile wall downgradient of the collection trench;
- Three ground water collection wells at various locations downgradient of the suspected source area (eastern property edge);
- Five ground water collection wells at various downgradient locations;
- *In situ* permeable reactive barriers (zero valent iron) to enhance arsenic precipitation; and
- Hanging barrier wall at various depths and lengths.

The purpose of these designs was to support the *in situ* source reduction air sparge system and:

- Capture or contain most of the particles traveling from the source areas and across the full extent of the observed ground water plume while minimizing the overall pumping rate and barrier wall dimensions;
- Minimize the disturbance to the current ground water flow pattern responsible for the on-going and effective on-site attenuation of the arsenic and boron plumes; and
- Fulfill the above criteria for both the historical ground water flow pattern that may have prevailed during the development of an off-site plume to the north of the site, and the current ground water flow pattern (no nearby off site pumping).

Remedial Design: Particle tracking utilizing MODPATH showed that the preferred remedial system entailed a series of five ground water extraction wells located downgradient of the suspected source area (**Figure 2**). The location and number of pumping wells were optimized to provide capture with the minimum overall pumping rate. These wells are 16.7 m deep (assuming ground elevation of +15 m) and screened from +13.7 m to –1.5 m. The combined pumping rate of the wells will be approximately 19 L per minute (5 gpm) to capture 99% of the impacted water (Table 1) without operation of the two off-site irrigation wells (east and west). If these wells were allowed to operate, the capture would be less complete (**Figure 3**).

TABLE 1. Comparative Analysis of Various Systems for Plume Containment.

Alternative Remedial System	Percent Captured [a]
Ground Water Collection System	99
Ground Water Collection System and Off- Site Pumping at East Well	74
Ground Water Collection System and Off-Site Pumping at West Well	98
Ground Water Collection System and Barrier Wall	99
Ground Water Collection System and Barrier Wall With Off-Site Pumping at East Well	88
Ground Water Collection System and Barrier Wall With Off-Site Pumping at West Well	98

[a] Refers to percentage of particles captured

CONCLUSIONS

The recently approved ground water MCL for arsenic (10 μg/L) has increased the need for effective, cost-efficient *in situ* strategies for managing arsenic. For the Dantlzer Site, a ground water model utilizing MODFLOW was constructed to simulate the ground water flow under various conditions. Based on field data, the model was calibrated under steady-state flow conditions and subsequently utilized to select an *in situ* remedial strategy alternative to hydraulic containment via pump-and-treat. This included:

- Air sparging for source reduction;
- Infiltration control (partial capping);
- Termination of nearby off-site irrigation wells;
- Ground water collection to manage ground water flow and enhance natural attenuation processes for arsenic (*i.e.*, hydraulically facilitated remediation by natural attenuation); and
- An engineered wetland to treat boron in extracted ground water prior to final off-site discharge.

The hydraulically facilitated RNA program represents a scientifically valid, safe, effective, and cost-efficient remedial option for arsenic in ground water, and was approved by the State of Florida for Site-wide application.

FIGURE 2. Five-well ground water collection system for hydraulically facilitated RNA (no off-site pumping of private irrigation wells).

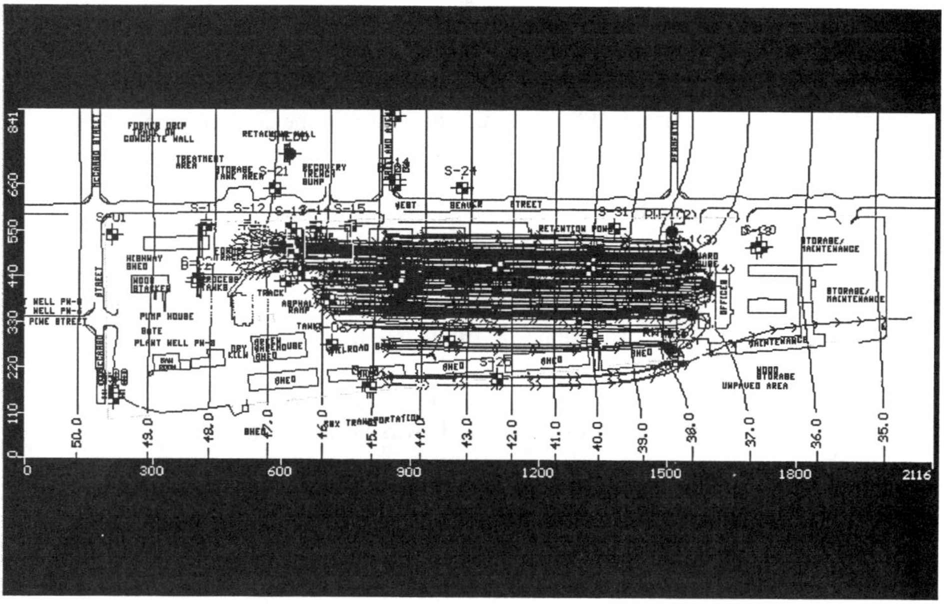

FIGURE 3. Five-well ground water collection system with off-site private irrigation well pumping at two locations.

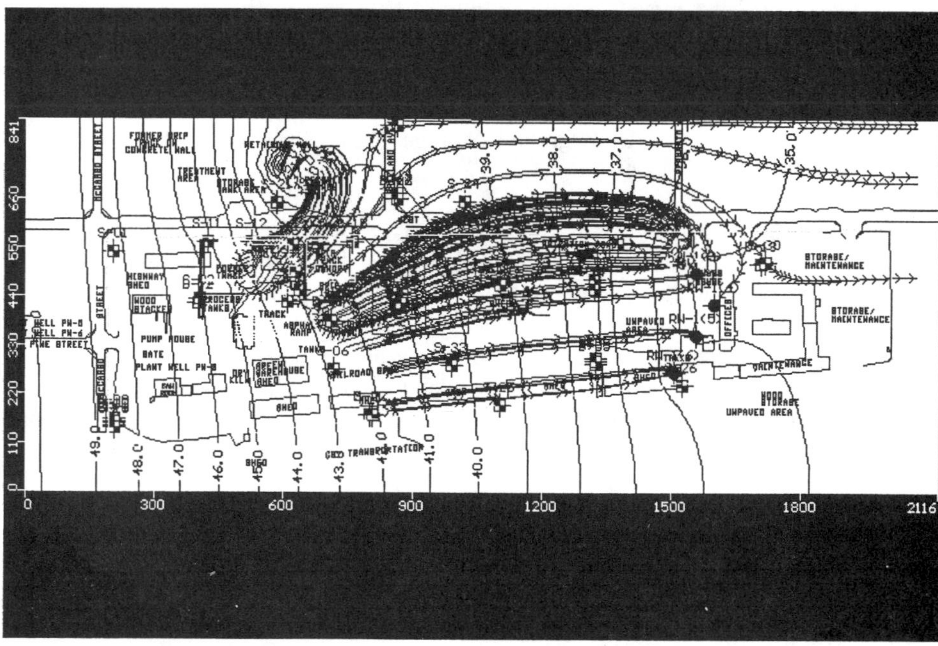

REFERENCES

Bhumbla, D. and R. Keefer. 1994. Arsenic mobilization and bioavailability in soils in *Arsenic in the Environment, Part I: Cycling and Characterization*, J. O. Nriagu, Ed. New York: John Wiley & Sons, p. 62-66.

Cullen, W. and K. Reimer. 1989. Arsenic speciation in the environment. *Chem. Rev.* 89: 713-764.

Daus, B., H. Weib, and R. Wennrich. 1998. Arsenic speciation in iron hydroxide precipitates. *Talanta* 46: 867-873.

Frey, M., D. Owen, Z. Chowdhury, R. Raucher, and M. Edwards. 1998. Cost to utilities of a lower MCL for arsenic. *J. Am. Water Works Assoc.* 90**:** 89-102.

Healy, H.G., 1982. Surficial and Intermediate Aquifers, Part 4: *in* Bernard J. Franks, ed., Principal Aquifers in Florida. USGS WRI Report 82-255i.

Krause, R.E., and R.B. Randolph, 1989. Hydrology of the Floridian Aquifer System in Southeast Georgia and Adjacent Parts of Florida and South Carolina. USGS Professional Paper 1403-D.

Miller, J.A., 1986. Hydrogeologic Framework of the Floridian Aquifer Systems in Florida and Parts of Georgia, Alabama, and South Carolina. USGS Professional Paper 1403-B.

Smith, A., Hopenhayn-Rich, C., M. Bates, H. Goeden, I. Hertz-Picciotto, H. Duggan, R. Wood, M. Kosnett, and M. Smith. 1992. Cancer risks from arsenic in drinking water. *Environ. Health Perspect.* 97: 259-267.

U.S. EPA. 1999. Use of Monitored Natural Attenuation at Superfund, RCRA Corrective Action, and Underground Storage Tank Sites. OSWER Directive 9200.4-17P. April 21, 1999.

U.S. EPA. 2001. Final Rule – National Primary Drinking Water Regulations; Arsenic and Clarifications to Compliance and New Source Contaminants Monitoring. 40 CFR Parts, 9, 141 and 142. Pages 6975-7066.

Welch, A., D. Westjohn, D. Helsel, and R. Wanty. 2000. Arsenic in Ground Water of the United States: Occurrence and Geochemistry. Ground Water 38:589-604.

MICROBIAL POPULATIONS AND MOISTURE CONTENT IN A CONVENTIONALLY OPERATED MUNICIPAL SOLID WASTE LANDFILL

Wendy J. Davis-Hoover (U.S. EPA, Cincinnati, OH ,USA), ***David A. Carson*** (U.S. EPA, Cincinnati, OH ,USA) David J. Slomczynski (U. of Cincinnati, Cincinnati, OH, USA), David Feldhake (U. of Cincinnati, Cincinnati, OH ,USA), Edith Holder (U of Cincinnati, Cincinnati, OH, USA), Gary R. Hater (Waste Management Inc., Cincinnati, OH, USA), Roger B. Green (Waste Management Inc., Cincinnati, OH, USA), John F. Martin (U.S. EPA, Cincinnati, OH, USA) and Rathi G. Kavanaugh (U. of Cincinnati, Cincinnati, OH, USA),

ABSTRACT: Landfills are not sterile environments. In fact it can be argued that they are the ultimate reactors for biodegradation as they contain nutrients, bacteria, and various redox conditions which, then, change over time. Enhancement of the landfill environment to optimize the rates of biodegradation and to ensure more rapid stabilization of the waste mass and availability of land fill volume needs to be developed. Research has progressed from laboratory to pilot to full-scale landfill studies. As part of a Cooperative Research And Development Agreement, Waste Management, Inc. and the United States Environmental Protection Agency's Office of Research and Development are studying bioremediation in full scale municipal solid waste landfills operated conventionally and as bioreactors over time and developing methods to monitor the degradation to ensure optimum operation. The microbial enumeration of harvested wastes landfilled for approximately five years conventionally in Unit 5, cell 1, coring 23 in the Outer Loop Landfill in Louisville, KY has been studied. The results of these analyses indicate that there is no statistically significant correlation between moisture content (47 to 65%) of the wastes and depth in the landfill (r^2 = 0.027), numbers of total aerobic (r^2 = 0.085), total anaerobic (r^2 = 0.179), sulfate reducing (r^2 = 0.185) and denitrifying bacteria (r^2 = 0.048). There is also no statistically significant correlation between the numbers of sulfate reducing and denitrifying bacteria (r^2 = 0.018). After addition of nitrified leachate to the landfill cell, it will be determined if this retrofit bioreactor demonstrates increased moisture content and numbers of bacteria.

INTRODUCTION.

Landfill material is composed of complex and heterogenous substances. The decomposition of this heterogenous waste involves the activity of many groups of bacteria. The decomposition process and the microbes involved have been studied by several researchers (Barlaz et al., 1989; Bogner, 1990; Donnelly, 1983; Donnelly and Scarpino, 1984; Jones et al., 1983). In a conventional landfill aerobic conditions initially prevail giving rise to aerobic decomposition. With the exhaustion and utilization of oxygen, anaerobic decomposition then occurs. There is a shift in the activity of functional groups of microorganisms involved in decomposition based on the availability of oxygen and nutrients necessary for their function.

Studies have shown that the decomposition of refuse is enhanced by recirculating the leachate generated (Barlaz et al., 1992, Pohland and Kim, 1999). Research in recent years has thus focused on the conversion of conventional landfills to "active" bioreactors by recirculation of leachate for faster stabilization. Some of the cells in the Outer Loop Landfill in Louisville, KY are in the process of being converted to a landfill bioreactor system for more investigation.

In June of 2000, Waste Management Inc. and U.S. EPA took waste samples from Unit 5, cell 1, coring 23 in the Outer Loop Landfill in Louisville, KY. This cell is in the process of being converted to a landfill bioreactor but is currently conventional. The samples received were analyzed for their microbiological and chemical characteristics. This analysis will represent a pre-conversion (to an "active" landfill bioreactor) survey of microbial and chemical characteristics for the landfill. The waste has been buried for about five years.

Objective. The objective of this paper is to survey microbiological populations in solid waste from a municipal landfill and compare them to the moisture content of the waste sample. It is expected that there is a direct relationship between moisture content and numbers of bacteria. A relationship between depth of the sample in a landfill and moisture content of the waste is not expected.

MATERIALS AND METHODS.

Sample Collection. Waste samples were collected by Waste Management Inc. (after augering into the waste mass with an about 1 meter (3 ft.) diameter bucket auger) and the US EPA by placing them in ziplock bags under nitrogen gas. Waste was collected from about 1 to 21 meters (3 to 63 ft). The first sets of samples were collected from Unit 5 at the Outer Loop Landfill, Louisville, Kentucky in June of 2000. Future sampling events are planned. Chain of custody was performed. The weight of each sample was taken upon arrival into the US EPA Center Hill Laboratory.

Sample Preparation. The waste was transferred to a mixer in one batch. A 1:1 ratio of a potassium phosphate buffer was added to the sample and blended for 60 seconds under nitrogen gas in a chemical hood. The resultant liquid was extracted and filtered using a conical shaped, stainless steel filtration device under nitrogen. Dry weights were determined on the solids. The filtrate was stored under nitrogen gas in polypropylene bottles and used for the microbial and chemical analyses.

Sample Analysis. The ***total anaerobic bacterial*** counts were determined using anaerobe agar (DIFCO) plates incubated at 37° C in GasPak jars after plating and diluting in an anaerobic chamber. Plate counts of ***total aerobic heterotrophic bacteria*** were determined using R2A agar (DIFCO) and a spiral plater (Biotec CASBA™). The plates were incubated at 37° C. The optimum incubation time was determined by running validation experiments.

Sulfate reducing bacteria were enumerated by MPN (Most Probable Numbers) techniques using a BIOMEK 2000 workstation. Microtitre plates (11 dilutions with 8 wells per dilution) inoculated with the landfill extract were incubated in a GasPak under anaerobic conditions. Incubation time was experimentally determined. The medium was as follows: yeast extract 1.0g, sodium lactate 4.0g, NH_4Cl 0.5g, KH_2PO_4 1.0 g, $MgSO_4 \bullet 7H_2O$ 0.2g, $CaCl_2 \bullet 2H_2O$ 0.1g, $FeSO_4 \bullet 7H_2O$ 0.1g, Na_2SO_4 0.5g (*Manual of Methods in Microbiology*, 1981).

Denitrifying bacteria were determined by MPN using a BIOMEK 2000 workstation and incubated in GasPak jars at 35° C. Incubation time was experimentally determined. The presence or absence of nitrate and nitrite was used to score the wells for denitrification. The medium used was sodium acetate 6.56g, added to Nitrate mineral salts medium (Atlas and Parks., 1997) $MgSO_4 \bullet 7H_2O$ 1.0g, KNO_3 1.0g, $Na_2HPO_4 \bullet 12H_2O$ 0.717g, KH_2PO_4 0.272g, $CaCl_2 \bullet 6H_2O$ 0.2g, ferric ammonium EDTA 4.0 mg, Trace element solution [Disodium EDTA 0.5g, $FeSO_4 \bullet 7H_2O$ 0.2g, H_3BO_3 0.03g, $CoCl_2 \bullet 6H_2O$ 0.02 g, $ZnSO_4 \bullet 7H_2O$ 0.01g, $MnCl_2 \bullet 4H_2O$ 3.0 mg, $Na_2MoO_4 \bullet 2H_2O$ 3.0mg, $NiCl_2 \bullet 6H_2O$ 2.0mg, $CaCl_2 \bullet 2H_2O$ 1.0 mg (Atlas and Parks., 1997)] 0.5mL, in 1 liter of the solution.

Dry weights of the solids were obtained by drying them at 103-105°C until two consecutive constant weight readings were obtained (based on, method SM 2540, *Standard Methods for the Examination of Water and Wastewater*, 1989).

RESULTS AND DISCUSSION.

Figure 1 shows that there is no correlation between the moisture content of the

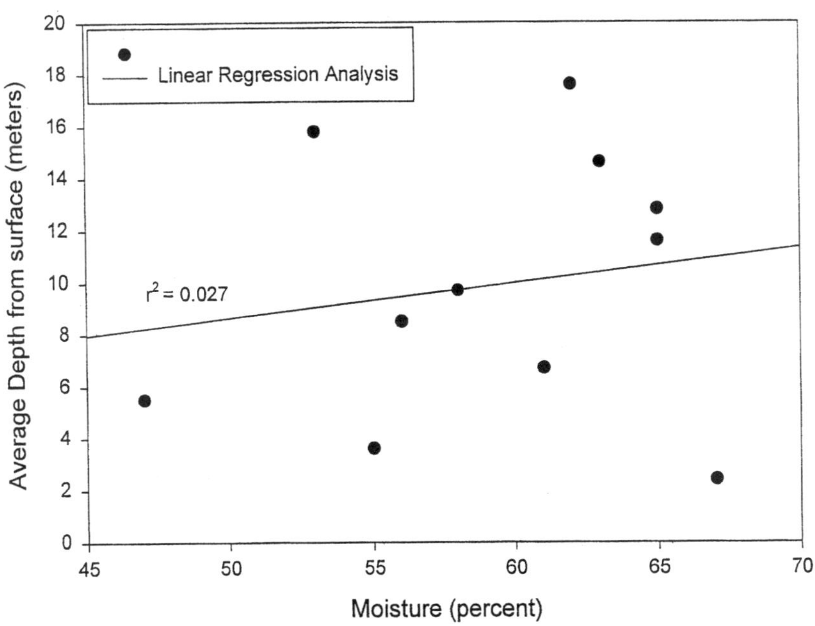

FIGURE 1. Percent moisture versus average depth in landfill

waste samples from a conventional landfill, and the depth of sampling. The coefficient of determination is 0.027.

Figure 2 shows the relationship of moisture content of the samples to numbers of total aerobic and anaerobic bacteria. In this conventionally operated landfill cell there is no statistically significant correlation between moisture content and aerobic bacteria ($r^2 = 0.085$) and anaerobic bacteria ($r^2 = 0.179$) .

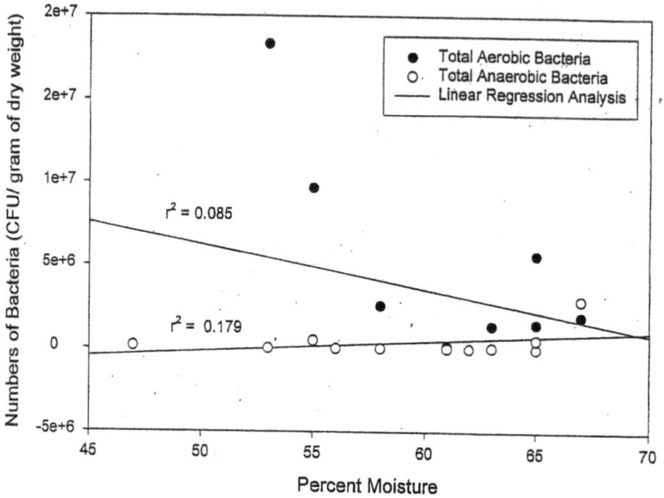

FIGURE 2. Percent moisture versus total aerobic and anaerobic bacteria

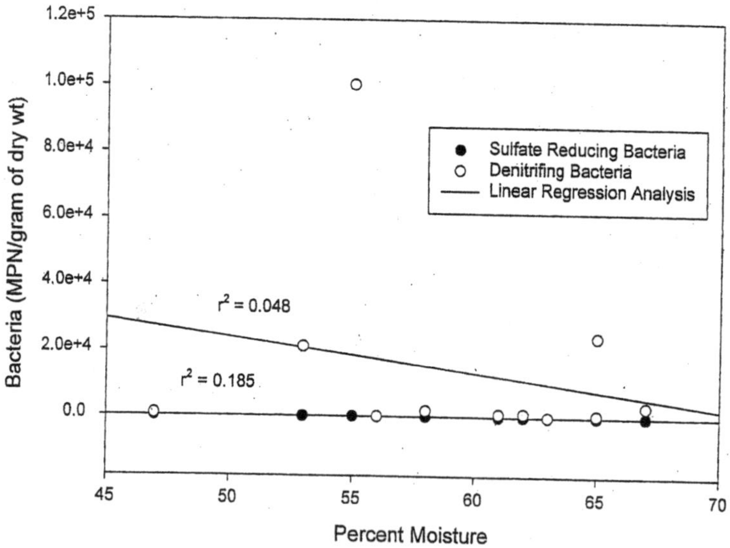

FIGURE 3. Percent moisture versus sulfate reducing and denitrifying bacteria

Figure 3 shows the relationship between the numbers of sulfate reducing bacteria and denitrifying bacteria and moisture content. Figure 4 shows the relationship between the numbers of the two groups of bacteria. This conventionally operated landfill cell shows no statistically significant correlation between moisture and denitrifying bacteria or sulfate reducing bacteria. (r^2 = 0.185). There is no statistically significant correlation between the sulfate reducing and denitrifying bacteria in this cell.

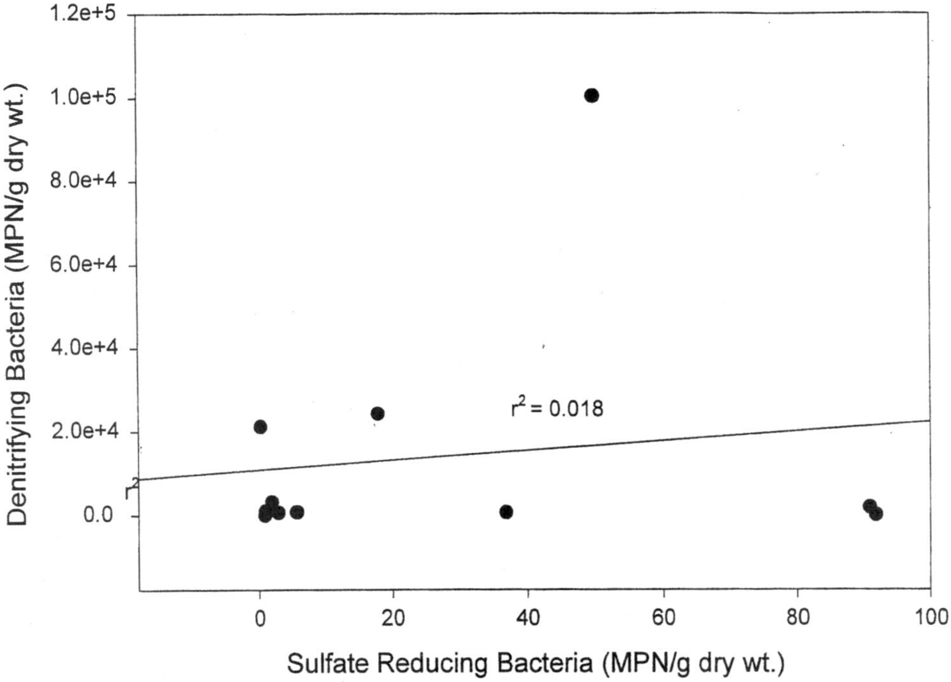

FIGURE 4. Sulfate reducing and denitrifying bacteria

CONCLUSIONS.

At the observed range of moisture content (47 to 65%), in this conventionally operated five year old cell of a landfill, there is no statistically significant correlation between moisture content (47 to 65%) of the solid waste samples and depth of the sampling point in the landfill (r^2 = 0.027). Also no statistically significant correlation was found between moisture content and numbers of total aerobic (r^2 = 0.085), total anaerobic (r^2 = 0.179), sulfate reducing (r^2 = 0.185) and denitrifying bacteria (r^2 = 0.048). There is also no statistically significant correlation between the numbers of sulfate reducing and denitrifying bacteria (r^2 = 0.018). This cell will be modified by adding nitrified leachate. Future studies will determine if there are any correlations between the moisture content of the waste and the microbial counts change.

ACKNOWLEDGMENTS.

The authors would like to thank J. L. Goetz and L, H. Wetzel for their assistance in sample collection and analysis and A. C. Chance, G. E. Kesterman and J. K. Jackson for sample preparation and analysis.

REFERENCES

Atlas., R.M and Parks, L.C.(Eds.).1997. *Handbook of Microbiological Media* . CRC press.

Barlaz, M.A., Ham, R.K., and Schaefer, D.M.. 1992. Microbial, Chemical and Methane Production Characteristics of Anaerobically Decomposed Refuse with and without Leachate Recycling." *Waste Management and Research 10*:257-267.

Barlaz, M.A., Schaefer, D.M., and Ham, R.K.. 1989. "Bacterial Population Development and Chemical Characteristics of Refuse Decomposition in a Simulated Sanitary Landfill". *Appl. Environ. Microbiol. 55*: 55-65.

Bogner, J.E., 1990. "Controlled Study of Landfill Biodegradation Rates Using Modified BMP Assays. " *Waste Management and Research 8*:329-352.

Donnelly, J. A. . 1983. " Isolation, Characterization and Identification of Microorganisms from Laboratory and Full-Scale Landfills." Ph.D. Dissertations, University of Cincinnati, Cincinnati, OH.

Donnelly , J. A., and P.V. Scarpino. 1984. *Isolation and Characterization and Identification of Microorganisms from Laboratory and Full Scale Landfills.* EPA Project summary. EPA 600/S2-84-119. U.S. Environmental Protection Agency. Washington, D.C.

Jones, K.L., Rees, J.F., and Grainger, J.M.. 1983. "Methane Generation and Microbial Activity in a Domestic Refuse Landfill Site." *Eur J Appl Microbiol Biotechnol 18*:242-245.

Manual of Methods in Microbiology 1981. American Society for Microbiology

Pohland, F.G., and Kim, J.C. 1999. "In situ Anaerobic Treatment of Leachate in Landfill Bioreactor." *Wat. Sci. Tech 40*(8): 203-210.

Standard Methods for the Examination of Water and Wastewater 1989. American Public Health Association and American Water Works Association and Water Pollution Control Federation

THE EFFECTS OF PERMANGANATE OXIDATION ON SUBSURFACE MICROBIAL POPULATIONS

Julia Klens, Dirk Pohlmann, Shirley Scarborough, Duane Graves, Ph.D
(IT Corporation, Knoxville, Tennessee, USA)

ABSTRACT: This paper describes the application of potassium permanganate at a site impacted with TCE and the initial observed effects on the subsurface aerobic and anaerobic microbial populations. In order to determine the effect of permanganate on subsurface microbial populations, soil and groundwater samples were collected from a site prior to, during, and after the application of potassium permanganate. The microbial populations in the soil and groundwater at each stage of the chemical oxidation process were examined. The general classes of microbial populations that are typically examined during a natural attenuation evaluation were measured. These included aerobic heterotrophs, anaerobic heterotrophs, iron reducing, nitrate reducing, sulfate reducing and methanogenic microbial populations. The objective of this test was to measure the effect of potassium permanganate ($KMNO_4$) on microbial viability and the potential for recovery.

INTRODUCTION

Monitored natural attenuation is used as an effective remedial technology at sites where source control measures have been implemented. Elimination of the contributing source of contamination at a site is critical to the effective use of monitored natural attenuation for site remediation. In-situ chemical oxidation with potassium permanganate or sodium permanganate is an emerging technology for the treatment of organic contaminants, especially chlorinated ethenes. Chemical oxidation is a popular technology because of the capability to destroy high concentrations of contaminants in-situ while generating relatively harmless byproducts. Permanganate is an effective oxidizing agent and is more stable and persists longer in the subsurface than stronger oxidizers. Coupling chemical oxidation with intrinsic bioremediation is attractive because oxidation can be used to reduce or remove source areas while intrinsic biodegradation can be used to treat residuals. The effectiveness of this approach relies on the ability of indigenous microbes to withstand exposure to permanganate.

According to the Office of Solid Waste and Emergency Response (OSWER) Directive, "monitored natural attenuation" is a term that refers to the use of natural attenuation processes to achieve site-specific remedial objectives within a time frame that is acceptable compared to the time frame offered by more active remedial technologies. At the majority of sites where monitored natural attenuation (MNA) is proposed as a remedial technology it is used as just one part of the total remedial strategy, either simultaneously or as a follow up measure. Typically the best use of monitored natural attenuation is at sites where source control measures have been implemented. Elimination of the contributing source of contamination at a site is critical to effective use of monitored natural attenuation.

Monitored natural attenuation is a remedial technology that could be used as a follow-up measure after the application of permanganate at a site. The biodegradative component of MNA is usually the most important natural attenuation process because intrinsic biodegradation is often the only means by which contaminant mass is actually destroyed. Therefore examining and understanding the effect of permanganate oxidation on subsurface microbial populations to ensure that a permanganate treated source area or site might then be fully remediated through the use of MNA is very important. Combining intrinsic biodegradation with permanganate treatment is a very attractive concept provided that indigenous microbes persist and resume biodegradative activity following chemical oxidation. However the effect of permanganate application on subsurface microbes has not been determined. Since permanganate oxidizes the subsurface, the activity and viability of anaerobic microbes may be effected. The oxidant may also be detrimental to aerobic microbes.

BACKGROUND

A chemical oxidation pilot test using potassium permanganate was conducted for the treatment of trichloroethene (TCE) and *cis*-1,2-dichloroethene (cis-1,2-DCE). The pilot test consisted of injecting approximately 11,000 gallons of 0.7 percent potassium permanganate solution into the subsurface in two batches over a two day time period. Prior to injection, immediately after injection, and at three and six month intervals, several groundwater quality parameters were measured including dissolved oxygen (DO), oxidation reduction potential (ORP), temperature, conductivity, and pH. Also prior to injection and immediately after injection soil and groundwater samples were collected, preserved on ice at 4 degrees Celsius and shipped overnight for microbial analyses. Groundwater samples only were collected at the three and six month intervals.

Because permanganate oxidizes the subsurface, changing the oxidation reduction potential of the soil and groundwater to a level likely to disrupt the anaerobic microbial processes that typically dominate biodegradation of chlorinated ethenes, it was important to measure and track ORP values in the groundwater in the treated area over time. Figure 1 illustrates the ORP values measured at the site over a six month period. Figure 2 illustrates the typical ORP values associated with respiration pathways. The effects of permanganate on other groundwater quality parameters critical to microbial activity were also examined. DO was lightly elevated in the treated groundwater; however, pH, specific conductivity, and temperature were not effected.

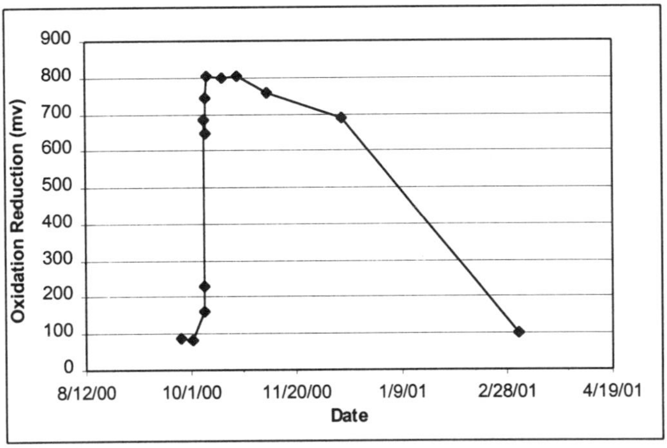

FIGURE 1. Site oxidation/reduction potential measurements over time.

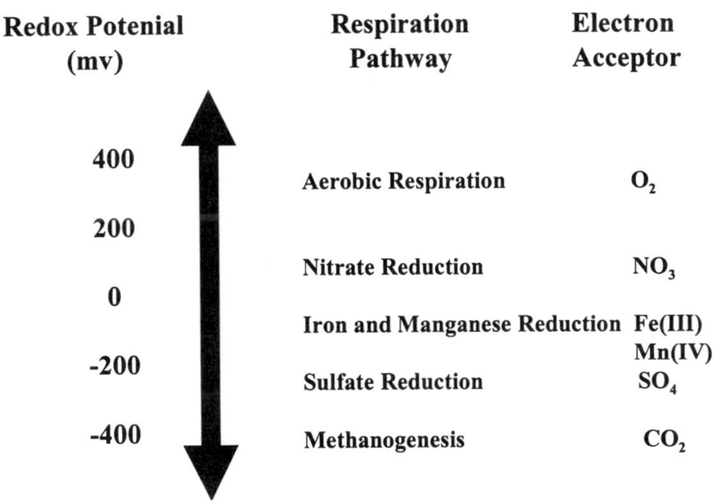

FIGURE 2. Electron acceptor for common microbial respiration pathways and approximate oxidation/reduction potential where each pathway occurs. (Atlas and Bartha, 1981)

MATERIALS AND METHODS

Site soil and groundwater samples were shipped to the laboratory within 24 hours of collection. The samples were analyzed for aerobic heterotrophic microbial populations, anaerobic heterotrophic microbial populations, nitrate reducing, iron reducing, and sulfate reducing microbial populations and methanogens. The aerobic heterotrophic microbial populations of the samples were measured using Standard Method 9215 C Spread Plate Method. Specific media designed for the selective growth of each reducing microbial population were used to determine population numbers. The most probable number (MPN) dilution technique was used for estimating the number of anaerobic microorganisms in the site soil and groundwater samples.

The media used for the anaerobic heterotrophs did not contain electron acceptors. The anaerobic microbes grown on this media represents other anaerobic microbes than the specific nitrate reducing, iron reducing, sulfate reducing and methanogen microorganisms.

RESULTS AND DISCUSSION

Results from the initial data from this study indicate that the naturally occurring subsurface microbial populations appeared to rebound after treatment with a 0.7 percent solution of potassium permanganate. The measured ORP value of the groundwater prior to permanganate treatment was less than 100 millivolts. Initial microbial enumerations indicated viable populations of aerobic and anaerobic microorganisms existed at the site. The first post treatment samples were collected two weeks after injection. These results indicated viable populations of anaerobic heterotrophs, nitrate reducers, sulfate reducers and methanogens were still present in the soil. Microbial enumerations of the groundwater from the same sampling point showed only a low population of methanogens still present. The ORP value of the groundwater two weeks post treatment was greater than 800 millivolts. Soil and groundwater that has effectively been saturated with potassium permanganate is colored a deep purple. The groundwater and soil collected from the site were deep purple as a result of the permanganate treatment and determined fully saturated.

Only groundwater was collected from the site at the three and six month intervals. Data from the three-month interval indicated that the nitrate reducing population was beginning to recover. No other viable microbial populations were identified at the three-month interval. Although methanogens were identified at low populations two weeks post injection, the three month post injection groundwater sample did not indicate the presence of viable methanogens. This variability is most likely due to experimental error in the methane analysis when dealing with such low counts. The ORP value of the groundwater at this sampling point was approximately 700 millivolts and the color of the groundwater was still purple. Six months after treatment with potassium permanganate the ORP value of the groundwater had declined to approximately 100 millivolts and the distinctive purple color was gone, the groundwater was clear. The population of aerobic heterotrophs at this time point had rebounded to 2,500,000 CFU/mL. Anaerobic data is not yet available from the six-month time point.

Aerobic Heterotrophic Microbial Enumerations. The pretreatment soil contained a microbial population of 140,000 colony forming units per gram (CFU/g) of dry soil. The pretreatment water sample contained 170 colony forming units per milliliter (CFU/ml) of groundwater. The immediate post treatment soil sample contained less than 100 CFU/g. Six months after potassium permanganate injection the heterotrophic microbial population in groundwater had rebounded to 2,500,000 CFU/ml.

Anaerobic Heterotrophic Population. The initial count for the anaerobic heterotrophic population was greater than 1100 CFU/g in the pretreatment soil and greater than 23 CFU/ml in the pretreatment groundwater. The immediate post treatment soil had a population of 240 CFU/g and the post treatment groundwater was less than 3 CFU/ml. Approximately 3 months after the permanganate treatment, the anaerobic heterotrophic population was still less than 3 CFU/ml in the groundwater.

Nitrate Reducing Population. The initial count for the nitrate reducing population was 46,000 CFU/g in the pretreatment soil and greater than 4,600 CFU/ml in the pretreatment groundwater. The immediate post treatment soil had a population of 23 CFU/g and the post treatment groundwater was less than 3 CFU/ml. Approximately 3 months after the permanganate treatment, the nitrate reducing population had recovered to 23 CFU/ml in the groundwater.

Iron Reducing Population. All of the samples, soil and groundwater, pretreatment and post treatment had an iron reducing population of less than 3 CFU per gram of soil or milliliter of groundwater. These microorganisms were not prevalent in the subsurface at the site.

Sulfate Reducing Population. The initial count for the sulfate reducing population was 1,100 CFU/g in the pretreatment soil and less than 3 CFU/ml in the pretreatment groundwater. The immediate post treatment soil had a population of 240 CFU/g and the post treatment groundwater was less than 3 CFU/ml. Approximately 3 months after the permanganate treatment, the sulfate reducing population was still less than 3 CFU/ml in the groundwater.

Methanogenic Population. The initial count for the methanogenic population was 23 CFU/g in the pretreatment soil and 21 CFU/ml in the pretreatment groundwater. The immediate post treatment soil and groundwater had populations of greater than 11 CFU/g and ml respectively. However, groundwater sampled 3 months after treatment was less than 3 CFU/ml.

Table 1 contains the microbial enumerations for each population at each time point to date.

The data to date indicate that application of a 0.7 percent potassium permanganate solution to the subsurface has an immediate effect on microbial populations and activity. Although the microbial populations declined, nitrate and sulfate reducing populations as well as methanogens persisted in the soil. Preliminary results indicate a rebound in the nitrate reducing population after three months in groundwater and in the aerobic heterotrophic population after six months. A final sampling event is scheduled for 9 months after $KMNO_4$ application. All of the microbial populations will be measured at this final time point in both soil and groundwater.

TABLE 1. Microbial enumerations and time point for each population measured.

Population	Pretreatment	Post (2 Weeks)	Post (3 months)	Post (6 months)
Aerobic Heterotrophs (soil)	140,000	<100	Not sampled	Not sampled
Aerobic Heterotrophs (groundwater)	170	Not Measured	Not Measured	2,500,000
Anaerobic Heterotrophs (soil)	>1100	240	Not Sampled	Not Sampled
Anaerobic Heterotrophs (groundwater)	23	<3	<3	<3
Nitrate Reducers (soil)	46,000	23	Not Sampled	Not Sampled
Nitrate Reducers (groundwater)	4,600	<3	23	Not available
Iron Reducers (soil)	<3	<3	Not Sampled	Not Sampled
Iron Reducers (groundwater)	<3	<3	<3	Not available
Sulfate Reducers (soil)	1,100	240	Not Sampled	Not Sampled
Sulfate Reducers (groundwater)	<3	<3	<3	Not available
Methanogenic Population (soil)	23	>11	Not Sampled	Not Sampled
Methanogenic Population (groundwater)	21	>11	<3	Not available

Soil is reported in CFU/g
Water is reported in CFU/ml

Available data at this interim point in the evaluation of permanganate application on subsurface microbes suggest a short term impact on microbial population density. However, treatment of the subsurface did not sterilize the soil and microbial populations appeared to recover after the oxidant dissipated. The preliminary results from this evaluation suggest that permanganate oxidation and natural biodegradation can be effectively coupled for site restoration provided other factors, such as inorganic nutrients, available respiratory substrates, or biodegradable organic carbon do not limit biodegradation following oxidation.

REFERENCES

Atlas, R.M., and R. Bartha, 1981. *Microbial Ecology: Fundamentals and Applications*, Addison-Wesley Publishing Co., Reading.

INCREASES OF ARSENIC IN GROUNDWATER DUE TO NATURAL DEGRADATION PROCESSES

James R. Chatham (BP Exploration, Anchorage, Alaska)
Paul Bitter (URS Corporation, Salt Lake City, Utah)
Mark Loucks (USAF, Salt Lake City, Utah)

ABSTRACT: Highly variable pH, arsenic, and manganese levels occur in shallow groundwater adjacent to a stormwater runoff holding system on Hill Air Force Base (Utah). Statistical and geochemical analyses of the data suggest that the elevated metal concentrations reflect biodegradation of natural organic carbon that infiltrates the pond sediment, resulting in metal oxide dissolution and a release of naturally occurring arsenic and manganese to the shallow groundwater. The elevated metal concentrations appear to be maintained via groundwater supersaturation with respect to manganese carbonate, and arsenic desorption at pH levels exceeding 7.5. After sampling and geochemical modeling were performed, the state agency and U.S. Environmental Protection Agency (USEPA) removed the pond from regulation as a National Priority List (NPL) site.

INTRODUCTION

Bamberger Pond is a stormwater runoff holding system on Hill Air Force Base, 30 miles north of Salt Lake City (Utah). The system consists of two parallel, unlined basins 50 to 100 feet wide and 600 to 800 feet long, that are connected with a culvert drain. The basins are approximately 8 to 10 feet deep with native vegetation (cattails, aquatic grasses) along the margins; natural deposits of organic debris (leaves, grass, twigs) are present in the pond sediment.

The pond is sited in unconsolidated sand and silty-sand strata. Groundwater beneath Bamberger Pond exists under unconfined conditions and ranges in depth from 10 feet below ground level (bgl) to 50 feet bgl. The regional groundwater flow is to the southwest. A clay-rich stratum at approximately 50 feet bgl restricts shallow groundwater recharge to the shallower sands. During rain events the pond fills rapidly and the groundwater is mounded due to infiltration through the pond. The magnitude of the mound influence on groundwater flow is based on the water level in the pond (URS, 2000).

Since 1989, highly variable pH, arsenic (As), and manganese (Mn) levels have been measured in shallow groundwater near the pond. Large variations in the concentrations with As and Mn occurred spatially between adjacent monitoring wells, and temporally within individual wells. In some instances, the metal concentrations were elevated above the maximum contaminant level (MCL) of 0.050 mg/L for both metals. In an attempt to determine the source of the As and Mn, groundwater monitoring and subsequent hydrologic and chemical investigations were performed at the site beginning in 1992. The metals in the past were presumed to be associated with activities in nearby buildings, including a paint shop and a former vehicle maintenance and repair facility. No source of anthropogenic contamination was identified during the investigations (URS, 2000).

SAMPLING AND ANALYTICAL METHODS

Quarterly- and monthly-based groundwater sampling events were performed in 1998 and spring, 1999 to determine if the metal variability and elevated concentrations were seasonally affected or associated with rainfall events. Samples of Bamberger Pond surface water and pond sediment were also collected and analyzed to supplement the source investigation. In 1998, an expanded suite of chemical analytes (e.g. redox, sulfide, alkalinity, iron, major cations and anions, and nutrients) was included in the quarterly monitoring program. The additional analytes were included to aid in identifying geochemical processes that were suspected to be active in the pond and shallow groundwater system.

Groundwater samples were collected using a low-flow submersible pump from 17 wells up-gradient and down-gradient of the pond The fluid pH, redox, temperature, dissolved oxygen, and conductivity were measured with a flow-through electrode system prior to atmospheric exposure. Conventional USEPA sampling, analytical, and quality control methods were followed (URS, 2000).

ANALYTICAL AND STATISTICAL RESULTS

Selected results from the expanded 1998 and 1999 monitoring program are summarized in Tables 1 and 2. Non-detectable levels of gasoline-range organic (GRO; Method 8015) concentrations, and very low levels of unidentified organic compounds (UIO; <0.76 mg/kg; from EPA Method 8015) and total benzene, toluene, ethylbenzene and xylenes (BTEX) (<0.250 mg/kg; Method 8020) occur in the pond sediment (Table 1). The low BTEX levels are presumed to reflect surface runoff from adjacent roads and parking lots. In contrast, the concentrations of total organic carbon (TOC; Method 9016) exceed 3.0% in some samples. The concentrations of As and Mn in the pond sediment are lower than local background levels, and the As and Mn concentrations in the pond water are well below the MCL and typical of shallow surface water.

TABLE 1. Summary of sediment data from 1998 and Spring, 1999 monitoring program.

Concentration (mg/kg)	N	Minimum	Maximum	Mean	Background Mean
BTEX[1]	30	0.023	0.224	0.070	NA
Gasoline (GRO)	30	ND	ND	ND	NA
UIO	30	0.350	0.761	0.604	NA
TOC shallow[2]	14	2,010	30,400	15,251	230
TOC deep[3]	16	1,450	70,000	13,485	230
As	30	2.3	9.4	4.8	7.5
Fe	30	3050	16,800	-	NA
Mn	30	23.5	187	73.7	200

[1]See text for acronyms and EPA method definition
[2]TOC shallow represents total organic carbon at 0.5 ft depth below the pond bottom.
[3]TOC deep represents total organic carbon at 3.0 ft depth below the pond bottom.
N = Number of samples; NA = Not Analyzed; ND = Not Detected

Several groundwater parameters listed on Table 2 contain minimum and maximum concentrations that range from two to over four orders of magnitude. The solubility of each of the three metals shown on Table 2 are greatly affected by variations in pH and redox potential (Schnoor, 1996; Stumm and Morgan, 1996). Redox levels in the shallow groundwater also are highly variable and range over 400 millivolts (mv). Hence, these measurements suggest that the shallow groundwater chemistry at Bamberger Pond is dominated by strong gradients in redox potentials (Eh) and pH, which subsequently affect the levels of As, Fe, and Mn.

TABLE 2. Summary of pond and groundwater data from 1998 and Spring, 1999 monitoring program.

Concentration (mg/L)	N	Minimum	Maximum	Mean	Std. Dev.
As (pond[1])	5	0.0012	0.0024	-	NA
Mn (pond[1])	5	0.020	0.039	-	NA
pH (gw[1])	107	7.02	9.44	8.02	0.636
Redox (mv; gw[2])	107	-205	225	65.6	82.3
As (gw[2])	105	<0.001	0.18	0.037	0.041
Fe (gw[2])	44	0.003	9.5	0.448	1.465
Mn (gw[2])	103	<0.0005	1.7	0.104	0.231
Ca (gw[2])	44	3.52	155	55.0	39.2

[1]pond = Pond surface water sample
[2]gw = Groundwater sample

The 1998-1999 data were statistically evaluated for significant trends. No significant difference in conductivity or Mn levels was observed at any location during the rainy periods. Arsenic concentrations were slightly higher during the rain events (80% probability level) immediately down-gradient (<100 ft) and up-gradient of the pond. During the rain events, groundwater pH was significantly lower down-gradient. A non-parametric Kruskal-Wallace ANOVA procedure was performed to evaluate seasonal differences in the groundwater chemistry. No significant seasonal differences were identified for As, Mn, pH, or Eh.. However, highly significant differences (95% probability level) were identified between individual wells for pH, Eh, conductivity, alkalinity, total organic carbon (TOC), As, Fe, Mn, calcium (Ca), magnesium (Mg), and potassium (K).

Three spatially consistent well groupings were identified through the Kruskal-Wallis mean ranks. A background (BKGD) grouping was observed that includes those wells that are greater than 250 feet and/or upgradient from the pond. Groundwater samples from these wells were found to have relatively low pH, As, Mn, and TOC concentrations and relatively high SO_4 concentrations.

Two geochemical zones were apparent in groundwater nearer to the pond. Wells that are generally down-gradient and within 100 feet of the pond (proximal; DWN-PROX) contained intermediate to elevated concentrations of As, and Mn, relatively high pH and TOC, and strongly depleted SO_4 levels. Wells down-gradient and distal to the pond but within its influence (DWN-DSTL; approximately 100 - 250 feet away) contained the highest pH, As, and Mn levels.

Summary statistics for each spatial grouping were recalculated by pooling the quarterly analyses from individual wells in each respective group. The distributions are graphically displayed as box plot diagrams in Figures 1 and 2.

Evidence of mixing between the surface (pond) water and shallow groundwater was evident in the TOC and Ca plots on Figures 1a and 1b, respectively. The mean concentration of TOC nearest the pond (DWN-PROX) was approximately three-fold higher than TOC levels in the background wells (3.28 vs. 1.03 mg/L, respectively), which is a significant difference at the 95% probability level. This statistic implies that TOC is infiltrating into the groundwater at Bamberger Pond. The Ca trend is similar to that of Mg, K, SO_4, and Cl, as referenced in Figure 1b. This distribution reflects simple dilution of the shallow aquifer by infiltrating pond water, and specific ion depletions by other geochemical mechanisms not addressed in this report.

The concentrations of As, pH, Fe, Mn, and Na follow yet another distribution (Figure 2). These parameters are moderately elevated in groundwater nearest the pond and highly elevated in the more distal wells. The mean pH of the distal class is 1.3 pH units higher than background. The mean arsenic concentration of the DWN-PROX class exceeds background by a factor of five (0.015 vs. 0.003 mg/L, respectively); whereas, the DWN-DSTL class mean (0.070 mg/L) is about 25 times higher than background. A similar trend is evident in the mean Mn concentrations of the proximal and distal groups, which exceed background levels by factors of 26 and 44 times, respectively (0.002 mg/L vs. 0.052 and 0.087 mg/L).

DISCUSSION

The analytical and statistical results portray a complex geochemical system in shallow sediment near the pond. The data indicate that TOC is being delivered to the shallow groundwater via infiltration from the pond, and that biodegradation of the natural organic carbon is creating an anaerobic (low Eh) environment.

Iron- and Mn-oxide dissolution occurs as bacteria utilize the oxides as electron acceptors. Elevated concentrations of Fe, Mn, and the associated As are subsequently released to solution. These trends were statistically supported by a factor analysis procedure. Although this general reaction sequence has been observed in numerous settings (Schnoor, 1996), the elevated pH and Mn levels, and the As association with these trends deserve further attention.

Geochemical modeling was performed to evaluate the data further. The Fe system at Bamberger Pond plays an important role in limiting the redox and pH of the background samples. Elevated levels of Fe[II] near the pond appear to occur in response to biodegradation of natural TOC, which results in Fe[III]-oxide dissolution. Cummings et al. (1999) determined that Fe-oxide reducing bacteria effectively dissolve oxides in metal-contaminated sediment in Idaho. Concurrent with oxide dissolution, the bacteria release elevated levels of As[V] to the pore water. This observation lends support to the mechanisms discussed above.

Manganese concentrations in most groundwater samples down-gradient of the pond are controlled by Mn-carbonate (rhodochrosite) solubility. Samples with

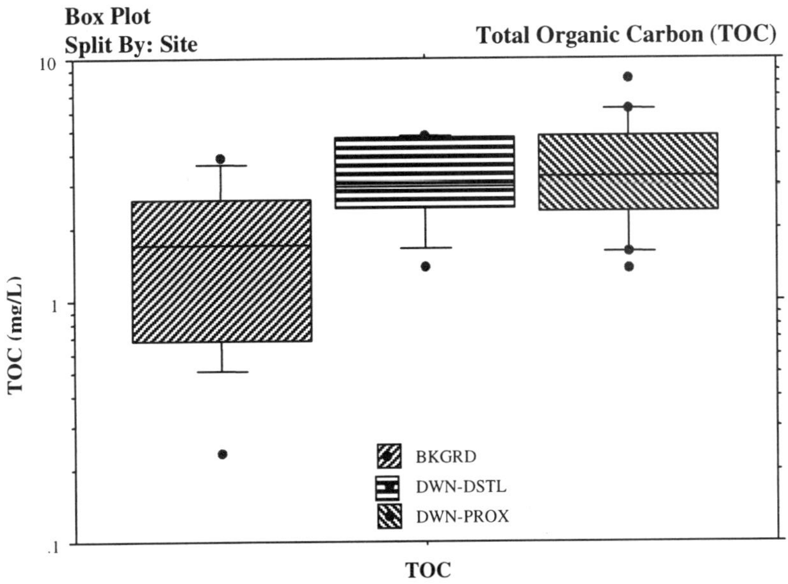

FIGURE 1a. Arithmetic mean of TOC in Bamberger Pond vicinity.

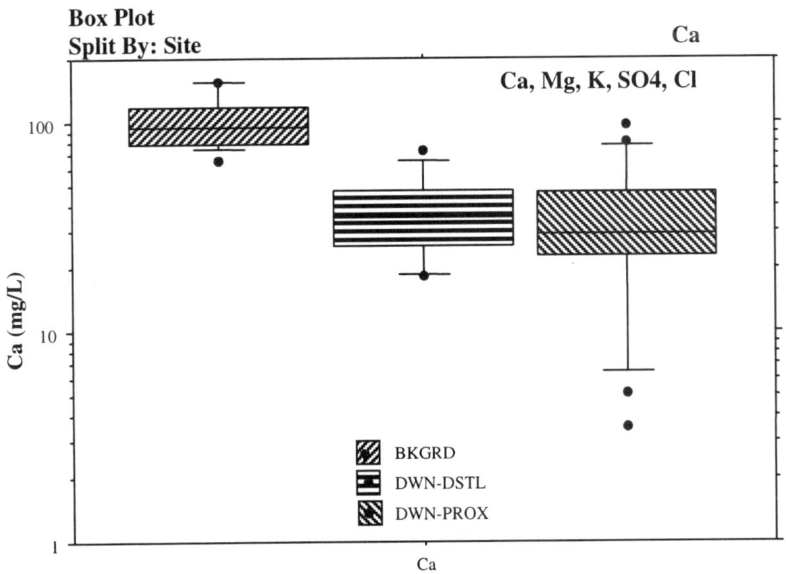

FIGURE 1b. Arithmetic mean of Ca in Bamberger Pond vicinity.

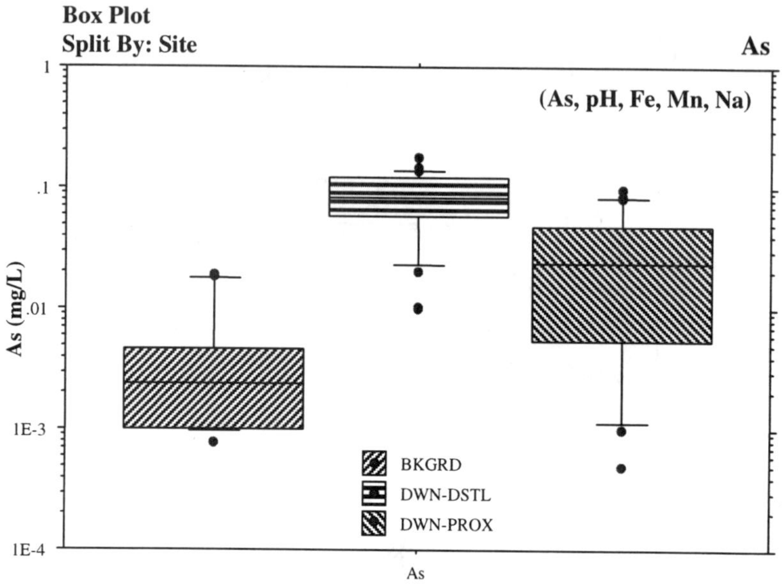

FIGURE 2. Arithmetic mean of As in Bamberger Pond vicinity.

pH levels exceeding 9.0 are positioned along the Mn^{+2}-carbonate-oxide triple point. An Eh-pH diagram with the combined phase boundaries from the Fe and Mn systems is plotted on Figure 3 for a low dissolved iron concentration of 0.005 mg/L representative of the groundwater quality. The background data (indicated by a "☆" on Figure 3) are constrained by the rhodochrosite phase boundary at a pH of 7.5, and by the $Fe(OH)3_{am}$ stability field. The down-gradient samples (indicated by a "◇" on Figure 3) plot along the Mn^{2+}-rhodochrosite stability field at successively higher pH values. Although several of the down-gradient samples occur in equilibrium with rhodochrosite, many others are supersaturated with respect to the carbonate phase. This trend towards supersaturation reflects the slow precipitation kinetics of the transition metal carbonates and some oxides, and is responsible for the elevated levels of dissolved Mn.

Recent studies have shown that As solubility in oxidizing environments is typically controlled by sorption on Fe-oxide minerals (Manning et al., 1998; Schnoor, 1996; Waychunas et al., 1993). To assess the affect of As sorption on shallow sediment at Bamberger Pond, a theoretical concentration of sorbed As was calculated with a geochemical model. Groundwater and sediment Fe concentrations similar to background conditions were assumed. The simulated concentration of sorbed As was 11.3 mg/kg, well within the range of local background levels (2.4 – 14.3 mg/kg). The results suggest that As in background sediment at Bamberger Pond is naturally associated with Fe-oxide phases.

Additional simulations over an increasing pH range of 7.0 to 9.5 indicated that As desorption is initiated at a pH of 7.5 and progresses with increasing pH. Other simulations suggested that the elevated pH values (approaching 9.5) are due

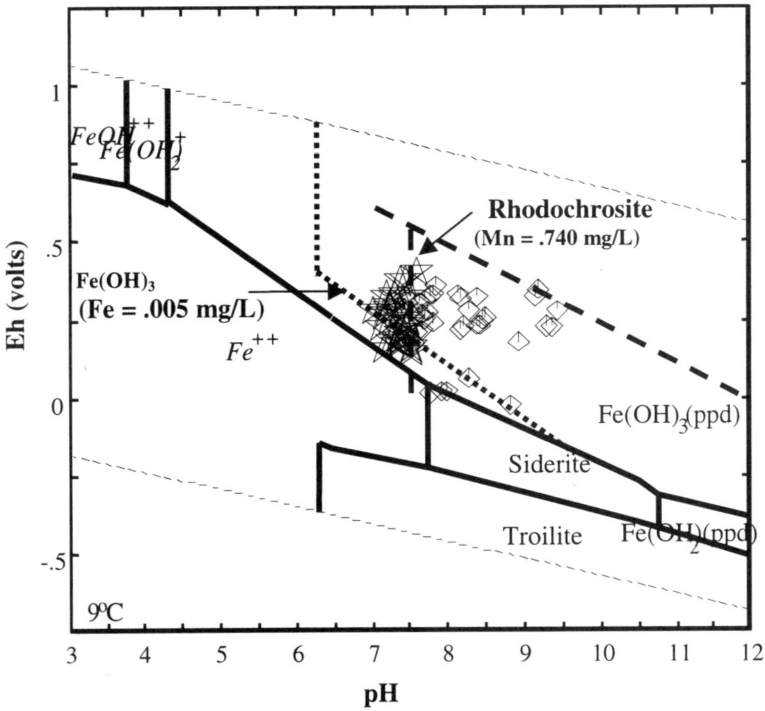

FIGURE 3. Redox potential vs. pH stability diagram for iron oxides, Fe $(OH)_x$, in the presence of manganese (Mn)

to evaporation of the shallow groundwater. Hence, the elevated levels of As are attributed to oxide dissolution and desorption at elevated pH levels.

In conclusion, the combination of statistical analyses and geochemical modeling summarized in this report, provided a sufficient basis for recommending cessation of any further investigation of the causes and sources of elevated levels of As and Mn at Bamberger Pond. Based on this analysis the regulatory community accepted the recommendation.

REFERENCES

Cummings, D. E., F. Caccavo, S. Fendorf, and R. F. Rosenzweig, 1999. "Arsenic Mobilization by the Dissimalatory Fe(III)-Reducing Bacterium Shewanell alga BrY"; in *J. Environ. Sci. Technol. 33*:723-729.

Manning, B. A., S. E. Fendorf, and S. Goldberg, 1998. "Surface Structures and Stability of Arsenic (III) on Goethite: Spectroscopic Evidence for Inner-Sphere Complexes"; in *J. Environ. Sci. Technol. 32*:2383-2388.

Schnoor, J. L., 1996. *Environmental Modeling*. Wiley, New York; 682 pp.

Stumm, W., and J. J. Morgan, 1996. *Aquatic Chemistry*, Wiley, New York, 1022 pp.

URS, 2000. *Fate and Analysis of Arsenic and Manganese in the Vicinity of Bamberger Pond*; *Vol. 2*. USAF Contract No. F42650-92-D-0007; Delivery No. 5099. Study performed for Hill AFB, Salt Lake City, Utah.

Waychunas, G. A., B. A. Rea, C. C. Fuller, and J. A. Davis, 1993. "Surface Chemistry of Ferrihydrite: Part I. EXAFS Studies of the Geometry of Coprecipitated and Adsorbed Arsenate"; in *Geochem. Cosmochim. Acta. 57*:2251-2269.

2001 AUTHOR INDEX

This index contains names, affiliations, and volume/page citations for all authors who contributed to the ten-volume proceedings of the Sixth International In Situ and On-Site Bioremediation Symposium (San Diego, California, June 4-7, 2001). Ordering information is provided on the back cover of this book.

The citations reference the ten volumes as follows:

6(1): Magar, V.S., J.T. Gibbs, K.T. O'Reilly, M.R. Hyman, and A. Leeson (Eds.), *Bioremediation of MTBE, Alcohols, and Ethers.* Battelle Press, Columbus, OH, 2001. 249 pp.

6(2): Leeson, A., M.E. Kelley, H.S. Rifai, and V.S. Magar (Eds.), *Natural Attenuation of Environmental Contaminants.* Battelle Press, Columbus, OH, 2001. 307 pp.

6(3): Magar, V.S., G. Johnson, S.K. Ong, and A. Leeson (Eds.), *Bioremediation of Energetics, Phenolics, and Polycyclic Aromatic Hydrocarbons.* Battelle Press, Columbus, OH, 2001. 313 pp.

6(4): Magar, V.S., T.M. Vogel, C.M. Aelion, and A. Leeson (Eds.), *Innovative Methods in Support of Bioremediation.* Battelle Press, Columbus, OH, 2001. 197 pp.

6(5): Leeson, A., E.A. Foote, M.K. Banks, and V.S. Magar (Eds.), *Phytoremediation, Wetlands, and Sediments.* Battelle Press, Columbus, OH, 2001. 383 pp.

6(6): Magar, V.S., F.M. von Fahnestock, and A. Leeson (Eds.), *Ex Situ Biological Treatment Technologies.* Battelle Press, Columbus, OH, 2001. 423 pp.

6(7): Magar, V.S., D.E. Fennell, J.J. Morse, B.C. Alleman, and A. Leeson (Eds.), *Anaerobic Degradation of Chlorinated Solvents.* Battelle Press, Columbus, OH, 2001. 387 pp.

6(8): Leeson, A., B.C. Alleman, P.J. Alvarez, and V.S. Magar (Eds.), *Bioaugmentation, Biobarriers, and Biogeochemistry.* Battelle Press, Columbus, OH, 2001. 255 pp.

6(9): Leeson, A., B.M. Peyton, J.L. Means, and V.S. Magar (Eds.), *Bioremediation of Inorganic Compounds.* Battelle Press, Columbus, OH, 2001. 377 pp.

6(10): Leeson, A., P.C. Johnson, R.E. Hinchee, L. Semprini, and V.S. Magar (Eds.), *In Situ Aeration and Aerobic Remediation.* Battelle Press, Columbus, OH, 2001. 391 pp.

KEYWORD INDEX

This index contains keyword terms assigned to the articles in the ten-volume proceedings of the Sixth International In Situ and On-Site Bioremediation Symposium (San Diego, California, June 4-7, 2001). Ordering information is provided on the back cover of this book.

In assigning the terms that appear in this index, no attempt was made to reference all subjects addressed. Instead, terms were assigned to each article to reflect the primary topics covered by that article. Authors' suggestions were taken into consideration and expanded or revised as necessary. The citations reference the ten volumes as follows:

6(1): Magar, V.S., J.T. Gibbs, K.T. O'Reilly, M.R. Hyman, and A. Leeson (Eds.), *Bioremediation of MTBE, Alcohols, and Ethers*. Battelle Press, Columbus, OH, 2001. 249 pp.

6(2): Leeson, A., M.E. Kelley, H.S. Rifai, and V.S. Magar (Eds.), *Natural Attenuation of Environmental Contaminants*. Battelle Press, Columbus, OH, 2001. 307 pp.

6(3): Magar, V.S., G. Johnson, S.K. Ong, and A. Leeson (Eds.), *Bioremediation of Energetics, Phenolics, and Polycyclic Aromatic Hydrocarbons*. Battelle Press, Columbus, OH, 2001. 313 pp.

6(4): Magar, V.S., T.M. Vogel, C.M. Aelion, and A. Leeson (Eds.), *Innovative Methods in Support of Bioremediation*. Battelle Press, Columbus, OH, 2001. 197 pp.

6(5): Leeson, A., E.A. Foote, M.K. Banks, and V.S. Magar (Eds.), *Phytoremediation, Wetlands, and Sediments*. Battelle Press, Columbus, OH, 2001. 383 pp.

6(6): Magar, V.S., F.M. von Fahnestock, and A. Leeson (Eds.), *Ex Situ Biological Treatment Technologies*. Battelle Press, Columbus, OH, 2001. 423 pp.

6(7): Magar, V.S., D.E. Fennell, J.J. Morse, B.C. Alleman, and A. Leeson (Eds.), *Anaerobic Degradation of Chlorinated Solvents*. Battelle Press, Columbus, OH, 2001. 387 pp.

6(8): Leeson, A., B.C. Alleman, P.J. Alvarez, and V.S. Magar (Eds.), *Bioaugmentation, Biobarriers, and Biogeochemistry*. Battelle Press, Columbus, OH, 2001. 255 pp.

6(9): Leeson, A., B.M. Peyton, J.L. Means, and V.S. Magar (Eds.), *Bioremediation of Inorganic Compounds*. Battelle Press, Columbus, OH, 2001. 377 pp.

6(10): Leeson, A., P.C. Johnson, R.E. Hinchee, L. Semprini, and V.S. Magar (Eds.), *In Situ Aeration and Aerobic Remediation*. Battelle Press, Columbus, OH, 2001. 391 pp.

A

B

C

D

E

F

G

H

I

J

K

L

M

N

O

R

S

T

U

V

W

X

Y

Z